Mesoscale Atmospheric Dispersion

Advances in Air Pollution

Objectives

The last decade has shown an increase in public and government concern around environmental issues due to air pollution, in particular that generated by man-made processes associated with the lifestyle of modern society.

Atmospheric pollution consists of the adverse effects of a variety of contaminants emitted into the atmosphere by natural and anthropogenic processes. These phenomena are complicated because an apparently inert contaminant can be transformed by chemical reaction into an adverse one, during its transport through the atmosphere.

Nowadays, the physical and chemical processes undertaken during an air pollution episode are to a large extent understood. The modelling of such processes has shown a constant and remarkable growth in the last three decades. Also, the development of efficient and economical monitoring devices has been very successful in recent years. Modelling and monitoring studies constitute only an initial activity. The results provide useful information for possible future implementation of emission regulations and control strategies.

The definition of efficient control strategies cannot be achieved without a good and clear knowledge of the complete pollution process, *i.e.* emission, transport and transformation. Pollution is widespread throughout the world and elimination of the risks to human health is of paramount importance.

Volumes in the series cover areas of current interest or active research in air pollution and include contributions by leaders in the field. Topics for the series consist of: air pollution assessment, management, monitoring and modelling; aerosols and particles; chemical transformation modelling; chemistry of air pollution; comparison of modelling with experiment; data analysis and observation; emission inventories; fluid mechanics for air pollution; global studies; health problems; indoor air pollution; laboratory studies; process studies; regulatory bodies; turbulence modelling at small and meso scales; urban air pollution, and urban and suburban transport emission.

G. Passerini
Universita di Ancona
Italy

T. Tirabassi
Institute FISBAT-CNR
Italy

R. Rosset
Laboratoire d'Aerologie
France

Mesoscale Atmospheric Dispersion

Edited by:

Z. Boybeyi
Center for Atmospheric Physics
Science Applications International Corporation, USA

WIT PRESS Southampton, Boston

Edited by Dr Z. Boybeyi

Published by

WIT Press
Ashurst Lodge, Ashurst, Southampton, SO40 7AA, UK
Tel: 44 (0) 238 029 3223; Fax: 44 (0) 238 029 2853
E-Mail: witpress@witpress.com
http://www.witpress.com

For USA, Canada and Mexico

Computational Mechanics Inc
25 Bridge Street, Billerica, MA 01821, USA
Tel: 978 667 5841; Fax: 978 667 7582
E-Mail: info@compmech.com
US site: http://www.compmech.com

British Library Cataloguing-in-Publication Data

A Catalogue record for this book is available
from the British Library

ISBN: 1-85312-732-9
ISSN: 1369-5886

Library of Congress Catalog Card Number: 99-68744

The texts of the papers in this volume were set individually by the authors or under their supervision.

Printed in Great Britain by IBT Global Ltd, London.

Thanks are due to United States Geological Survey (USGS) for the figure which appears on the front cover of this book.

Contents

Preface

Early studies of atmospheric dispersion focused on local dispersion only due to civilian concerns over the effects of industrial air pollutants on public health and military concerns over the impact of chemical and biological agents on the battlefield. In the case of local dispersion, dilution occurs largely due to turbulent diffusion. Therefore, both the mean transport wind and turbulence may be considered steady and horizontally homogeneous.

In the last few decades, however, there has been increasing recognition and interest in evaluating air pollution problems over mesoscale and synoptic-scale. In particular, the most serious problems of our atmospheric environment on mesoscale, such as urban air pollution and the behavior of urban scale plumes downwind of large industrialized cities, regional haze, photochemical smog, acidic precipitation, accidental toxic gas and radioactive releases, and stratospheric ozone depletion have received tremendous public attention.

In the case of mesoscale dispersion, the approximation of steady and horizontally homogeneous flow no longer holds. Instead, a large number of mesoscale flow modes with a wide range of temporal and spatial scales, constitutes the primary mechanism of dilution. Mesoscale atmospheric dispersion is, therefore, an even more complex phenomenon than local scale atmospheric dispersion due to the additional physical processes that may play a role and to the greater number of characteristic time and space scales.

This monograph presents an organized collection of invited review papers covering the topic of mesoscale atmospheric dispersion. The organization of the book in a sequence of chapters follows a review of the basic concepts, theories, and models of mesoscale atmospheric dispersion and the related mesoscale atmospheric systems affecting transport, diffusion, transformation, and removal of air pollutants. In dealing with a problem so complex as mesoscale atmospheric dispersion, the cooperation of experts in many different areas is required. Each chapter was authored by a leading scientist in the field and was written in such a way that it provides the reader with an organized and consistent discussion on the field of mesoscale atmospheric dispersion. Each chapter is independent and treats a single part of the broad subject.

Finally, I would like to thank to my colleagues (authors of each chapter) for their valuable contributions to this monograph. I would also like to thank to Dr. David P. Bacon and Dr. Ananthakrishna Sarma for their support and help in the editing effort of this book. We hope that the subject material of this book will be of interest and use to researchers as well as students intending to pursue studies in the field of mesoscale meteorology, mesoscale dispersion, atmospheric boundary layer, and air pollution meteorology.

Zafer Boybeyi
Editor

About the Authors

S. Pal Arya

Pal Arya is currently a Professor of meteorology in the Department of Marine, Earth and Atmospheric Sciences, North Carolina State University (NCSU), Raleigh, North Carolina. He has been a faculty member at NCSU since August 1976. Before that he was on the faculty of the Department of Atmospheric Sciences, University of Washington, Seattle, Washington, from 1969 to 1976. He received his Ph.D. in fluid mechanics from Colorado State University in 1968.

Professor Arya has taught courses in micrometeorology, environmental fluid mechanics, planetary boundary layer, air-sea interaction, air pollution meteorology, atmospheric turbulence, and atmospheric dispersion, including some at Wessex Institute of Technology. He has advised many undergraduate and graduate students in meteorology and atmospheric sciences. He has been conducting research in the areas of environmental fluid mechanics, micrometeorology, atmospheric boundary layer, turbulence, diffusion and air pollution meteorology for more than thirty years. He has received many research grants and cooperative agreements and published more than ninety papers, book chapters and two textbooks. Professor Arya is a Fellow of American Meteorological Society, a Fellow of American Association for the Advancement of Science, and a member of American Geophysical Union. He is also a member of the honor society Phi Kappa Phi.

David P. Bacon

David Bacon is trained as both an experimental and theoretical physicist.. He holds S.B degrees in both Mathematics and Physics from the Massachusetts Institute of Technology and a Ph.D. in Physics from Dartmouth College. Dr. Bacon has an extensive and broad research background ranging from experimental and theoretical plasma physics to theoretical hydrodynamics to atmospheric science.

For the past fifteen years, Dr. Bacon has been developing new numerical methods for atmospheric simulation. As Director of the Center for Atmospheric Physics of Science Applications International Corporation, Dr. Bacon leads a team of experienced scientists with expertise in numerical methods, computational fluid dynamics, atmospheric dynamics, thermodynamics, chemistry, and microphysics, and aerosol and gas dispersion. For the past seven years, Dr. Bacon has lead the development of the Operational Multiscale Environment model with Grid Adaptivity (OMEGA), the highest resolution operational atmospheric dispersion model currently available. Dr. Bacon is an Affiliated Professor at the Institute for Computational Science and Informatics of George Mason University and the author of numerous papers and presentations.

Zafer Boybeyi

Zafer Boybeyi has extensive research experience in both mesoscale modeling and air pollution dispersion modeling. He has obtained a B.S. degree in Meteorological Engineering from Istanbul Technical University, Istanbul, Turkey and a M.S. degree in Meteorology from San Jose State University. His interest in mesoscale modeling led him to pursue the subject further at North Carolina State University where he obtained his Ph.D. in Atmospheric Sciences. He has authored and co-authored several international journal articles, conference proceedings, book chapters, and technical reports dealing with various aspects of atmospheric sciences.

Dr. Boybeyi currently works at Science Application International Corporation (SAIC), where he conducts research in the areas of numerical modeling of mesoscale physical and dynamical processes, numerical weather prediction methods and analysis, planetary boundary layer physics, land-surface processes, data assimilation techniques, cloud physics, aerosol transport and diffusion, dispersion modeling techniques, aerosol chemistry and scavenging processes. Dr. Boybeyi is also an Affiliated Professor at the Institute for Computational Science and Informatics of George Mason University.

Michael J. Brown

Michael Brown specializes in atmospheric turbulence, boundary layer mechanics, and air pollution dispersion modeling. While at Los Alamos National Laboratory he has worked on air quality projects in Mexico City, Chile, Los Angeles, and El Paso/Ciudad Juarez. Currently he is working on concentration fluctuation modeling and computational fluid dynamics modeling of flow around buildings. Michael Brown earned Ph.D. and M.S. degrees in atmospheric science from North Carolina State University and a B.S. from the University of California at Davis in the same field. He has also worked at the Unites States Environmental Protection Agency and the Royal Netherlands Meteorological Institute performing air quality research.

Daewon W. Byun

Daewon Byun has extensive research experience in atmospheric turbulence and diffusion, planetary boundary layer, and photochemical air quality modeling. He has obtained a B.S. degree in Meteorology from Seoul National University, Korea, and M.S. and Ph.D. in Marine, Earth, and Atmospheric Sciences from North Carolina State University. He has worked on the several Eulerian photochemical modeling projects, such as the Acid Deposition and Oxidant Model (ADOM), the Regional Acid Deposition Model (RADM), and the U.S. Environmental Protection Agency (EPA) Models-3 Community Multiscale Air Quality (CMAQ) modeling system. He worked at the ENSR (formerly ERT) consulting company and Computer Sciences Corporation before joining National Oceanic and Atmospheric Administration (NOAA), on assignment to EPA, in 1992. Since, he has been serving as the science team leader for the Models-3/CMAQ development project.

Dr. Byun has authored and co-authored many journal articles, conference proceedings, and project reports. His current research spans a wide spectrum of interests, such as software engineering of comprehensive environmental models, numerical algorithms for atmospheric transport and transformations, treatment of sensitivity and uncertainty, model evaluation, and environmental policy decision making processes.

Jeremy M. Hales

Jeremy Hales has been a principal of Envair since 1990, an unincorporated association dedicated to carrying out contract research and offering consulting services in the environmental and earth sciences. Areas of focus include technical consulting in air quality and atmospheric sciences; coordination of research-oriented technical studies, including large, combined monitoring, modeling, and data analysis programs; meteorological/pollution model development and application; and interpretive analysis of air-chemistry, cloud-physics, and pollutant-removal processes. Prior to joining Envair, Dr. Hales worked at Battelle Pacific Northwest National Laboratory for 23 years as a scientist and manager of the Atmospheric Sciences Department. While at the laboratory, Dr. Hales led the acquisition, development, and deployment of the DC-3 and Gulfstream-1 (G-1) research aircraft. He has been an active team member in adapting and deploying advanced analytical equipment for aircraft use and has led several major field studies using the aircraft.

Dr. Hales is currently involved in a variety of technical activities, including serving as Management Coordinator for the North American Research Strategy for Tropospheric Ozone (NARSTO) program, conducting local, regional, and global modeling studies in atmospheric chemistry, developing advanced numerical techniques for atmospheric models. Dr. Hales is the co-developer of the PLUVIUS reactive-storm models, and the original team leader for development of the GChM global atmospheric chemistry model. Dr. Hales is the author of over one hundred journal articles, book chapters, and technical reports dealing with various aspects of atmospheric sciences and engineering.

Michael D. Moran

Michael Moran is a research scientist in Toronto, Ontario with the Meteorological Service of Canada (MSC), a part of Environment Canada, the Canadian national environmental agency. His research focus at MSC has been regional-scale air quality modeling, including numerical studies of acid rain and particulate matter. He holds a B.Math. from the University of Waterloo, an M.Sc. from the Florida State University, and a Ph.D. from the

Colorado State University, where the topic of his Ph.D. dissertation was the numerical modeling of mesoscale atmospheric dispersion.

Devdutta S. Niyogi

Devdutta Niyogi is the Assistant State Climatologist for North Carolina State, and a Meteorologist at the Department of Horticultural Science in North Carolina State University, in-charge of a statewide agrometeorological observational network called AgNet. Devdutta Niyogi obtained his bachelor of Civil Engineering in 1994, from University of Poona, India, concentrating on air pollution models. He was the Rajiv Gandhi Summer fellow for Atmospheric Sciences at the Indian Institute of Sciences in 1994, and same year he joined the graduate program in Atmospheric Sciences at the North Carolina State University. He obtained his M.S. in 1996, with his thesis being on Dynamic Interactions in Soil Vegetation Atmosphere Transfer (SVAT) process. He obtained his Ph.D. from NCSU in October2000 with his research concentrating on carbon and soil moisture assimilation and terrestrial ecological processes in mesoscale models.

Dr Niyogi has been involved with various field and collaborative projects principally through NSF, DOE, and UNEP. He has published or communicated over 25 research papers in various international journals. His areas of research include surface-atmosphere exchanges, stomatal resistance parameterizations, SVAT schemes, applied climatology, air pollution, earth system sciences, observational and numerical analysis of mesoscale processes, and applied statistics. He is a member of the AMS committee on Agriculture and Forest Meteorology, and is a reviewer for publications related to land surface processes for different atmospheric journals.

Sethu Raman

Sethu Raman is a Professor of Atmospheric and Marine Sciences at the Department of Marine, Earth, and Atmospheric Sciences and the Director of the State Climate Office, a center for climate environment interactions, at the North Carolina State University. Dr. Raman obtained his Ph.D. from Colorado State University in 1972, and resumed worked with the

North Carolina State University working on land air sea interactions and mesoscale processes.

He has published over 150 peer-reviewed papers, has edited books and journals on atmospheric and environmental processes. In addition to this, he has coordinated several national and international experimental programs through NSF, DOE, and similar other agencies. At NCSU, he has chaired or co-chaired many M.S., and Ph.D. committees, in addition to active participation in several university and national committees. His list of awards includes the Alumni Extension Award for 1998, and the Outstanding Extension Award at NCSU for 1997. He is a fellow of the AMS and a member / reviewer of various journals and committees / panels through AMS, NSF, and other funding agencies

Ananthakrishna Sarma

Ananthakrishna Sarma is a senior research scientist at the Center for Atmospheric Physics of Science Applications International Corporation, which he joined in 1986. He is part of a team of scientists who have designed and built the Operational Multiscale Environment model with Grid Adaptivity (OMEGA). OMEGA is a novel atmospheric prediction tool built upon the concept of unstructured adaptive grids which gives it the unique capability to adapt its underlying grid to terrain and other static features as well as to the evolving weather.

Dr. Sarma received his M.S. in Meteorology from the S.D. School of Mines and Technology in 1978 and his Ph.D. in Atmospheric Sciences from the Colorado State University in 1986 where he studied the formation of sulfate in convective clouds through heterogeneous chemical processes. He has worked for the last two decades in the field of numerical modeling and has studied a variety of atmospheric processes ranging from natural convective clouds to highly perturbed atmospheres associated with rocket exhaust clouds, nuclear fireballs and large area fires. His research interests include atmospheric dynamics, cloud microphysics and atmospheric chemistry. He is a member of the American Meteorological Society. He has served as a member of the AMS committee for atmospheric chemistry and as a reviewer for atmospheric chemistry related proposals to the U.S. Department of Energy. He is also an affiliate Professor at the George Mason University.

Ian Sykes

Ian Sykes has a BA (Mathematics) from Cambridge University, 1973, and Ph.D. (Applied Mathematics) from Imperial College, University of London, 1979. He joined the UK Meteorological Office in 1973, and worked in the Boundary Layer Research Branch on turbulence modeling and flow over topography. In 1980, Dr. Sykes moved to New Jersey to join Aeronautical Research Associates of Princeton (ARAP), which later became part of the Titan Corporation. Since 1980, he has worked principally on atmospheric dispersion and modeling turbulent concentration fluctuations using second-order turbulence closure and Large-Eddy Simulation (LES). Dr. Sykes has authored and co-authored several publications, many for international journals and conference proceedings.

Chapter 1

A brief overview of mesoscale atmospheric features

Zafer Boybeyi
Center for Atmospheric Physics
Science Applications International Corporation
1710 SAIC Drive
McLean, VA 22102, USA

Abstract

Mesoscale atmospheric flows possess various mesoscale time and space scales resulting from the diurnal cycle, the atmospheric inertial mode, thermal and mechanical forcing due to mesoscale terrain inhomogeneities, and downscale energy transfer from the synoptic-scale to the mesoscale due to nonlinear flow interactions. The transport and diffusion of atmospheric pollutants can be affected by this wide range of flow scales through variations in the mean transport wind, differential advection due to vertical and horizontal wind shear, and vertical mixing. It is, therefore, important to understand these mesoscale space-time flow scales when the mesoscale dispersion processes are studied.

1 Introduction

Since atmospheric motions occur over a wide range of scales from a few millimeters (e.g., turbulence) to thousands of kilometers (e.g., planetary waves), all of these flow scales can affect the transport and diffusion of atmospheric pollutants to some degree. Therefore, one typically desires to capture not only the development and evolution of small-scale flow features but also their interaction with and influence upon the larger-scale flow features.

Over mesoscale distances, plumes and/or puffs are not only relatively localized with respect to their environment, but are also forced on scales larger than their own. In the atmosphere, mesoscale flow dynamics are quite variable and complex. Mesoscale energy is distributed over a variety of flow modes, including mean mesoscale circulations and mesoscale eddies. Every atmospheric dynamical process operating on scales larger than molecular dissipation and smaller than the latitudinal variation of the Coriolis force can play a role in mesoscale circulations and hence dispersion. Mesoscale flows may be hydrostatic or nonhydrostatic. Nonhydrostatic motions may contain significant features on scales ranging from several meters to several tens of kilometers with time scales of minutes to many hours. Hydrostatic motions, in which the nonhydrostatic motions are embedded, have motion scales orders of magnitude larger than the nonhydrostatic motions.

For example, thermal and mechanical forcing due to mesoscale terrain inhomogeneities can generate externally forced mesoscale flow inhomogeneities such as upslope precipitation events, lee cyclogenesis, drylines, land and sea breezes, mountain-valley winds, and urban heat island circulations. The atmospheric inertial mode can generate internally generated circulations such as fronts and jet streaks, buoyant instability, synoptic instability, Kelvin-Helmholtz instability waves, horizontal vortices, quasi-stationary convective events, and mesoscale structure of hurricanes. Moist processes can generate mesoscale convective systems such as cumulus-scale convection, extratropical squall lines, mesoscale convective complexes, mesoscale cellular convection, tropical cyclones, mesoscale rainbands, tornadoes, and tornadogenesis. Synoptic-scale wave-wave interactions may produce higher wavenumber circulations or flow features. This wide range of mesoscale circulations and their associated mesoscale vertical ascents and descents can have a significant impact on the dispersion of pollutants.

In addition, density stratification in the atmosphere plays an important role in both flow dynamics and dispersion of pollutants. The presence of wind shear further complicates the flow dynamics and the dispersion of pollutants. Many flow instabilities occur partly as a result of the presence of wind shear, which may enhance turbulent diffusion by increasing turbulence intensities. Large-scale baroclinicity and large-scale vertical motions affect the local structure and the dynamics of the planetary boundary layer. Latitude affects both the inertial period and diurnal forcing. Time of day affects the dynamics of the boundary layer, while time of year affects the diurnal forcing.

In summary, a large number of mesoscale flow modes active on a wide range of temporal and spatial scales, constitute the primary mechanism for dispersion on this scale. It is, therefore, first necessary to understand the behavior of atmospheric flows on these scales, in order to understand the dispersion processes.

2 Definition of mesoscale

The highly variable and complex phenomena of atmospheric flow can be classified according to the physical scales of apparently coherent structures that appear generally or intermittently within the flow (e.g., synoptic-scale, mesoscale, and local-scale). This scale classification might suggest that the atmosphere is somehow quantized and the scales are discrete. In reality, of course, the spectrum of atmospheric motions is smooth and continuous between the limits imposed by the mean free path of molecules and the circumference of the Earth.

For example, the spectra of wind speed near the ground for two different meteorological conditions are shown in Figure 1. These spectra are referred to as energy spectra, because they have units of kinetic energy per unit mass. The spectrum indicated by a solid line shows a large peak near a period of 4 days, a small peak near a period of 1 min, and a reduced energy region in between these two peaks. The reduced energy region indicates the existence of a mesoscale gap. The portion of the spectrum to the right of the small peak is called the inertial subrange (Kolmogorov [1]). Theoretically, energy in the inertial subrange is neither created nor destroyed but transferred to higher and higher frequencies until it is dissipated by friction into heat.

On the contrary, the second spectrum indicated by a dashed line does not show the existence of the mesoscale gap. This implies that energy is produced in this range by mesoscale motions or that energy produced on a larger scale is broken down into this scale. Note that the peaks in the spectra occur at frequencies having the greatest contribution to the overall variance, and correspond to production of turbulent kinetic energy.

The existence of such a mesoscale gap should be viewed as a near-surface feature resulting from boundary effects of the Earth's surface. The mesoscale gap has significant implications for (a) the predictability of large-scale atmospheric motions, (b) the parameterization of small-scale motions in terms of the characteristics of the large-scale motions, (c) the statistical stability of atmospheric turbulence, and hence (d) the predictability of mesoscale transport and diffusion processes.

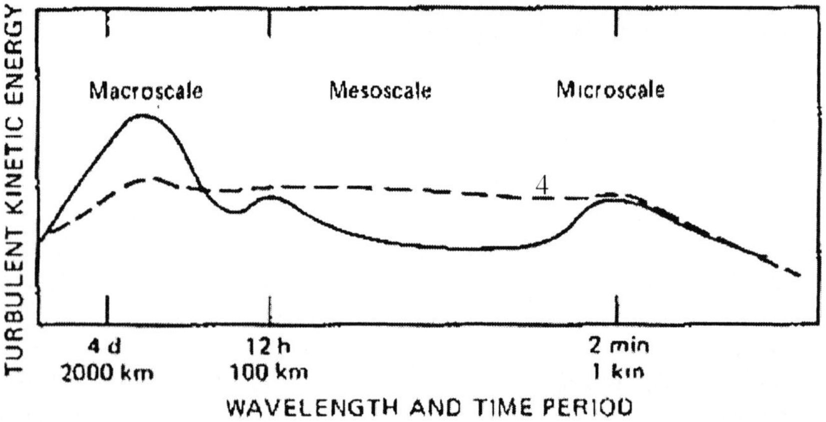

Figure 1: Spectra of wind speed near the ground for two different synoptic conditions. (from D. Randerson, Atmospheric Science and Power Production, USDOE Report TIC-27601 [2]).

For example, consider a cloud of pollutant released during periods with meteorological conditions corresponding to the two spectra presented in Figure 1. Initially the two clouds would grow at the same rate by eddies, since the two time series have similar magnitudes for energy up to a scale of 1 km. As the clouds grow to mesoscale sizes, the cloud released during series 2 indicated by a dashed line would grow faster than the cloud released during series 1 indicated by a solid line, since more energy is available at these frequencies. As the clouds continue to grow to larger and larger sizes, their growth rates will be determined by the amount of energy at that scale. Differences in growth rates of the two clouds are, therefore, proportional to differences in the energy spectra.

As a somewhat arbitrary middle region of the atmospheric energy spectrum, the mesoscale can then be defined as that part of the spectrum between the synoptic-scale and local-scale. Further, the mesoscale can also be classified into three spatial decades as suggested by Orlanski [3]: the meso-γ scale (2-20 km), the meso-β scale (20-200 km), and the meso-α scale (200-2000 km). Although somewhat arbitrary, Orlanski's geometric definition has been widely adopted. Over these mesoscale distances, ubiquitous energy transfer processes and the special events occur locally and intermittently as a result of topographic forcing and mesoscale instabilities. Finally, the mesoscale atmospheric dispersion can be defined as the transport and diffusion of air pollutants over horizontal distances between 2 and 2000 km with time periods ranging from 1 to 48 hours.

3 Mesoscale flow scales

The atmosphere is characterized by phenomena whose space and time scales cover a very wide range (Table 1). The space scales of these features are determined by their typical size or wavelength, and the time scales by their typical lifetime or period. Over mesoscale travel times and distances, a large number of mesoscale flow modes are possible with a wide range of attendant temporal and spatial scales. The implication, of course, is that mesoscale flow scales and flow structures must be considered and characterized, when mesoscale atmospheric dispersion is studied (a detailed discussion on mesoscale atmospheric dispersion is given by Moran [4]).

Table 1: Typical time and space scales for micro and mesoscale. (after Orlanski [3]).

	1 month – 1 day	1 day – 1 hour	1 hour – 1 minute	1 minute – 1 sec	
2000 km	Fronts Hurricanes				Meso- α Scale
200 km		Squall lines Inertial waves Cloud clusters Orography			Meso- β Scale
20 km		IGW CAT Urban effect Thunderstorms			Meso- γ Scale
2 km			Tornadoes Convection		Micro- α Scale
200 m			Dust devils Thermals Wakes		Micro- β Scale
20 m				Plumes Turbulence Roughness	Micro- γ Scale
2 m					

3.1 Mesoscale time scales

The influence of the Earth's surface is effectively limited to the lowest 10 km of the atmosphere in a layer called the troposphere. Over time periods of about one day this influence is restricted to a very much shallower zone known as the planetary boundary layer. The planetary boundary layer contains typically about ten per cent of the overlying mass of air. This layer is particularly characterized by well developed turbulence generated by frictional drag as the atmosphere moves across the rough and rigid surface of the Earth, and by the density stratification of the air which results from differences in temperature of the surface and the air.

These differences in temperature arise over land primarily in the course of the daily cycle of radiative heating and cooling of the ground, but they may also arise from overflowing of air from warmer or cooler regions of the Earth. The boundary layer receives much of its heat and all of its water through this turbulence process. Undoubtedly, effective dispersion of atmospheric pollutants released into the atmosphere near the ground depends largely on mixing processes on a variety of scales. This mixing is a direct consequence of turbulent and convective motions generated in the planetary boundary layer.

In the aerodynamic sense, the boundary layer is simply the layer from which momentum is extracted and transferred downward to overcome surface friction. This layer has important variations in depth as well as in mixing efficiency, arising from the control on vertical transport by the density stratification. The height of the boundary layer is not constant with time; it depends upon the strength of the surface generated mixing. By day, when the Earth's surface is heated by the sun, there is an upward transfer of heat into the cooler atmosphere. This vigorous thermal mixing enables the boundary layer depth to extend to about 1 to 2 km. Above the planetary boundary layer, a stably stratified free atmosphere usually forms in which turbulent motion is suppressed. Conversely by night, when the Earth's surface cools more rapidly than the atmosphere, there is a downward transfer of heat. This tends to suppress mixing, and the boundary layer depth is confined to a shallower layer than in the daytime.

Within the planetary boundary layer, these changes in vertical stability and vertical mixing resulting from the diurnal heating cycle modify the wind speed, wind direction, and turbulence profiles even over flat and uniform terrain. One striking example of this is the frequent formation of a low-level jet over land on clear nights. Although the timing of the formation and subsequent erosion of nocturnal jets is controlled by the diurnal cycle, the jet itself can also be explained as an inertial oscillation. Thus, two natural mesoscale time scales need to be considered for travel times of more than one hour: the 24-hour diurnal period and the latitude-dependent inertial period $2\pi/f$, where f is Coriolis parameter. The inertial period may be longer than, equal to, or shorter than the diurnal period depending on the latitude.

For thermally forced mesoscale circulations, the natural circulation time scale is the forcing time scale, that is, the duration of heating or cooling. This time scale may vary from 1 to 24 hours depending upon latitude and season. Short-duration forcing is less likely to generate a mesoscale atmospheric response than longer duration forcing. However, some shallow circulations such as drainage winds can form in as little as one or two hours. In contrast, mesoscale circulations resulting from the mechanical forcing of stably stratified flows by topographic features will have the time scale of the larger scale flow even though they are mesoscale in spatial extent. Still other mesoscale time scales will be associated with internally forced mesoscale phenomena such as fronts, jet streaks and latent heat driven phenomena (e.g., mesoscale convective complexes or mesoscale rainbands). These time scales will often be comparable to or larger than the inertial or diurnal periods, but they can also be shorter, especially if internal gravity waves are important.

3.2 Mesoscale space scales

Atmospheric flows are seldom steady over mesoscale time periods. Therefore, they are not usually horizontally homogeneous over mesoscale distances. Many mesoscale circulations are atmospheric responses to externally imposed boundary forcing such as mountains, plateaus, and valleys. The presence of such topographic features can result in mesoscale flow phenomena such as mountain waves, lee waves, vortex shedding, upstream blocking, channeled flow, slope winds, and mountain-valley circulations. Variations in other terrain properties may also affect atmospheric flows. Horizontal inhomogeneities in roughness length or surface temperature can generate internal boundary layers, and mesoscale atmospheric circulations such as land-sea breezes and urban heat island circulations.

These various surface-forced mesoscale circulations possess a number of natural spatial scales. In the case of mechanically-forced circulations, the horizontal and vertical scales will depend on obstacle height h, obstacle width l, horizontal wind speed V, vertical stratification represented by the Brunt-Vaisala frequency N, and density scale height $H = \rho_o(\partial\rho_0/\partial z)^{-1}$, where $\rho_0(z)$ is the base state density. In general, the vertical scale of mechanically-forced mesoscale circulations ranges between the obstacle height and the density scale height, while the horizontal scale lies between the horizontal scale of the obstacle and the Rossby radius of deformation based on obstacle height Nh/f. In the case of thermal forcing, thermally-forced mesoscale circulations generally tend to be shallow so that the planetary boundary layer height, z_i, is a relevant vertical scale within the Rossby radius of deformation Nz_i/f as the limiting scale. However, differences in surface heating due to surface inhomogeneities may also trigger moist convection and mesoscale convective clouds. In this case, convection tends to be cumulus scale and may develop throughout the depth of the troposphere.

Spatial variations in atmospheric fields may also arise due to internal flow dynamics and instabilities. Mesoscale circulations can be generated by synoptic-scale dynamical or thermal forcing. Examples of such circulation systems include frontal and dryline circulations, density currents, squall lines, jet streaks, and mesoscale gravity waves. In addition to geostrophic adjustment (and flow over topography), internal gravity waves with horizontal wavelengths ranging from several hundred meters to several hundred kilometers can be generated by Kelvin-Helmholtz instability, frontal surges, convective activity, and restratification processes. Quasi-horizontal vertical modes may be generated by three-dimensional turbulence decay and upscale energy transfer in a stratified fluid. Synoptic-scale baroclinicity can also modify vertical profiles of mean wind and turbulence within the planetary boundary layer by imposing additional vertical shear. Mesoscale or synoptic-scale vertical motions resulting from internal flow dynamics may concentrate pollutants near the Earth's surface or lift them out of the planetary boundary layer. Clearly, mesoscale space scales can be quite complicated, and can directly influence mesoscale dispersion (see for more information Moran [4]).

4 Scale interaction

"Scale interaction" is one of the most popular phrases in the meteorological literature. In planetary-scale meteorology, scale interaction generally refers to the interactions between the zonal flow and a fairly limited set of waves that are quantized by the circumference of the Earth. In turbulence theory, it means the interactions among a continuous spectrum of eddies of all sizes. In mesoscale meteorology, scale interaction is the limited set of interactions among discrete scales such as the meso-γ, the meso-β, and the meso-α.

In a very simple system consisting of a slowly varying mean flow with a very weak perturbation upon it, the main scale interaction can be considered to be the influence of the mean flow upon the perturbation. As the perturbation becomes more substantial, it exerts an increasing influence on the mean flow. This generates other scales of motion as a result of secondary instabilities. Hence, the scale interactions become more and more numerous and the general degree of disorder becomes greater. In a highly nonlinear system such as fully turbulent flow, the scale interaction becomes very complex and numerous. Therefore, an explicit description of their interaction becomes problematic. One then may resort to stochastic treatments of the interactions. Perhaps the most difficult flows to deal with are ones for which the degree of disorder is too great to allow explicit treatment of the dynamics but too small to allow a simple stochastic treatment.

The question of the degree of disorder of mesoscale motions is a serious problem confronting mesoscale meteorologists. Can the mesoscale be treated as a finite number of waves interacting with each other and the mean flow? Or can mesoscale be considered as a continuous spectrum of eddies? According to Lilly [5], the mesoscale is a region of disordered waves and eddies serving to shunt energy between the cumulus and cyclone scales. Mesoscale motions may also occur locally and intermittently to convert potential energy to kinetic energy on this scale. If we assume that this is a qualitatively correct view, then we should ascertain the physical nature of the intermittent organized mesoscale motions and aim to determine the way in which energy, heat, and momentum are transferred by both the intermittent ordered motions and the more continuous disordered motions.

5 A brief overview of mesoscale circulations

After Atkinson [6] and Ray [7], mesoscale circulation systems can be divided into three broad types: (1) terrain-induced mesoscale circulations, (2) internally generated mesoscale circulations, and (3) mesoscale convective systems. Here, we will only briefly discuss the mesoscale systems and circulations that have profound affects on the transport and diffusion of pollutants. A comprehensive review of atmospheric circulations and their influences on air pollution can also be found in Arya [8].

5.1 Terrain-induced mesoscale circulations

The atmosphere has only one natural boundary, namely the Earth's surface, where the atmosphere exchanges momentum and energy with the Earth. Two types of mesoscale atmospheric circulations are constantly generated by the interaction between the atmosphere and the Earth. Most of these circulations are associated with fixed geographical features and do not generally move far from their point of origin. They are the result of influences on the airflow by different geographical and topographical configurations, such as sea and land, city and surrounding rural area, and mountain and valley.

The first type of terrain-induced mesoscale circulations is a "thermally forced circulation". This type of atmospheric circulation is generated mainly due to differential surface heating. The forcing of mesoscale systems by horizontal gradients of surface heating is conceptually one of the most straightforward physical processes in the atmosphere. The classical thermally forced mesoscale systems include the sea and land breezes, urban heat island circulations, and mountain and valley winds. Less commonly considered types of thermally forced mesoscale flows include those due to horizontal contrasts in ground wetness, snow and vegetation coverage, and to spatial variations in cloud cover. The horizontal and vertical extents of these circulations are intimately related to the time of day, month or season, and synoptic conditions.

The second type of circulation is a "mechanically forced circulation". This type of atmospheric circulation is generated mainly due to the interaction between the stably stratified atmospheric flow and topographic obstacles. Atmospheric circulations resulting from airflow over hills, valleys, and mountains exist over the whole range of atmospheric scales of motion. The classical mechanically forced mesoscale systems include lee waves, upslope and downslope winds, and mountain and valley wakes. The horizontal and vertical extents of these circulations are strongly related to the height, slope, and shape of the forcing obstacle (topography) and the ambient atmospheric conditions, such as winds and atmospheric stability.

5.1.1 Sea and land breezes

A sharp thermal contrast often occurs at the coastline due to large differences in the thermal properties of land and sea surfaces. Because of the large heat capacity of water, the sea surface temperature does not change by more than 2^0C between day and night. The land, on the other hand, heats up more rapidly during the day and cools at night, especially under clear-sky and fair-weather conditions with weaker large-scale atmospheric flow. A diurnal range of 10^0C to 20^0C in surface air temperatures over land is typically observed in coastal areas. Consequently, large temperature differences occur between land and sea across the coastlines. These temperature contrasts induce mesoscale thermal circulations, known as sea and land breezes.

During the mid-morning hours of a clear summer day with light winds, the land heats up much more quickly than the adjacent sea water, and this heating generates a shallow thermal low over land. As warm air rises over land and cool air sinks over coastal water, this brings colder air from the sea toward the land near the surface and warmer air from the land to the sea at higher levels. Hence, this leads to the development of sea breeze circulation. As the sun rises, the sea breeze penetrates inland and increases its depth, and speed. By mid afternoon, the sea breeze reaches its greatest development, and as the sun sinks by late afternoon it begins to weaken.

Frequently, at night after a sea breeze has dissipated, a weak land breeze circulation forms where the land is cooler than the water surface. This is more likely to occur during the winter season, at which time the land breeze reaches its maximum development. The land breeze is usually much weaker than the sea breeze (Boybeyi and Raman [9]).

Similar thermal circulations also develop in the coastal areas of large lakes. The daytime breeze from the lake is called the lake breeze, while the nighttime breeze from the land is called the land breeze. Inland penetration of lake breezes is usually not more than a few kilometers and their depths are also smaller as compared with sea breezes. But even these smaller thermal circulations have important influences on local environments around large lakes.

Sea and land breezes are important for mesoscale dispersion processes because of the associated changes in atmospheric stability, turbulence and transport patterns. Moreover their almost daily occurrence at many seaside locations during the warmer seasons results in significant differences in diffusion climatology over rather short distances. They sometimes appear to be associated with a large suppression of plume rise associated with industrial sources. Strong fumigation usually occurs at the sea breeze front, which can bring high concentrations down to ground level several tens of kilometers downwind of the source.

5.1.2 Urban heat island circulation

On a much smaller scale, urban circulations are similar to sea and land breezes in that it is the differential heating and cooling between the rural and urban areas that generates and sustains the urban heat island circulation. Over the cities and suburban areas, such alterations as asphalting, buildings, and the removal of vegetation usually alter the surface heat budget, and therefore the intensity of heat flux to the air. Direct thermal circulations over large metropolitan areas are associated with this differential heating between the city and the surrounding rural areas, when there is a weak large-scale flow.

The urban heat island circulation is essentially caused by the rising of warm air over the city and the sinking of cold air over the surrounding rural area. The near-surface air from rural surrounding areas converges toward the thermal low that forms over the city, while at upper levels the air diverges away from the city. This circulation may extend up to the base of the lowest inversion, which may then acquire a dome shape with the maximum mixing height at the city center.

The difference between the urban and rural air temperatures is called the urban heat island intensity. The maximum difference generally occurs at night. The urban rural temperature differences during the day are much smaller. The intensity of an urban heat island depends on many factors, such as the size and energy consumption of the city, geographical location, time of day, month or season, and synoptic conditions (the ambient atmospheric stability and wind speed). The maximum intensity for a given city occurs on clear and calm nights, a few hours after sunset, during the winter season.

The urban heat island circulations have consequences also for mesoscale dispersion. These consequences can be summarized as an increase in vertical diffusion, mixing depth, and the probability of convective precipitation, especially in summer, and difficulty in prescribing appropriate wind speed and turbulence intensity. Since mixing is generally restricted to the mixed layer below the inversion, pollutants (such as dust, smoke, and haze) from urban emissions tend to accumulate during stagnant conditions. Therefore, the air under the city's dust dome on low wind speed days tend to become increasingly polluted. The dust particles generally accumulate over the city during the evenings and clear away during the daytime because of vigorous mixing. As a result, during the night, the heat released by the ground and building surfaces is trapped within this dust dome causing the temperature to rise further. The effect is further aggravated by the heat released by the particulate matter forming the so-called dust haze (Boybeyi et al. [10]).

5.1.3 Mountain-valley winds

In a region with irregular terrain, local wind patterns can develop due to the diurnal heating and cooling of mountain slopes. A larger diurnal temperature variation usually occurs at the ground, so that during the day the higher terrain becomes an elevated heat source, whereas at night it is an elevated heat sink. In response to this diurnal temperature variation, thermal circulations often develop along mountain slopes. These circulations are known as mountain and valley winds and are most pronounced on clear summer nights and days when the large-scale gradient winds are weak.

During the night, the mountain slopes cool more quickly than the valley floor. The air adjacent to the slopes cools through conduction and turbulent mixing. This air would be cooler than the air at the same elevation over the valley floor. Thus, gravitational flow begins and the cooler dense air flows down the slope. Such air movement is referred to as a downslope wind or mountain wind. These winds are usually very shallow (2 to 20 m) and have velocities of the order of 1 to 5 m/s. Continued downslope winds fill the valley with cool air. The downslope mountain winds begin shortly after sunset, strengthen throughout the night, and reach their peak strength just before sunrise.

During the day, solar radiation warms the mountain slopes or valley walls, which in turn warm the air in contact with them. Due to convective mixing, air gets heated up to several hundred meters above the sloping surface. This heated air, being less dense than the air at the same elevation above the valley floor, rises as an upslope wind and is called a valley wind. The valley winds begin early in the morning shortly after the beginning of sunshine on the slope, reach a maximum speed by midday, and die down before sunset. Valley winds tend to be deeper than mountain winds, because the heating of the ground by the sun is mixed upward effectively by turbulent heat fluxes. At night, radiational cooling predominates if the winds are light and the resultant perturbation flow field is more shallow.

In general, mountain-valley winds are best developed when large-scale gradient winds are weak. Under such conditions, air pollutants will follow the local circulation once they are trapped in it. For example, during the night, strong downslope winds cause efficient dispersion of pollutants from sources on their top and downwind bases. The presence of valley inversion, on the other hand, can trap the pollutants and lead to high concentrations. This can be hazardous for people, animals, and plant life on the slopes.

5.1.4 Lee waves

A commonly observed phenomenon in stably stratified flows over hills and other obstacles are lee waves, which are produced by disturbance of the stratified flow by an obstacle in the vertical direction. When an air parcel is forced by the obstacle to rise, a restoring force due to gravity tends to bring it back to its original (equilibrium) level. In fact, due to its downwind momentum or inertia force, the parcel tends to overshoot the equilibrium level and is then further subjected to the buoyancy force to raise it back toward the original level. In other words, particles of air experience a vertical oscillation in a stratified airstream.

Primarily, the occurrence of mountain lee waves depends on atmospheric stability and ambient wind speed. A strong stability at levels comparable to the mountain height is a necessary requirement for lee wave development. However, the characteristics of the lee waves are strongly controlled by the winds. For example, in light winds, perpendicular to the mountain, the flow is characterized by a smooth shallow solitary wave over the ridge. With somewhat stronger winds, a larger amplitude solitary wave usually forms with a standing eddy of recirculating flow underneath on the lee side of the mountain. With still stronger winds increasing with height, rotors can form near the surface under the crest of large amplitude lee waves. But a more favorable condition for the rotor development is a strong wind maximum extending through a limited vertical depth comparable with the height of the mountain, followed by decreasing wind aloft.

These large amplitude lee waves can produce several weather phenomena that significantly affect human activity and therefore require the attention of the weather forecaster. For example, the strong downslope winds observed along the lee slopes of mountain barriers are usually associated with large amplitude waves. Dangerous regions of clear air turbulence are also produced by these waves. Distinctive orographic clouds can be formed in the wave crests by these waves. If the air is close to saturation, and is lifted to higher levels by flow in the wave crests, the saturation level can be reduced to below the local concentration. This produces water droplets (condensation), and hence forms local clouds. This condensation process occurs very rapidly, in a time scale of a few seconds, and the associated cloud pattern is the primary means of visualizing waves produced by flow over topography in the atmosphere. When the air again descends below the level for condensation, evaporation of the water droplets may occur as rapidly as their formation.

Flow separation can also occur on the lee slope of the hill under certain conditions. This separation of flow from the hill surface is actually induced by lee waves and their associated rotor-like vortices near the surface. These wake flows are characterized by weak and variable winds and high turbulence intensities. They are typically confined to downwind distances of a few hill heights. At large distances, the wake expands in both the lateral and vertical directions, while the hill-induced flow perturbations decay with increasing distance from the hill. Eventually, the wake relaxes back to the undisturbed approach flow and may not be distinguished as a separate entity. Since flow reverses and recirculates in the region of flow separation, pollutants from any source in that region would be trapped, resulting in very high concentrations on the lee-side slope and base of the hill with a relatively slow escape to the main flow outside. At higher levels, lee waves may break and produce turbulence to cause efficient mixing.

5.2 Internally-generated mesoscale circulations

Atmospheric motion is largely a result of the balance between inertia, pressure gradient, Coriolis force, and frictional force. Since a simple geostrophic balance (like that of large-scale flows) does not exist in mesoscale atmospheric flows, the atmospheric inertia mode becomes one of the important responsible mechanisms for the development of mesoscale circulations such as gravity waves, fronts and jet streaks, buoyant instability, synoptic instability, Kelvin-Helmholtz instability waves, horizontal roll vortices, quasi-stationary convective events, and mesoscale structure of hurricanes. Here, only a brief discussion on gravity waves, mesoscale instabilities and horizontal roll vortices will be given because of their importance for mesoscale atmospheric dispersion.

5.2.1 Gravity waves

Over mesoscale travel times and distances, density stratification in the atmosphere plays an important role in both flow dynamics and dispersion of pollutants. For example, a large number of processes (such as flow over orography, rising convective elements, shear instability, and frontal systems) can generate internal gravity waves due to the presence of density stratification. Atmospheric gravity waves are the simplest and most fundamental motions on the mesoscale, and can only exist when the atmosphere is stably stratified (so that a fluid parcel displaced vertically will undergo buoyancy oscillations). Gravity waves essentially propagate mesoscale free oscillations. In a fluid which has no upper boundary, such as the atmosphere, gravity waves may propagate vertically as well as horizontally.

Gravity waves transport momentum and energy from one region of the atmosphere to another. The so-called spectral gap in the mesoscale at tropospheric levels may reflect this ability of gravity waves. Gravity waves are also the predominant cause of the clear-air turbulence encountered by high-altitude aircraft flying in stable air. Some of this clear-air turbulence is generated locally, but some is generated from below by convective storms, which impulsively force oscillations of the stable environment as they develop. Finally, gravity waves trigger mesoscale instabilities that lead to the development of severe weather that subsequently strongly modifies the ambient environment. They also interact strongly with one another, and, in the process, drive major spectral energy transfer in the atmosphere.

The gravity waves can influence the atmospheric dispersion of pollutants in several ways. For example, their presence may result in mesoscale fluctuations in wind speed and direction depending upon the direction of propagation of the wave relative to the mean wind. Such periodic variations in wind direction may cause plume meander. These waves can also trigger flow instabilities such as Kelvin-Helmholtz instability and convective instability. These mesoscale flow instabilities produce isolated patches of turbulence, even in the interior of stably stratified fluids. Thus, gravity waves can also influence the dispersion of pollutants by increasing turbulence intensity levels.

5.2.2 Mesoscale instabilities

The presence of wind shear further complicates the flow dynamics and the dispersion of pollutants over mesoscale travel times and distances. Many flow instabilities occur partly as a result of the presence of wind shear. These mesoscale instabilities influence the dispersion of pollutants by increasing turbulence intensity levels through the flow instabilities. For example, three kinds of instability can grow in a horizontally uniform flow with vertical wind shear and/or buoyancy.

These instabilities occur over scales of tens to thousands of meters and produce much of the small-scale turbulence observed in the troposphere.

The first type of instability is a pure buoyant instability. Adiabatic oscillations of a fluid parcel about its equilibrium level in a stably stratified atmosphere are referred to as buoyant instability. The second type of instability is the inertial-buoyancy type called symmetric instability. Symmetric instability, in its moist form, occurs over scales of tens to hundreds of kilometers and may be the immediate cause of many of the rain and snow bands usually associated with warm and occluded fronts. The third type of instability is the shearing type instability. Most clear air turbulence and even some of the turbulent motions close to the ground are the result of small-scale wavelike undulations, which develop spontaneously when the vertical shear of the horizontal flow exceeds some critical value. The process responsible for the development of these waves is known as shear instability or Kelvin-Helmholtz instability.

Furthermore, wind shear and/or buoyancy play an important role in generation of atmospheric boundary layer turbulence. For example, at night under stable conditions, the formation of a nocturnal surface inversion due to long-wave radiational cooling decouples the nocturnal boundary layer from the remainder of the well-mixed daytime planetary boundary layer. In the nocturnal boundary layer, the wind speed decreases and the wind direction backs due to the new force balance between the pressure gradient force, the Coriolis force, and a weakened friction force. These changes in the wind direction and speed may result in pollutants at different levels being advected at different speeds (speed shear) or in different directions (directional shear). When morning arrives and the unstable daytime planetary boundary layer begins to grow, the shear-distorted pollutants mix in the vertical. Vertical turbulent mixing acts to reduce vertical wind shear and shear enhanced dispersion. Nevertheless, vertical shear will still be present, mainly due to surface friction and baroclinicity, and its interaction with vertical diffusion results in simultaneous mixing.

5.2.3 Horizontal roll vortices

During conditions of combined surface heating and strong winds, weak horizontal helical circulations can form in the boundary layer. These circulations are called horizontal roll vortices and consist of clockwise and counterclockwise pairs of helices with their major axes aligned almost parallel with the mean wind. The depth of these rolls equals the mixed layer depth, and the ratio of lateral to vertical dimensions for a roll pair is about 3 to 1. Tangential velocities around the helices are usually less than 1 m/s, although faster velocities occasionally occur. However, these velocities sweep other convective plumes and thermals into rows, creating along-wind rows of strong updraft. If sufficient moisture is present, cloud streets can form along these updraft rows. Rolls are frequently observed during cold-air advection over warmer bodies of water, and are strongly associated with air-mass modification. Rolls are also common in the low-level jet ahead of cold fronts, and can occur between pairs of closed isobars of warm season anticyclones.

Horizontal roll vortices have an appreciable influence on a pollution plume. Of particular interest is the case in which a pollution plume is forced to the ground by descending air. Horizontal roll vortices do not generally occur unless the mean wind speed is approximately 10 m/s. These strong winds would tend to dilute pollutants so that, over an extended period, the effect of a horizontal roll circulation would not be detrimental to the overall air quality of a point some distance downwind from the source. This hypothetical receptor downwind from the source could, however, receive high concentrations over short periods of time. Stationary horizontal roll vortices with respect to a source could cause a significant reduction in the overall air quality at a downwind receptor for extended periods.

5.3 Mesoscale convective systems

Mesoscale convective systems are formed more or less randomly from convection over heated land and water surfaces, or they spawn from larger storms and weather systems. Consequently, they may move with the large-scale flow and the mother storm to which they are attached. These moving mesoscale systems include frontal circulations, individual convective clouds, extratropical squall lines, mesoscale convective complexes, tropical cyclones, mesoscale rainbands, and tornadoes.

In general, mesoscale convective systems have a very dramatic effect on the boundary layer and its contents. They represent a very real breakdown of the normal structure. Usually, a strong exchange of air between the boundary layer and the free atmosphere takes place in the presence of these systems. This implies not only a rapid loss of boundary layer pollution to other levels, but also removal of the soluble pollutants by scavenging processes.

In other word, pollutants can be vented out of the mixed layer by deep penetrative convection, as well as by upward motion associated with mesoscale convective systems. The vertical transport and mixing in the free atmosphere above the planetary boundary layer are confined to rather limited areas of upward motion and turbulence associated with the convective systems. At the same time, the pollutants can be scavenged by clouds and rain and subsequently wet deposited on the ground. Thus, synoptically induced mesoscale systems serve more to remove the pollutants from the atmosphere and to dilute their concentrations.

On the other hand, large-scale synoptic systems such as fronts, cyclones, and anticyclones with synoptically-induced mesoscale systems attached to them can also effect the dispersion of pollutants released in the lower part of the atmosphere, at time scales of the order of a day or longer. These large-scale synoptic systems further tend to spread the pollutants over the much greater horizontal extents of their circulations.

5.3.1 Convective clouds

It is only over the relatively warm and humid regions of the earth that convective clouds commonly grow and assume major importance as rain producers. The primary energy driving a convective cloud development is the buoyancy experienced by updrafts as latent heat is released as vapor condenses to form cloud droplets, supercooled drops freeze, and vapor deposits on ice crystals. The buoyancy, which is determined by the temperature excess of an updraft relative to its environment multiplied by the acceleration due to gravity, is a local measure of the acceleration of the updraft. The actual magnitude of the updraft strength at any height in the atmosphere is largely determined by the integrated buoyancy that an updraft experiences as it ascends from cloud base to a given height in the atmosphere. The integrated buoyancy through the atmosphere is called convective available potential energy. In general, the greater the convective available potential energy, the stronger the updrafts in a convective cloud.

In the initial stage, a convective cloud consists entirely of a warm, buoyant plume of rising air. The updraft velocity increases rapidly with height within the cloud and there is considerable entrainment through the lateral cloud boundaries. The top of the cloud moves upward with a velocity of the order of 10 m/s. Because of the large updraft velocities, supercooled raindrops may be present well above the freezing level.

In the mature stage, the convective clouds are characterized by a vigorous downdraft circulation, which coincides with the region of heaviest rain. This downdraft circulation is initiated by the downward frictional drag force induced by the drops. Dry environmental air entrained into the downdraft circulation and unsaturated air below the cloud base are cooled by evaporation of falling precipitation. In some cases the resulting evaporative cooling is capable of greatly enhancing the negative buoyancy of the downdraft circulation. In the mature stage, supercooled raindrops still exist well above the freezing level in the downdraft. The maximum upward vertical velocities are in the middle of the cloud, with detrainment above that level. The top of the cloud approaches the tropopause and begins to spread out.

As precipitation develops throughout the cloud, the downdraft circulation gradually becomes more extensive until in the dissipating stage it occupies virtually the entire cloud. Deprived of a source of supersaturated updraft air, cloud droplets can no longer grow and as a consequence precipitation soon ceases and the cloud evaporates.

When pollutants are emitted at the bottom of the mixed layer they are quickly drawn by the convergence zones into the rising updraft. Pollutants emitted above the surface layer move downward on average because downdrafts comprise a larger area than updrafts, and because the updraft air is quickly recycled back down. Eventually, the pollutants become uniformly mixed in the vertical regardless of their source altitude, as expected in the mixed layer. In this case, pollutants transported up from the surface by thermals are trapped beneath the capping inversion. Certain convective clouds can break though the stable environment zone to vent pollutants and moisture out of the boundary layer into the free atmosphere.

5.3.2 Frontal circulations

Synoptic-scale fronts may be defined as sloping zones of pronounced transition in thermal and wind fields. Consequently, they are characterized by a combination of relatively large horizontal temperature gradients, static stability, absolute vorticity (horizontal wind shear), and vertical wind shear. They are also directly related to a horizontal temperature gradient averaged over a depth in the atmosphere (e.g., between 1000-850mb, 1000-500mb). Such a mean temperature gradient is referred to as thickness and results directly from the hydrostatic assumption.

When the horizontal thickness gradient associated with a front has been steady for a long period of time, the wind field closely approximates gradient wind balance. By definition, these fronts are not mesoscale features. When the thickness distribution changes with time, however, significant nongradient winds are generated as the adjustment toward a new balanced state begins. Williams [11], for example, has suggested that cold frontogenesis is created from the large-scale baroclinic field by the divergent component of the wind. Such frontal circulations are mesoscale and are an integral component in frontogenesis and cyclogenesis. These mesoscale frontal circulations are closely related to the convective bands imbedded in stratiform cloud systems.

5.3.3 Squall lines

In midlatitude cyclones and along synoptic-scale fronts, cumulonimbus towers are often arranged in long lines in which adjacent elements are so close together that this alignment is regarded as a squall line. Squall lines are highly variable in space and transient in time. Precipitation along a squall line is usually not uniformly distributed, but occurs in well-organized mesoscale sized bands of heavier rain. The squall lines are often accompanied by gusty surface winds and move with a speed characteristics of the winds in the middle troposphere.

Squall lines often form in association with synoptic weather features such as cold and warm fronts, dry lines (Homan and Vincent [12]), and tropical waves (Fortuna [13]), although they typically travel at a greater speed than these larger-scale weather phenomena. Squall lines may also be initiated through differential temperature gradients in the boundary layer interacting with the synoptic flow in an analogous manner to that causing sea and land breezes (Sun and Ogura [14]).

The vast majority of squall lines, however, form usually in a convectively unstable environment in which there is considerable vertical wind shear between the low-level flow and upper troposphere, and their appearance is strongly influenced by factors such as the occurrence of early morning cloudiness. In such an environment, convective precipitation associated with squall lines persists for long periods of time in their mature stage without destructive interference between updrafts and downdrafts.

5.3.4 Mesoscale convective complexes

A type of large, long-lived convective weather system that frequently occurs over the middle latitudes of the Unites States was identified and discussed by Maddox [15]. These systems are defined as mesoscale convective complexes (MCCs) and appear to originate over and downstream of the Rocky Mountains and to propagate from west to east. MCCs often produce very heavy rains and widespread damaging winds at night during the summer.

Maddox [16] presented composite analyses of the large-scale settings attending 10 MCCs. The important aspect of the analyses is that many MCCs result from mergers and interactions between groups of storms that develop in different locations. Some of the MCCs are initially squall lines that gradually acquire MCC characteristics as they persist and grow in size. This often happens on the southern, or trailing portion of squall lines where the associated (or triggering) cold fronts often weaken and become slow moving. As the warm season progresses, the favored region of MCC occurrence shifts slowly northward so that by July and August MCCs primarily affect the north-central states. Regardless of the time of year, MCCs are relatively rare along the east coast and west of the Rocky Mountains.

6 Summary

Mesoscale atmospheric flows possess various mesoscale time and space scales resulting from the diurnal cycle, the atmospheric inertial mode, thermal and mechanical forcing due to mesoscale terrain inhomogeneities, and downscale energy transfer from the synoptic-scale to the mesoscale due to nonlinear flow interactions. Even for horizontally homogeneous flows over flat and uniform terrain, mesoscale frequencies such as the diurnal heating cycle and formation of a nocturnal low-level jet will usually be present.

This wide range of mesoscale circulations and their associated mesoscale vertical ascents and descents can have a significant impact on the dispersion of pollutants. The transport and diffusion of atmospheric pollutants can be affected through variations in the mean transport wind, differential advection due to vertical and horizontal wind shear, and vertical mixing. In addition, large-scale baroclinicity and large-scale vertical motions affect the local structure and the dynamics of the planetary boundary layer. Latitude affects both the inertial period and diurnal forcing. Time of day affects the dynamics of the boundary layer, while time of year affects the diurnal forcing.

Therefore, these mesoscale space-time flow scales should be represented accurately when the mesoscale dispersion processes are studied. Otherwise, significant components of mesoscale transport and diffusion may be neglected. Thus, knowledge of mesoscale systems is necessary for the air pollution meteorologist.

References

[1] Kolmogorov, A. N., The local scale structure of turbulence in incompressible viscous fluids for very large Reynolds numbers. *C. R.. Acad.. Sci. URSS*, **30**, pp. 301-305, 1941.

[2] Randerson, D., Atmospheric Science and Power Production, USDOE Report TIC-27601. Available from National Technical Information Service, Springfield, VA, 1984.

[3] Orlanski, I., A rational subdivision of scales for atmospheric processes. *Bull. Amer. Meteor. Soc.*, **56**, pp. 527-530, 1975.

[4] Moran, M. D., *Numerical Modeling of Mesoscale Atmospheric Dispersion*. Ph.D. dissertation, Colorado State University, Fort Collins, Colorado, pp. 758, 1992.

[5] Lilly, D. K., Stratified turbulence and mesoscale variability of the atmosphere. *J. Atmos. Sci.*, **40**, pp. 749-761, 1983.

[6] Atkinson, B. W., *Meso-scale Atmospheric Circulations*. Academic Press, New York, pp. 495, 1981.

[7] Ray, P. S., *Mesoscale Meteorology and Forecasting*. American Meteorological Society, Boston, pp. 793, 1986.

[8] Arya, S. P., *Air Pollution Meteorology and Dispersion*. Oxford University Press, New York, pp. 310, 1999.

[9] Boybeyi, Z. and Raman, S., A three-dimensional numerical sensitivity study of convection over the Florida peninsula. *Boundary Layer Meteor.,* **60**, pp. 325-359, 1992.

[10] Boybeyi, Z., Raman, S. and Zannetti, P., A numerical investigation of the possible role of local meteorology in the Bhopal gas accident. *Atmospheric Environment*, **29***,* pp. 479-496, 1995.

[11] Williams, R. T., Atmospheric frontogenesis: A numerical experiment. *J. Atmos. Sci.*, **29**, pp. 3-10, 1967.

[12] Homan, J. H. and Vincent, D. G., Mesoscale analysis of surface variables during the severe storm outbreak of April 10-11, 1979. *Mon. Wea. Rev.*, **111**, pp. 1122-1130, 1983.

[13] Fortuna, M., Properties of African squall lines inferred from time-lapse satellite imagery. *Mon. Wea. Rev.,* **108**, pp. 153-168, 1980.

[14] Sun, W.-Y. and Ogura, Y., Boundary layer forcing as a possible trigger to a squall line formation. *J. Atmos. Sci.*, **36**, pp. 235-254, 1979.

[15] Maddox, R. A., Mesoscale convective complexes. *Bull. Amer. Meteor. Soc.*, **61**, pp. 1374-1387, 1980.

[16] Maddox, R. A., Large-scale meteorological conditions associated with mid latitude mesoscale convective complexes. *Mon. Wea. Rev.*, **111**, pp. 1475-1493, 1983.

Chapter 2

Basic aspects of mesoscale atmospheric dispersion

Michael D. Moran
Air Quality Research Branch, Meteorological Service of Canada
Downsview, Ontario, M3H 5T4, Canada

Abstract

If pollutant sources are strong, numerous, or long-lasting enough, or if the pollutants themselves are sufficiently toxic or active in some way, then pollutant concentrations may still be of concern even after mesoscale dilution. The multi-country radionuclide contamination experienced during the 1986 Chernobyl disaster is one example. Others include regional-scale acidic deposition, volcanic plume dispersion, photochemical smog, airborne pathogen dispersal, and heavy-metal and pesticide accumulation in remote areas.

Mesoscale atmospheric dispersion (MAD) is even more complicated than shorter-range dispersion because the mean wind field can seldom be considered steady or horizontally homogeneous over mesoscale time and space scales. Wind shear also plays a much more important role on the mesoscale; horizontal dispersion can be enhanced and even dominated by vertical wind shear through either the simultaneous or the delayed interaction of horizontal differential advection and vertical mixing over one or more diurnal periods. In addition, concepts as fundamental as turbulence and diffusion are situational, not absolute, and must be reconsidered when applied to MAD. Important characteristic time scales for MAD include the release, sampling, and travel times and the diurnal and inertial periods.

1 Introduction

Atmospheric dispersion is the irreversible process by which a cloud of tracer or pollutant expands, mixes, and is diluted by atmospheric motions acting on the cloud (unless otherwise noted, the word 'cloud' will be used herein to refer to a cloud of tracer or pollutant material and not to a cloud of water droplets or ice crystals). It may be thought of as the result of two component processes, *advection* and *diffusion*. The advective or transport component can in turn be further divided into two subcomponents: (a) *whole-body advection*, which determines the trajectory of the center of mass of the tracer cloud; and (b) *differential advection*, in which the tracer cloud is deformed and even fragmented as a result of horizontal and vertical shear (Pielke et al. [1]). Oceanographers studying oceanic dispersion often refer to differential advection and diffusion as *stirring* and *mixing*, respectively (e.g., Eckart [2]).

Differential advection (or stirring) increases the areal extent of the interfacial region between the cloud and the surrounding atmosphere. Figure 1 shows an example of two-dimensional synoptic-scale stirring. Diffusion (or mixing) then acts irreversibly on the concentration gradients in the interfacial region, resulting in weaker gradients and lower concentration maxima (e.g., Garrett [3]). Such a separation of dispersion into two separate processes is somewhat artificial, however; for example, the *interaction* of differential advection with turbulent diffusion results in enhanced diffusion or mixing. Csanady [4, p. 137] suggested the term *turbulent dispersion* to distinguish such augmented diffusion from "pure" turbulent diffusion. Following Csanady and Pielke et al., the term *turbulent diffusion* will be used in this chapter to refer to pollutant dispersal due to unresolved or subgrid-scale motions, and the term *turbulent dispersion* will be used to denote dispersal due to both turbulent diffusion and differential advection.

Diffusion is itself a macroscopic concept, a statistical description of unresolved, small-scale advective transport (e.g., Hinze [6, Sec. 1.9]). *Molecular* diffusion results from the aggregate movements of individual molecules. Like other quasi-continuous macroscopic

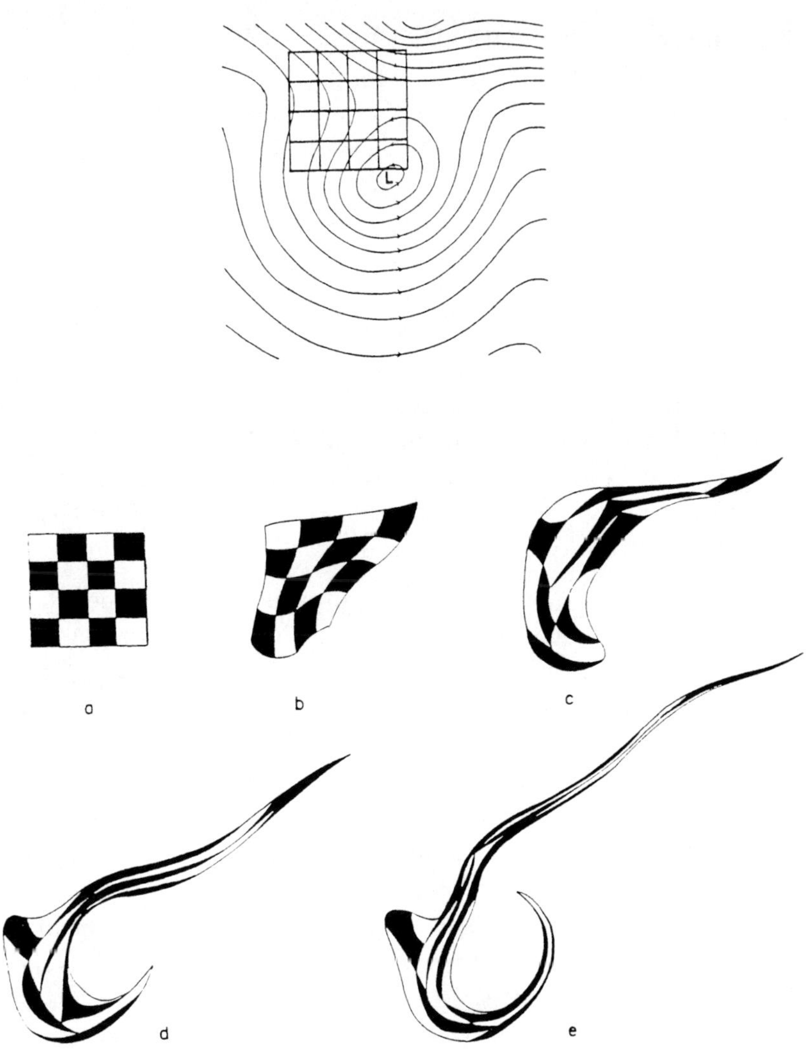

Figure 1: Horizontal deformation of a large-scale parcel by two-dimensional incompressible flow at 500 mb as predicted by a barotropic numerical model: (a) 0 h; (b) 6 h; (c) 12 h; (d) 24 h; and (e) 36 h. The initial streamline pattern is shown at the top. The sides of the shaded grid squares are 300 km long (from Welander [5]).

fluid properties of statistical mechanics such as pressure, density, and temperature, molecular diffusion is defined based on an *average* over a large number of molecules in a small volume and, like pressure, density, and temperature, it can be measured directly without knowledge of the details of molecular motions. Much of our present theory of turbulence is based on an analogy to statistical mechanics and the kinetic theory of gases. *Turbulent* diffusion thus corresponds to the mean advective transport due to the aggregate contribution of numerous individual turbulent velocity fluctuations. As in the molecular case, the details of individual turbulent velocity fluctuations are not known; rather, our description is confined to only a few of the statistical properties of the turbulent fluctuations. The distinction between the processes of advection and turbulent diffusion thus depends critically upon the specification of the *averaging operator* and the *averaging scale* (see Section 4).

By far the majority of studies of atmospheric dispersion to date have been concerned with the *local* scale, that is, dispersion within a few kilometers of a source. Atmospheric turbulence as defined for averaging periods on the order of 30–60 minutes is the dominant dispersion mechanism over these distances. More recently, a number of studies have dealt with the *long-range* transport and diffusion of air pollutants over regional-scale, synoptic-scale, and even global-scale distances. Synoptic-scale "eddies" such as extratropical cyclones constitute the primary mechanism for dispersion on these scales (e.g., Kao [7]; Gifford et al. [8]; Maryon and Buckland [9]). Mesoscale atmospheric dispersion (MAD) falls between these two regimes. Small-scale turbulent diffusion contributes to MAD but so too do mesoscale temporal and spatial variations in the mean wind field caused by changing surface characteristics, diurnal forcing, and large-scale atmospheric dynamics (see previous chapter).

A fundamental problem arises, however, in trying to apply the theoretical and analytical framework developed for the study of small-scale atmospheric turbulence and diffusion (e.g., Frenkiel [10]; Batchelor [11]; Monin and Yaglom [12]; Tennekes and Lumley [13]; Csanady [4]; Pasquill and Smith [14]) to MAD. Practical applications of this framework usually assume steady, horizontally homogeneous mean flows and statistically stationary, horizontally homogeneous turbulence as defined for time and space scales on the order of 30–60 minutes and 5–20 km. Since atmospheric flows are seldom steady or horizontally homogeneous at larger time and space scales, and since the mesoscale energy spectrum differs from the small-scale energy spectrum (Section 7), it is not obvious how much of this theory and the implicit and explicit assumptions about short-range atmospheric dispersion that have proven useful will also hold on the mesoscale. Moreover, if small-scale turbulence contributes to small-scale turbulent diffusion, it is then natural to ask whether mesoscale "turbulence" also exists and, if so, whether it also contributes to MAD and whether it can be parameterized and modeled?

All of these issues suggest a need to return to first principles in considering MAD. For example, what are appropriate averaging operators and scales for MAD; what is a mesoscale flow realization; how is an ensemble average defined in an unsteady flow; what are the roles of release time, travel time, sampling time, and averaging time in MAD; and what are the differences between mesoscale relative diffusion and mesoscale absolute diffusion?

In the remainder of this chapter, I will present some examples and observations of MAD (Section 2), compare and contrast MAD with shorter-range dispersion (Section 3), review some basic concepts from the statistical theory of turbulence and diffusion as applied to the mesoscale (Sections 4 and 6), consider puffs and plume releases and absolute and relative diffusion on the mesoscale (Section 5), discuss the roles of characteristic time and space scales for mesoscale flows and dispersion (Section 6.3), consider mesoscale kinetic-energy (KE) spectra (Section 7), and describe the important contribution of shear effects to MAD (Section 8).

2 Examples of MAD

The smoke trails from cities have been observed by aviators to be hundreds of miles long. If aviators would also take note of the horizontal breadth of the trail at various

> *distances from the source, and of the speed of the mean wind, it might be possible to extract a measure of the horizontal diffusivity.*
>
> L.F. Richardson [15, p. 220]

Early studies of atmospheric dispersion after World War I focused almost exclusively on local dispersion, motivated by civilian concerns over the effects of industrial air pollutants on public health and by military concerns over the impact of smoke and poison gases on the battlefield. The conventional wisdom of the time held that since the atmosphere's capacity to dilute air pollutants is effectively limitless, maximum ground-level concentrations and fluctuations close to the pollutant source should be the main concern of the air pollution meteorologist. For pollutants transported over mesoscale or larger distances, it was argued, sufficient dilution will have taken place for there to be little practical need to know pollutant concentrations at these distances.

In the last few decades, however, there has been increasing recognition that in situations where (a) pollutant sources are either strong, numerous, or long-lasting enough or (b) where the pollutants themselves are sufficiently toxic or active in some way, e.g., bioaccumulative or catalytic, then pollutant concentrations or dosages even far from the source may be of concern. The multi-country radionuclide contamination experienced during the 1986 Chernobyl disaster in the Soviet Union is one recent example of mesoscale and synoptic-scale pollutant transport (e.g., Persson et al. [16]; Smith and Clark [17]). Earlier significant releases and transport of radionuclides occurred during the Windscale reactor accident in England in 1957 and during numerous atmospheric tests of nuclear devices in the 1950s (e.g., Crabtree [18]; Feely et al. [19]; List et al. [20]; Reiter [21]).

Regional-scale (i.e., meso-α-scale: 200–2000 km) acidic deposition, ozone, and haze episodes are the result of the transport and mixing over long distances of chemical compounds emitted by numerous large and small sources (e.g., Lyons [22]; NRC [23]; Pudykiewicz et al. [24]; Ackermann et al. [25]). Even remote regions of the planet without local sources of pollutants have been affected by air pollution in recent years. For example, present-day air pollution impacts in the Arctic and Antarctic include wintertime haze episodes, diminution of polar stratospheric ozone concentrations, and uptake and accumulation of pesticides and other chemicals by Arctic flora and fauna (e.g., Stolarski [26]; Trivett et al. [27]; Maenhaut et al. [28]; Barrie [29]). The behaviour of urban-scale plumes downwind of large, industrialized cities is another example of MAD (e.g., Aguado [30]; Lamb et al. [31]; Trainer et al. [32]).

Significant mesoscale transport of natural aerosols also occurs frequently as a result of volcanic eruptions, forest fires, and sandstorms (e.g., Crabtree and Kitchen [33]; Chung [34]; D'Almeida [35]; Segal et al. [36]; and Ferrare et al. [37]). Biological organisms, including pollens, seeds, bacteria, spores, viruses, and harmful insects such as locusts and spruce budworms, can also be transported over mesoscale or greater distances (e.g., Pedgley [38]). Some more references to other examples of MAD may be found in Moran and Pielke [39].

2.1 Observational studies

Observations of MAD underpin much of our current understanding of this phenomenon and have helped to motivate theoretical and numerical studies. A variety of Lagrangian marker systems can be considered to observe MAD, but basically these all fall into one of two main categories: *balloons* and *tracer gases.* Relatively few observations of MAD exist, however. There are a number of reasons for this, including (a) a lack of interest early on by both the scientific community and funding agencies, but more importantly, (b) a lack of suitable technologies for observing atmospheric dispersion over larger scales, and (c) the high costs and organizational and logistical demands associated with mounting a MAD field experiment. A well-designed MAD field program requires coordinated and extensive sensor networks deployed over a large area to measure both meteorological variables and tracer concentrations or marker locations.

Balloons have been used to study MAD in three ways: (a) in isolated releases as Lagrangian markers of pollutant trajectories for atmospheric flow over complex terrain or long distances (e.g., Angell et al. [40]; Pack et al. [41]; Clarke et al. [42]); (b) in sequential releases to provide one-particle diffusion estimates (e.g., Angell [43];

Thomas and Vogt [44]); and (c) in simultaneous or cluster releases to study relative diffusion (e.g., Angell and Pack [45]; Kao and al-Gain [46]; Er-El and Peskin [47]).

Tracer releases have the advantage over balloon releases of giving a more direct measurement of MAD. Measurements of the dispersion of tracer gases and aerosols over mesoscale distances have been made both for *controlled* releases in formal mesoscale tracer experiments and for *uncontrolled* releases from 'sources of opportunity'. The latter include large industrial point sources, urban plumes, major fires, volcanic eruptions, and accidental and intentional radionuclide releases (e.g., Peterson [48]; Randerson et al. [49]; Millán and Chung [50]; Gillani [51]; Carras and Williams [52, 53]; Hoff and Gallant [54]; Brown et al. [55]; Haagenson and Morris [56]; Bastable et al. [57]; Davies [58]; Randerson [59, 60, 61]; Limaye et al. [62]; Chung et al. [63]; Crabtree and Kitchen [33]; Persson et al. [16]; Wheeler [64]; Heffter [65]; List et al. [20]).

Formal mesoscale tracer experiments, on the other hand, are more effective than uncontrolled releases for studying MAD because of the control they give over the strength, timing, and length of the tracer release. Such control permits the design and deployment of the best possible network of fixed and mobile air samplers and the most appropriate sampling protocol given available resources, including (a) proper transport and storage of samples, (b) stringent quality assurance and quality control of samples, (c) calibration of instruments before, during, and after the experiment, and (d) laboratory checks and intercomparisons (e.g., Dabberdt and Dietz [66]; Ferber et al. [67]; Lagomarsino et al. [68]). Some examples of MAD tracer experiments include the Great Plains experiment (Ferber et al. [69]; Moran and Pielke [39]), the Cross-Appalachian Tracer Experiment (CAPTEX: Ferber et al. [67]), the Øresund experiment (Gryning [70]), and the Across North America Tracer Experiment (ANATEX: Draxler and Heffter [71]; Haagenson et al. [72]).

Heffter [65], Bauer [73], Gifford [74, 75, 76], Draxler [77], and Moran [78] have all summarized observations of boundary-layer, free-tropospheric, and stratospheric mesoscale dispersion over a wide range of scales and from a variety of experiments.

3 Mesoscale vs. short-range dispersion

In general, MAD is an even more complex phenomenon than short-range atmospheric dispersion due to the additional physical processes that may play a role and to the greater number of characteristic time and space scales (e.g., Pasquill [79]). Consider the following differences between MAD and smaller-scale dispersion.

3.1 Unsteady, inhomogeneous mean flow

In the case of near-source or local dispersion, dilution occurs due largely to turbulent diffusion. To a first approximation, both the mean transport wind and turbulence may be considered *steady* and *horizontally homogeneous* although they may vary in the vertical. Under such atmospheric conditions, simple, semi-empirical diffusion models such as Gaussian plume models and similarity-theory models have been applied with some success to small-scale dispersion in flat, uniform terrain and with less success to small-scale dispersion in complex terrain (e.g., Hanna et al. [80]; Pasquill and Smith [14]).

Over mesoscale travel times and distances, however, this approximation of steady, horizontally homogeneous flow no longer holds (e.g., Wendell [81]; Pasquill [79]; Atkinson [82]). Instead, as discussed in the previous chapter, a large number of mesoscale flow modes are possible, with a wide range of accompanying temporal and spatial scales.

3.2 Dominance of horizontal over vertical diffusion

The relative importance of vertical diffusion vs. horizontal diffusion differs between mesoscale dispersion and shorter-range dispersion. Over short distances, say 10 km or less, vertical and horizontal diffusion will be comparable in magnitude under neutral or unstable conditions. Once a tracer has mixed through the depth of the planetary boundary layer (PBL), however, additional dilution can only result from either further horizontal dispersion or from leakage out of the PBL due to such mechanisms as cloud venting, mesoscale venting by fronts or mesoscale convective systems, large-scale convergence, and surface deposition. Under stable conditions, vertical motions will be constrained by buoyancy forces whereas horizontal motions will not be so constrained. Horizontal diffusion will thus dominate both short-range dispersion in stably stratified flows and MAD.

3.3 Importance of shear-enhanced dispersion

Low-level vertical shear is nearly always present in the atmosphere as a result of surface friction and synoptic-scale horizontal temperature gradients. Mesoscale circulations driven by terrain or synoptic forcing may also produce vertical or horizontal shear and differential advection. Horizontal dispersion can be enhanced by vertical shear through either the simultaneous or delayed interaction of horizontal differential advection and vertical mixing. While shear-enhanced dispersion can usually be neglected for short travel distances, as described in Section 8, shear processes will significantly augment and often dominate horizontal diffusion for mesoscale travel distances.

3.4 Contribution of vertical advection

Short-range dispersion models frequently neglect vertical advection. Over mesoscale distances, however, the presence of mesoscale and larger-scale vertical motions associated with convergence and divergence in the horizontal wind field may have a significant impact on dispersion. Mean vertical velocity ($\overline{w}$) values on the synoptic scale can range from 1 $\mathrm{cm\,s^{-1}}$ downward in an intense anticyclone to a few $\mathrm{cm\,s^{-1}}$ upward in extratropical cyclones (e.g., Pasquill [79]). Much larger vertical velocities can occur in thunderstorms and mesoscale convective systems and along frontal surfaces. A vertical velocity of 1 $\mathrm{cm\,s^{-1}}$ will displace a parcel 864 m in 24 hours. Even net vertical transport due to active nonprecipitating convective clouds can exchange significant amounts of PBL and free-tropospheric air, a process known as *cloud venting* (e.g., Dickerson et al. [83]; Ching et al. [84]; Martner and Kropfli [85]; and Vukovich and Ching [86]). Studies of the impact of vertical motions on mesoscale transport and dispersion include Draxler [87], Haagenson et al. [88], Martin et al. [89], Segal et al. [90], Uliasz ([91], and Thompson et al. [92].

3.5 Multi-modal surface concentration patterns

Contrary to the common idealization for shorter-range dispersion (e.g., Gifford [93]; Hanna et al. [80]), multi-modal, non-Gaussian *cross-plume* concentration profiles are often observed at mesoscale travel distances. Carras and Williams [53] have presented plume cross-section plots for an isolated Australian smelter plume for downwind distances ranging from 230 to 1200 km (Figure 2). These plots show significant cross-plume mesoscale structure. What is especially interesting about the Mt Isa measurements is that transport has taken place across fairly flat, homogeneous terrain, suggesting that terrain-forced mesoscale circulations due to mesoscale physiographic inhomogeneities are not responsible for the cross-plume mesoscale structure. This case is discussed further in Section 8.

Another example includes the set of three-hour-average measurements of perfluorocarbon surface concentration made along an arc of samplers 600 km from the release site during the 1980 Great Plains mesoscale tracer experiment. Measurements along this sampler arc for each of the first two 3-hour observing periods showed three relative concentration maxima across the tracer puff (Ferber et al. [69]; Moran and Pielke [39]). Csanady [94] reported measurements of two cross-wind peaks even after travel over distances of only 30 km in some tracer trials carried out in rural Arkansas. He attributed these features to the effect of wind shear and to expected irregularities in individual tracer clouds.

Along-wind multi-modal surface concentrations have also been observed during MAD. In CAPTEX in 1983, some six-hour-average surface concentration fields and also time-composited maximum concentration fields showed two or more concentration relative maxima (Ferber et al. [67, Figs. 2, 4, 11)]. Multi-modal surface concentration patterns were also measured in ANATEX in 1987 (e.g., Draxler [95, Figs. 2–5]).

3.6 Larger Lagrangian integral time scale

Some measurements appear to suggest that the effective Lagrangian integral time scale for MAD is larger than for short-range dispersion.

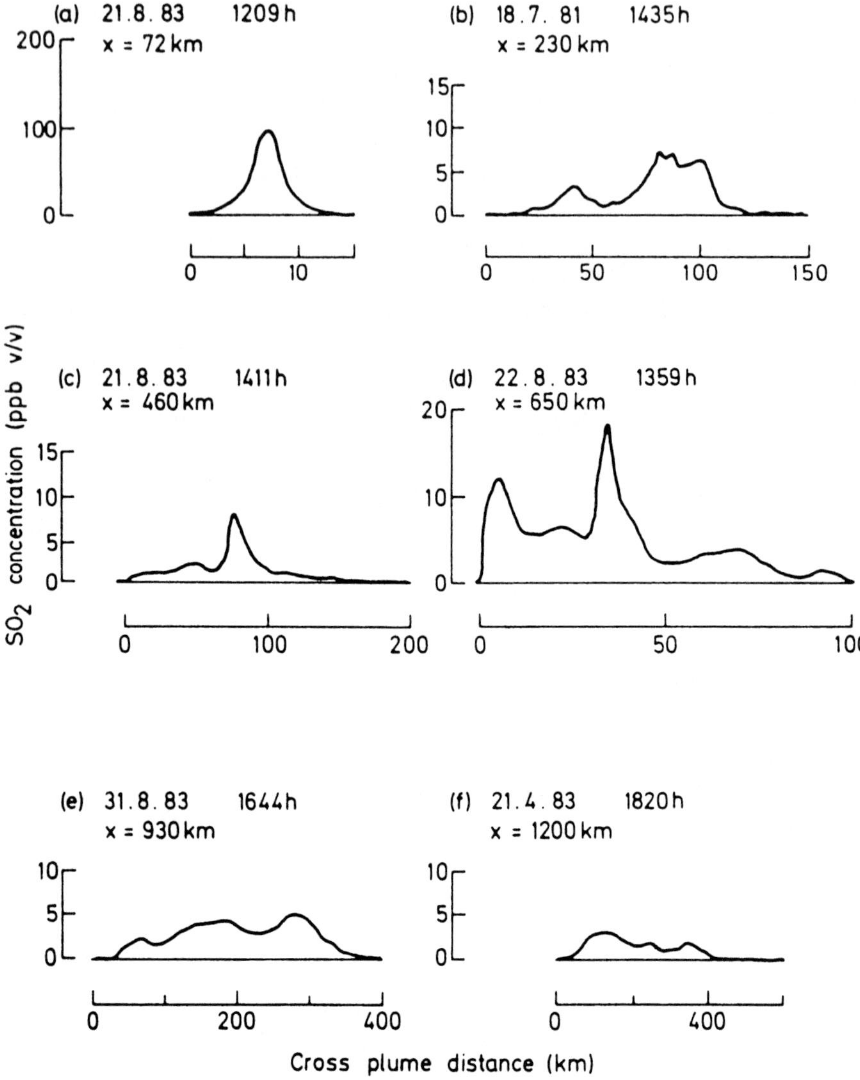

Figure 2: Variations of Mt Isa plume SO_2 cross-wind concentration distributions with distance from source as measured by aircraft (from Carras and Williams [53]).

Suppose that the u-component of velocity $u(t)$ can be decomposed into a mean and a turbulent component (see Section 4):

$$u = \bar{u} + u' \quad , \tag{1}$$

where the overbar denotes an average of some sort over some scale and the prime denotes a fluctuation from that average. For *statistically stationary* and *homogeneous* conditions (i.e., dependent upon separation but not position in time and in space), we can define the temporal autocorrelation function $R(\xi)$ of the u-component of velocity as

$$R(\xi) = \frac{\overline{u'(t)\, u'(t+\xi)}}{\overline{u'^2}} \quad , \tag{2}$$

where ξ is a time lag and $\overline{u'^2}$ is the variance. If the u-component is measured at a fixed point in space, then eqn (2) defines an Eulerian autocorrelation function R_E, whereas if the u-component is measured for a tagged fluid element moving with the flow, then eqn (2) defines a Lagrangian autocorrelation function R_L.

An integral time scale T_I may then be defined for the u-component of velocity by the integral

$$T_I = \int_0^\infty R(\xi)\, \mathrm{d}\xi \quad . \tag{3}$$

T_E is the Eulerian value of T_I (based on R_E) while T_L is its Lagrangian value (based on R_L). T_L is often interpreted as a measure of the longest time during which a tagged fluid element, once started, persists on average in moving in a given direction (e.g., Hinze [6]). It is thus an indication of the parcel's 'memory' of earlier motion.

One important consequence of Taylor's [96] statistical theory of diffusion (see Section 6) is that in *statistically stationary* and *homogeneous* turbulence, the cross-wind concentration standard deviation σ_y of the ensemble-mean plume from a continuous point source will initially grow linearly with travel time T, but then, as T becomes large compared to the Lagrangian integral time scale T_L, say greater than $5T_L$, σ_y will asymptotically approach a square-root dependence on T (Pasquill and Smith [14]). Hanna [97] reported average values for T_L of 60–90 s in the daytime boundary layer. Other estimates for a wider range of stability conditions are higher, in the 200–3000 s range (Gifford [98]). Thus, the parabolic $T^{1/2}$ regime for lateral plume spread could be approached for travel times of less than 5–10 minutes if the appropriate T_L is the PBL convective time scale up to travel times of four hours for the most stable conditions. For a mean wind speed of 10 km h^{-1}, this corresponds to travel distances in the 1–40 km range.

However, observations of a number of mesoscale plumes suggest that the linear time regime for lateral plume spread may persist out to mesoscale distances of 500 km or more (e.g., Pasquill [79]; Panofsky [99]; Clarke et al. [42]; Gifford [75, 76]; Pasquill and Smith [14]; Carras and Williams [100]), which suggests that T_L for MAD is larger than the above estimates by an order of magnitude. Figure 3 shows a composite of three different sets of mesoscale plume observations that illustrate this behaviour (bearing in mind the discussion in Section 6.2.2 about the equivalence at large travel times of absolute and relative diffusion). In addition, many regional-scale Lagrangian-puff dispersion models use Pack et al.'s [41] formula for lateral plume spread, $\sigma_y = 0.5T$, where σ_y is in units of meters and T is in seconds, for all mesoscale travel times (e.g., Heffter [101]; Carhart et al. [102]).

As discussed in Section 8, the augmentation of horizontal dispersion by the simultaneous interaction of vertical shear and vertical diffusion may be responsible for this continued linear dependence during the first few hours. At longer travel times, when the flow can no longer be considered steady, Gifford [75, 104, 98, 105] and Barr and Gifford [106] have argued on the basis of both observations and theory that the Coriolis-force time scale f^{-1} may be the intrinsic and appropriate T_L for MAD, with a value on the order of 10^4 s in mid-latitudes. That is, the Earth's rotation imposes an outer integral time scale for turbulence in atmospheric flows. McNider et al. [103, 107], on the other hand, have suggested that the combined influence of two mesoscale time scales, the diurnal period (24 h) and the inertial period ($2\pi f^{-1}$), can explain the continued linear dependence for lateral plume spread through a different mechanism – delayed shear enhancement – for which Taylor's theorem is not applicable

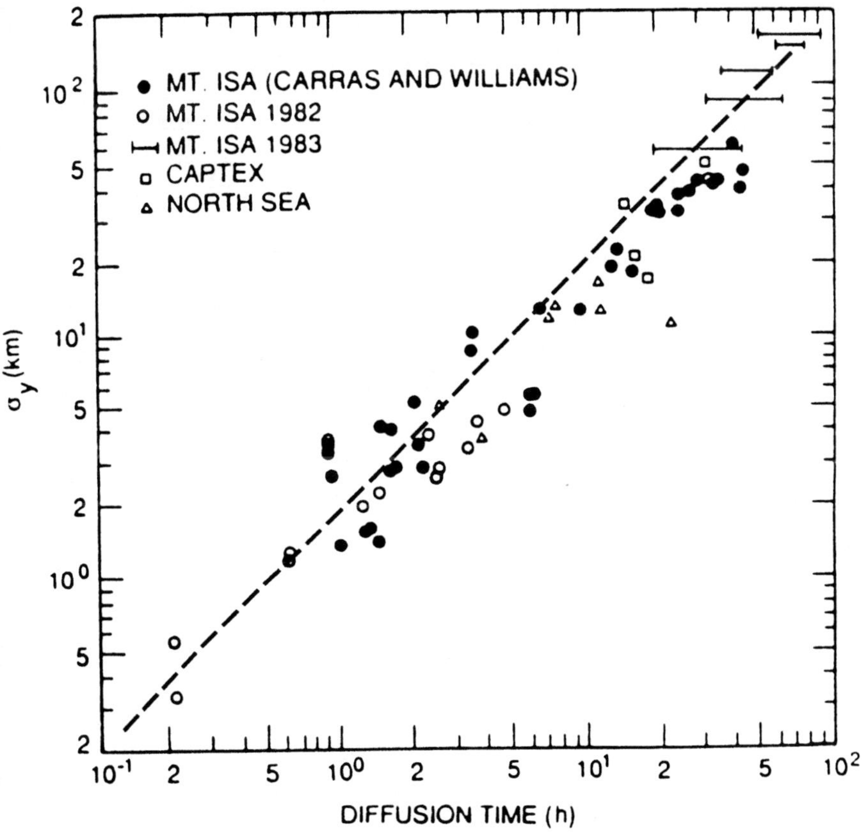

Figure 3: Composite of plume width observations for mesoscale travel times as compiled by Gifford [76] (adapted from McNider et al. [103]). Mt Isa observations were made by aircraft of an isolated sulphide smelter plume in northern Australia during several winters; CAPTEX observations of a perfluorocarbon tracer plume were made by a network of surface monitors in autumn, 1983; the North Sea observations of an SF_6-labelled English power plant plume over the North Sea were made by aircraft in 1980. The dashed line shows the relationship $\sigma_y = 0.5T$ proposed by Pack et al. [41].

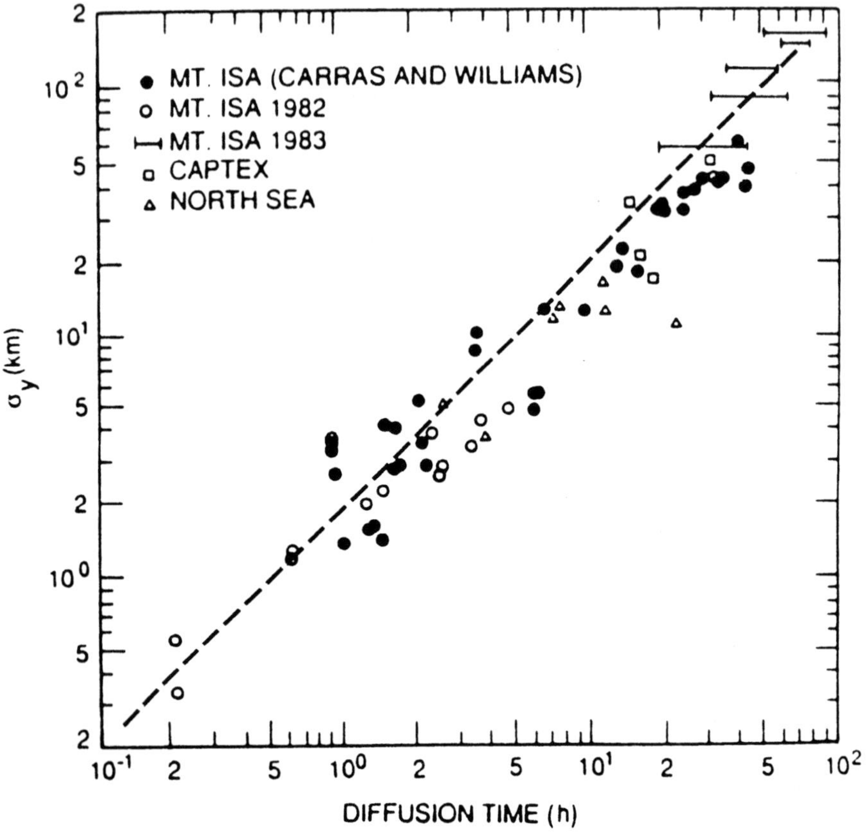

Figure 3: Composite of plume width observations for mesoscale travel times as compiled by Gifford [76] (adapted from McNider et al. [103]). Mt Isa observations were made by aircraft of an isolated sulphide smelter plume in northern Australia during several winters; CAPTEX observations of a perfluorocarbon tracer plume were made by a network of surface monitors in autumn, 1983; the North Sea observations of an SF_6-labelled English power plant plume over the North Sea were made by aircraft in 1980. The dashed line shows the relationship $\sigma_y = 0.5T$ proposed by Pack et al. [41].

(e.g., Onsager [113]; Lee [114]; Ogura [115]; Fjørtoft [116]; Kraichnan [117, 118]; Leith [119]; Lin [120]; Kraichnan and Montgomery [121]; Lilly [122]). One important finding is that the kind of energy cascade towards higher wavenumbers that occurs in three-dimensional flows does not occur in two-dimensional flows because vortex-tube stretching is not possible in two-dimensional flows. In fact, Tennekes [123] has distinguished two-dimensional turbulence from three-dimensional turbulence by referring to the former as *vorticity-advecting* chaos and to the latter as *vorticity-stretching* chaos. Batchelor [124] noted that the two basic properties of turbulence, *randomness* and *nonlinearity*, are present in two- as well as in three-dimensional turbulence. Gifford [125] and Gifford et al. [8], on the other hand, have argued that two-dimensional turbulence, unlike three-dimensional turbulence, conserves area and hence cannot contribute to relative diffusion (Section 5) or to pollutant dilution through the irreversible growth of pollutant-cloud volumes even though it causes rapid cloud deformation and distortion.

4.1.2 Geostrophic turbulence

Although the example of the atmospheric general circulation served to motivate many studies of two-dimensional turbulence in the 1950s and 1960s, Charney [126] pointed out in a much referenced paper that two-dimensional turbulence is not the best model for large-scale atmospheric energetics since baroclinic instability, an important large-scale energy source, requires three-dimensional motions. However, he then described a deeper similarity between three-dimensional *quasi-geostrophic* flows and purely two-dimensional flows. Quasi-geostrophic flows, like two-dimensional flows, are governed by the conservation of two scalar invariants. For quasi-geostrophic flows these invariants are pseudo-potential vorticity, a combination of potential vorticity and potential temperature, and total energy, the sum of kinetic energy and available potential energy. Next, Charney showed that the conservation of pseudo-potential vorticity forbids a downscale energy cascade under certain conditions. Once he had established this analogy, all other theorems applying to energy exchanges amongst spectral components in two-dimensional flows could then be shown to apply to three-dimensional quasi-geostrophic flows as well, including the k^{-3} downscale enstrophy cascade (see Section 7). Finally, Charney derived several additional properties of what he termed 'geostrophic turbulence', including the fact that in the k^{-3} enstrophy-cascade inertial range, the temperature variance will also exhibit a k^{-3} spectrum.

Observational evidence supporting Charney's description of geostrophic turbulence has been published by Baer [127], Chen and Wiin-Nielsen [128], Lambert [129], Boer and Shepherd [130], Nastrom and Gage [131], and Morita and Uryu [132], among others. Follow-on theoretical papers about geostrophic turbulence include those by Rhines [133], Herring [134], Holloway and Kristmannsson [135], McWilliams [136], and Gage and Nastrom [137].

4.1.3 Stratified turbulence

Stratified flows possess other important flow modes, including internal gravity waves (e.g., Hines [138]; Dewan [139]; Hooke [140]) and mesoscale vortical modes (e.g., Lilly [141]; Müller et al. [142]; Lelong and Riley [143]). Internal gravity waves (IGW) are mesoscale atmospheric free oscillations, the simplest and most fundamental motions on the mesoscale. They have very different transport and mixing properties from turbulence: for example, IGW can transport momentum and potential and kinetic energy but to first order cannot contribute to net mass or heat transport. As a result, estimates of transport, diffusion, or dissipation made from measured flow variances may be grossly incorrect if the wrong flow mode is assumed. IGW are also non-local, propagating, long-lived, coherent, weakly linear, and weakly dissipative whereas turbulence is local, non-propagating, short-lived, incoherent, strongly nonlinear, and strongly dissipative (Dewan [139]).

One important problem in the study of stably stratified flows as a consequence concerns the degree to which turbulent motions can be distinguished from IGW when both are present in a flow, as they generally will be and, moreover, when they interact and modify each other (Busch [144]). Stewart [145] considered this identification problem and concluded that a rigorous general solution seems impossible. However, he proposed several experimental criteria that would allow an approximate identification and separation. Other researchers have since contributed to this difficult problem (e.g., Axford [146]; Metcalf [147]; SethuRaman [148]; Hunt et al. [149]; Finnigan [150]).

Lumley [151, p. 101] commented that the difference between a random IGW field and turbulence is in one sense a semantic question that depends critically on our definition of 'turbulence': "if [turbulence] is taken to be any random, three-dimensional velocity field with a continuous spectrum, displaying spectral transfer and dissipation at high wave numbers, then the case of internal waves is also included". A similar consideration motivated Court [152] to propose the term 'undulance' to describe a fluid "in which particles (or 'parcels') oscillate more or less regularly, in contrast to turbulence, in which the motions are irregular, random, or chaotic".

4.1.4 Turbulent 'eddies'

It is also worth noting the confusion that can arise from the use of the term 'eddy' to describe turbulence structure in *both* physical and spectral space (e.g., Gifford [93, p. 86]). In discussions of turbulence spectra, 'eddy' is often used to denote a Fourier component with characteristic time and length scales. Consider, for example, Batchelor [153, p. 133]) on the subject: "Let it be said firmly and right at the beginning, that the word 'eddy' does not refer to any particular local distribution of velocity. It is merely a concise term for a component of motion with a certain length scale, i.e., an arbitrary flow pattern characterized by size alone." In contrast, Corrsin [154, p. 2136]) noted that the 'traditional' usage of the term "implies a spatially local blob of fluid, usually swirling".

To summarize, many atmospheric flow scales display turbulent behaviour. The geometry, kinematics, and dynamics of atmospheric turbulence can vary significantly due to buoyancy effects, and in stably-stratified fluids turbulence will often coexist with other flow modes, some of whose properties are similar. But for pollutant dispersion, as Batchelor and Townsend [155, p. 352] observed, "the most striking property of turbulent motion is its ability to disperse fluid particles that initially are grouped together". In fact, Gifford [105] suggested that a useful definition of turbulence might be motions that cause relative diffusion, and in these terms, two- and three-dimensional turbulence differ fundamentally.

4.2 Role of averaging

The various definitions of turbulence given above all included the use of averaging: in fact the characteristics of the averaging operator determine in part the partitioning between mean and turbulent components. For example, Stommel [156, p. 215] stated that "*the term 'velocity' is meaningless unless accompanied by some indication of the mode of averaging employed in its determination*", and he repeated this statement in discussing 'eddy viscosity'. Richardson [157, p. 2] observed that "... the motion of numerous particles is not naturally divided into a mean motion and a residuum; but that the mean is a mental artefact, and may be variously apprehended, so that the motion of the residuum is a compensatory artefact; and that therefore different types of mean may need to be compensated by different laws of diffusion". Thompson also noted that Reynolds' formal definition of turbulence depends on the mode of averaging assumed and concluded that turbulence "... is not a real and distinctive physical phenomenon, but is simply a formal and rather useful concept for dealing with fluid systems whose state is not completely known" (Thompson [158, p. 152]). In a somewhat similar vein, Lamb [159] has argued that since the continuous atmosphere has an infinite number of degrees of freedom, its state at any time cannot be determined completely. That is, for any finite set of observations, there is an infinite subset of atmospheric states that could fit the observations. Thus, rather than attempting to divide the atmospheric flow field into a deterministic mean component and a stochastic turbulent component, Lamb advocated that in regional-scale modelling, atmospheric flow should be considered as a single nondeterministic class of possible atmospheric states which he termed 'metulence'.

Nevertheless, the classical Reynolds treatment of three-dimensional turbulence is based upon the concept of a rapidly fluctuating turbulent component superimposed on a steady, or at least slowly varying, mean flow (Reynolds [160]). For example, consider eqn (1). By introducing averaging, we can concentrate on mean values and ignore many flow details just as individual molecular behaviour is ignored in the kinetic theory of gases. However, an averaging operator must first be chosen, bearing in mind that the use of different operators may lead to different decompositions of the flow and hence to different mean and turbulence fields and to different diffusion characteristics.

4.3 Types of averages

The four main types of averages are time averages, space averages, space-time averages, and ensemble (or statistical or probability) averages (e.g., Frenkiel [161]; Hinze [6]; Wyngaard [162]; Cotton and Anthes [163]). Kampé de Fériet showed in the 1930s that the derivation of the Reynolds-averaged equations of motions requires the averaging operator to satisfy the following Reynolds conditions (Kampé de Fériet [164, 165, 166]; Monin and Yaglom [12]):

$$\overline{f+g} = \overline{f}+\overline{g} \ , \tag{4}$$

$$\overline{\alpha f} = \alpha\overline{f} \ , \quad \alpha = \text{constant}, \tag{5}$$

$$\overline{\overline{f}g} = \overline{f}\,\overline{g} \ , \tag{6}$$

and

$$\overline{\partial f/\partial s} = \partial\overline{f}/\partial s \ , \quad \text{where } s = x_1, x_2, x_3, \text{ or } t. \tag{7}$$

It then follows that $\overline{\overline{f}} = \overline{f}$ and $\overline{f'} = 0$. Kampé de Fériet [166] also noted Oseen's observation in 1930 that neither a time average over a finite interval of time nor a space average over a finite volume of space exactly satisfies the Reynolds conditions. Monin and Yaglom [12] pointed out that it is eqn (6) that causes the problem, but they also noted that a time or space average can satisfy this condition *approximately* provided that the averaging interval is chosen to be large compared to the characteristic periods or length scales of f' but small compared to the characteristic periods or length scales of $\overline{f}$.

An ensemble average, on the other hand, can be shown to satisfy the Reynolds conditions exactly, where the ensemble average is the average of an infinite set of flow realizations associated with the same set of initial and external conditions (e.g., Monin and Yaglom [12, Sec. 3.2]). Accordingly, the modern theory of turbulence has adopted a statistical or probability-theory approach in which fields of fluid dynamical variables in a turbulent flow are assumed to be random fields and ensemble averages are used instead of time or space averages. One diffusion experiment then reduces to a single sample value or realization, and ensemble averages are taken over an infinite number of such independent realizations. No intrinsic time or space scales are involved.

4.4 Applications of different averages

These various types of averages are appropriate to different situations. Clearly, a true ensemble average taken over an infinite number of realizations has only theoretical utility, although a sample ensemble average can be calculated based on only a finite number of realizations. Time-based averages, on the other hand, are commonly used for measurements made over a sampling interval τ at a fixed point in space by *in situ* instruments such as hot-wire anemometers, instrumented towers, or tethered balloons. Space-based averages are more demanding for practical applications since they require knowledge of instantaneous variable values over one, two, or three spatial dimensions. Nevertheless, some measurements made by networks of samplers and by remote-sensing techniques such as photography, laser interferometry, radar, sodar, and lidar do yield space averages. Strictly speaking, moving *in situ* probes such as airborne, balloon-borne, or vehicle-mounted instruments give *space-time* averages but, if the probe velocity is large relative to the internal time scales of the flow, these averages will approximate instantaneous space averages. Obviously, measurements from sampler networks and remote-sensing instruments made over a finite sampling time will also yield space-time averages.

The fundamental average used in numerical models is also a space-time average, namely a grid-volume–time average taken in space over the volume of an individual grid cell and in time over a model time step (e.g., Pielke [167]). Lamb [159]) has noted that we can generate a system with any desired number of degrees of freedom from a continuous fluid flow with an infinite number of degrees of freedom by performing suitable space-time averaging. This is what is done in a finite-difference model, for example, when a continuous flow is represented by values at a finite number of discrete grid points. Mesoscale meteorological models frequently use two different time averages. The obvious one is the time average over the model time step, but a second, longer time average over 15–60 minutes is implicit in the use of ensemble-average-based turbulence parameterizations (e.g., Pielke [167]; Mahrt [168]). Cotton [169] argued that such 'generalized ensemble' averages are different from grid volume-time averages and in fact are the fundamental averaging operators actually used in mesoscale and larger-scale meteorological models.

It has been suggested that the horizontal grid scale of mesoscale meteorological models provides a natural demarcation between turbulence and the mean flow (e.g., Lewellen [170]). According to this view, turbulence consists of subgrid-scale flow motions while the mean flow corresponds to grid-resolved motions. Unfortunately, this appealing definition does not always coincide with common notions of turbulence. In some mesoscale models, larger PBL turbulence scales will be grid-resolvable, while in others, even some mean mesoscale circulations will be subgrid-scale in size. Moreover, the form of the model's turbulence parameterization scheme must be considered. As discussed by Wyngaard [162] and Lamb [171], boundary-layer turbulence models may be divided into two broad classes: (a) *ensemble-average* models and (b) *volume-average* models. Ensemble-average boundary-layer models parameterize *all* PBL turbulence; they are unable to generate large eddies even though such scales of motion may be grid-resolvable. Thus, equating turbulence to subgrid-scale motion is not consistent with the formulation of ensemble-average boundary-layer models.

The best known volume-average boundary-layer models are the large-eddy simulation (LES) models. In these *three-dimensional, nonhydrostatic* models with *inertial-range grid spacing* in *both* horizontal and vertical directions (all of the italicized characteristics must be present), subgrid-scale motions are parameterized but eddies larger than a few grid lengths in size are resolved explicitly. Most boundary-layer meteorologists would deem these large eddies to be 'turbulent' motions. As the size of the averaging volume increases, however, the fraction of eddies that can be resolved explicitly decreases. Mesoscale meteorological models with horizontal grid spacing of several kilometers or more will be incapable of resolving any PBL eddies. In fact, most volume-average mesoscale meteorological models employ an ensemble-average boundary-layer model (e.g., Mellor [109]).

Continuing upscale to horizontal grid spacing of tens to hundreds of kilometers, synoptic-scale meteorological models will be unable to resolve even terrain-forced meso-β-scale circulations such as sea breezes that a boundary-layer meteorologist would be likely to deem 'mean' motions. Thus, in the case of volume-average turbulence models, equating turbulence to subgrid-scale motions does provide an operational though sometimes nontraditional definition of turbulence.

4.5 Relationships between different averages

Although the use of ensemble averages and the techniques of modern mathematical probability theory provide a powerful tool with which to investigate turbulence theoretically, there is still a need to compare theoretical deductions with laboratory or atmospheric measurements and with numerical model calculations. From a practical viewpoint, then, it is necessary to invoke an ergodic hypothesis ("Of or relating to a process in which every sequence or sizeable sample is equally representative of the whole (as in regard to a statistical parameter)" (*Webster's New Collegiate Dictionary*, 1975)) of some sort to relate various kinds of averages and to determine how close empirical mean values obtained by physical measurement or numerical simulation are to the ensemble mean value (Hinze [6]; Panofsky and Dutton [172]).

Ensemble-average turbulence statistics can be approximated (a) as the ensemble average of a spatially and temporally local variable (i.e., an instantaneous point measurement) taken over a *finite* number of flow realizations, or (b) as the time average of a spatially local variable measured at a finite number of times in a single, statistically-stationary flow realization, or (c) as the space average of a temporally local variable measured at a finite number of locations in a single, statistically-homogeneous flow realization (e.g., Hinze [6]). In each case the accuracy of the sample estimate is improved by maximizing sample size (see eqn (8)). For example, turbulence measurements made at a fixed point in a wind tunnel or at a certain height on an instrumented tower are averaged over as long a time as is practical without violating the assumption of flow stationarity (typically 30–60 minutes) in order to obtain an estimate of the ensemble average at that point. Similarly, aircraft measurements of turbulence quantities are averaged over as long a spatial 'leg' as is possible without violating the assumption of spatial homogeneity (recall the comment about the characteristic scales of f' and $\overline{f}$ in Section 4.3). LES model results are usually averaged over both time and the full horizontal domain (e.g., Moeng and Wyngaard [173]; Hadfield et al. [174]; Walko et al. [175]).

Wyngaard [176] discussed the sampling times used to estimate ensemble averages in the atmospheric surface layer at a fixed point. He used the expression

$$\tau = 2T_I\overline{f'^2}/(a^2\overline{f}^2) \tag{8}$$

to examine sampling fluctuations, where τ is the sampling time required to determine the ensemble average $\overline{f}$ to an accuracy a, T_I is the integral time scale (cf. Section 3.6) of the time-dependent function f, and $\overline{f'^2}$ is the ensemble variance of f about $\overline{f}$. τ is assumed to be much larger than T_I (e.g., Tennekes and Lumley [13, p. 212]). To estimate the mean horizontal wind speed $\overline{u}$ to 1% accuracy (i.e., $a = 0.01$) for a typical value of relative turbulence intensity $\overline{u'^2}/\overline{u}^2$ of 0.05, τ should thus be equal to approximately $1000T_I$. For an integral time scale of 1 second, this corresponds to a sampling period of about 15 minutes, a typical value for surface-layer measurements. Wyngaard then showed, however, that averaging times for a given accuracy increase with the order of the statistical moment of interest and with height. For example, to measure the stress component $\overline{u'w'}$ with the same accuracy at 200 m as at 20 m, τ must be about 50 times larger.

In a similar vein, Mellor [109] has discussed the minimum sampling *distances* required to obtain ensemble-mean statistics from data in both the atmosphere and the ocean. If Λ is the turbulence integral length scale and if we accept a 10% error in our estimate, then for second-order moments, the measurement distance should be at least 100Λ, assuming Gaussian probability distributions. For a 1% error, this distance increases to $10,000\Lambda$. If U_t is a translation velocity relative to the turbulence field being observed, then a corresponding sampling time would be $100\Lambda/U_t$. If we further assume that Λ for the PBL is one-half the PBL height, then for a 1 km deep PBL the measurement distance should be at least 50 km for only a 10% error bound. For a translation velocity U_t of 10 $\mathrm{m\,s^{-1}}$, the corresponding measurement time would be about an hour and a half. For these same assumptions in a numerical model with square horizontal grid cells, each areal cell should be at least 5 km along a side for cell values to represent ensemble-average values. This is consistent with the results of Moeng and Wyngaard [173, p. 2325]), who suggested that the largest PBL-scale eddies will not be resolvable on a 5 km by 5 km grid. Wyngaard [176] also briefly considered spatial averages for fixed arrays of sensors, for aircraft measurements, and for remote sensing techniques.

The overall implication for MAD is that sampling fluctuations and representativeness will always be concerns at mesoscale time and space scales and that the ergodic hypothesis will not always apply at these scales.

4.6 Averaging and turbulent flow realizations

As already alluded to, one crucial difference between ensemble and volume averages is that ensemble averaging completely removes randomness whereas volume averaging does not (Wyngaard [162]; Mellor [109]; Cotton [169]). Since an ensemble average is in principle taken over an infinite number of realizations, it samples all possible eddy sizes, phases, and amplitudes so that all transient eddies are averaged out (e.g., Yule [177]). Stationary, surface-forced eddies may not be removed by ensemble averaging, but then these structures are by definition neither random nor transient. Grid-volume averages may, however, retain transient eddies or coherent structures larger than the averaging length employed since such fields represent single flow realizations.

Wyngaard [162, p. 71] has the following to say on this point:

> ... while the ensemble averaging operation removes all the turbulence, the volume averaging operator removes only those components of turbulence having spatial scales smaller than the averaging scales. In a statistically homogeneous, stationary, unbounded turbulent flow, the volume average converges to the ensemble average in the limit as the averaging volume increases. Thus one can think of the volume-averaged field as being the sum of the ensemble-averaged field and the large-scale turbulence.

Note that Wyngaard implicitly defines turbulence based on an ensemble average. If a volume-average definition is used, then what he refers to as large-scale turbulence must instead be interpreted as the resolved mean flow for the single flow *realization* being considered.

Under what conditions will volume-average model predictions closely resemble ensemble-average model predictions? If the volume-average model grid scale is much larger than the turbulence integral length scale and the flow is both statistically homogeneous and stationary, then volume-average and ensemble-average models become identical since both schemes will parameterize all turbulence. That is, ensemble-average turbulence models always parameterize all scales of turbulence while for very large averaging volumes, volume-average models will be unable to resolve any eddy-scale motions and hence must also parameterize all eddy effects. If the horizontal resolution is considerably coarser than the vertical resolution, as is usually the case in mesoscale meteorological models, then any resolved eddies will be unrealistically long and flat (Lilly [122]). Moreover, if the model is hydrostatic, it will be unable to produce turbulent eddy motions even for a small averaging volume. Similarly, two-dimensional models of any resolution will effectively be ensemble-average models since real boundary-layer turbulence is three-dimensional and hence must be entirely parameterized in these models. In fact, Deardorff [178, p. 429] states that "the use of a two-dimensional model to simulate three-dimensional flows is scarcely justifiable unless the problem is one in which continuous and strong forcing in two dimensions occurs".

In the case of MAD tracer experiments, a tracer release will be tracked for a period of days, which means that the flow realization and dispersion realization will also be several days in length. Even for MAD experiments with many tracer releases, such as ANATEX, the meteorological conditions during the course of each release are seldom similar enough to permit ensemble averaging over two or more releases (realizations). Thus, MAD generally has to be studied through the examination of individual realizations.

When such mesoscale tracer experiments are modelled, the assumption is made that the mesoscale flow over the period of the experiment is resolvable and deterministic and that only subgrid-scale diffusion needs to be parameterized (e.g., Moran and Pielke [179, 180]). Given the range of possible mesoscale flow modes discussed in Chapter 1 and in Section 4.1, this assumption may not always hold. For example, in the 1984 Øresund mesoscale dispersion experiment, the general synoptic conditions on two consecutive Øresund release days, June 4 and 5, were quite similar but ground-level concentrations measured on June 4 were one-third to one-half those measured on June 5 (Andrén [181, 182]). Andrén also noted that pibal observations on June 4 indicated the presence of a low-level jet in the lowest 100 m that was not observed on June 5 nor predicted by any of the numerical dispersion models that attempted this case. One possible contributing factor is inherent limitations in the initialization of mesoscale meteorological models. Model initialization is limited by the spatial and temporal resolution of the synoptic observing network, which implies that sub-synoptic flow features – in effect much of the mesoscale energy spectrum – will not be represented in model initial flow fields (e.g., Lamb and Hati [183]; Moran and Pielke [180]).

4.7 Nonstationarity

What about flows that cannot be considered statistically stationary? Hinze [6, p. 6] discussed the case of a turbulent flow with slow, nonturbulent variations. If the turbulence integral time scale is T_I and the time scale for the slow, nonturbulent variations is T_S, then the averaging time t_a should be chosen so that $T_I \ll t_a \ll T_S$. Problems arise, however, if T_S is itself small, since a time average based on a short time sample may be quite different from the corresponding ensemble average, resulting in a large sample variance (e.g., Tennekes and Lumley [13]; Sykes [184]). Such fluctuations can be reduced if averaging is instead carried out over multiple (nonstationary) realizations for identical initial and external conditions in an analogous manner to the conditional sampling and compositing techniques used to study coherent flow structures (e.g., Yule [177]; Elsberry [185]; Ruscher and Mahrt [186]; Narasimha and Kailas [187]). That is, a sample ensemble mean can be used in place of a sample time mean in some cases.

Nonstationarity is a particular problem in the atmosphere with its many time scales, including the diurnal cycle and cyclone–anticyclone lifetimes and passage times (Lagrangian and Eulerian time scales, respectively: cf. Section 3.6) (e.g., Frenkiel [10]; Csanady [4]). In the case of a convective boundary layer (CBL) with a depth z_i of 1200 m and a convective velocity scale w_* of 2 m s^{-1}, one large-eddy turnover time will take about 10 minutes. An averaging time of 30 minutes would thus sample three large-eddy turnover times but would still be considerably shorter than the significant variations over one diurnal period. On the other hand, Wyngaard [176] and Millán [188] noted that even three or four hours of measurement in a CBL may not be long enough to obtain a stable sample. Turbulence integral time scales will be smaller in neutral and stable boundary layers. Csanady [4, Sec. 3.13] argued that sub-hour PBL measurements have to be considered within the context of a geophysical turbulence continuum in which the 'mean wind' will itself depend on larger-scale eddies. He pointed out that micrometeorologists often reject data obtained during periods when the mean wind had 'shifted' midway through the observing period or showed an upward or downward trend or even complete fluctuation cycles.

How can ensemble-average turbulence parameterizations or models handle nonstationary conditions? If there is a significant separation between the time scale of turbulence and the time scale of the large-scale nonstationarity, and if the turbulence field adjusts quickly to the slower changes in the mean flow field, then the turbulence field will traverse a sequence of near-equilibrium states as the large-scale environment changes so that the time scale of the statistical properties of the turbulence field is equal to the time scale of the large-scale flow variations. This approach has been followed when using ensemble-average equations to simulate a complete PBL diurnal cycle (Wyngaard [176]). The turbulence time scales are assumed to be much smaller than the averaging time which is in turn much smaller than a diurnal period. Given this wide scale separation, diurnally-varying surface forcing can be assumed to carry the slowly-evolving PBL through a sequence of quasi-equilibrium states over the course of a day (e.g., Nieuwstadt and Brost [189]). In fact, Mellor and Yamada's [190] Level 2, Level 2.5, and Level 3 ensemble-average turbulence-closure schemes ignore most or all of the local time-rate-of-change terms in the governing equations because these time scales are much larger than the time scale of turbulence. Only after the morning transition, when the PBL is growing rapidly, does this simplification cause problems (e.g., Helfand and Labraga [191]). The overall success that ensemble-average turbulence-closure models have had in simulating the full PBL diurnal cycle supports the application of the quasi-equilibrium assumption to this case.

5 Types of releases and types of diffusion

Pollutant releases range in duration from instantaneous *puff*-type releases at one extreme to continuous, long-term *plume*-type releases at the other. Accordingly, *release time* t_r constitutes an important dispersion time scale since instantaneous releases and continuous releases must often be treated quite differently for purposes of analysis and simulation. In the intermediate case of a point release made over a finite time interval into a flow of mean velocity U, the longitudinal cloud length will be Ut_r (ignoring longitudinal diffusivity K_x).

There are also two basic forms of turbulent diffusion: (a) diffusion of a cloud about its center of mass; and (b) diffusion from a fixed, continuously emitting source (e.g., Batchelor [192, 153]). The former, which is often referred to as *relative* or *two-particle* diffusion, requires a Lagrangian analysis framework moving with the center of mass of the fluid element, while the latter, which is often referred to as *absolute* or *one-particle* diffusion, requires a fixed or Eulerian analysis framework. While puff releases are generally associated with relative diffusion and plume releases are most often associated with absolute diffusion, what actually distinguishes these two basic types of turbulent diffusion from each other is the analysis reference frame rather than the release type. (It is worth noting that Herterich and Hasselmann [193] and Sanderson and Okubo [194] have discussed a third, intermediate form of turbulent diffusion that they term 'patch diffusion' to handle the case where a finite tracer patch is neither small enough to move as a single particle (two-particle limit) nor large enough to be unaffected by eddy motions (one-particle limit).)

5.1 Puff releases

For tracer puffs or clouds, the natural frame of reference is a coordinate system that *moves* with the puff center of mass, i.e., a Lagrangian reference frame, and the puff characteristic of greatest interest is the *size* of the puff. The simplest representation of the diffusion of such a puff is the rate of separation of a *pair* of particles or fluid elements released simultaneously from the same source (Batchelor [153, 195]). For this reason, the growth and spreading of a puff about its center of mass is often referred to as *two-particle* or *relative* diffusion.

The manner in which a puff release disperses will be determined by the size of the puff relative to the spatial scale of the dominant motions of the flow field in which it is embedded. If the dominant elements of the flow field are large relative to the puff, then the puff will tend to be transported as a whole and will experience little growth and dilution due to mixing, although velocity gradients associated with these large flow scales may deform the puff (Fig. 4a). If the dominant flow field elements are comparable in size to the puff, then deformation, mixing, growth, and dilution of the puff will be rapid (Fig. 4b). If the dominant flow field elements are small relative to the puff, then the puff will experience only slow growth at its boundaries (Fig. 4c). Thus, a fundamental property of relative diffusion, first recognized by Richardson [196]), is that the rate of growth of a puff depends directly on the size of the puff. This means that the growth of a large cloud cannot be represented as the superposition of the growth of component elements.

In spectral terms, the spatial scale of a puff or cloud defines a high-pass filter: spatial spectral scales larger than the scale of the puff cannot contribute to puff growth (e.g., Smith and Hay [198]; Mikkelsen et al. [199]). The problem of defining a sampling time, and hence of specifying an operational definition of turbulence, which complicates the analysis of plume growth (see below), is thus avoided for puff growth since the sampling time is of necessity equal to the puff travel time. On the other hand, the characteristics of the 'high-pass filter' change in time as the puff grows (cf. eqn (22)).

Even flow structures on the scale of extratropical cyclones can act to 'stir' and deform a tracer puff if it is large enough, drawing the puff out into a long, narrow 'streak' or 'filament' and increasing the areal extent of concentration gradients (e.g., Maryon and Buckland [9]). Figure 1 shows an example of large-scale horizontal deformation due to the 500 hPa flow field. As in Fig. 4b, the puff length scale and the dominant eddy length scale are comparable so that deformation is both rapid and pronounced.

5.2 Plume releases

A plume release is an extended continuous release of a pollutant from a fixed source. The resulting downstream plume remains connected to the source. Thus, the natural reference frame for a plume release is a coordinate system that is *fixed* with respect to the source (e.g., Batchelor [192, 153]). In such an Eulerian reference frame, it is easy to vary the period of time over which the resulting pollutant plume is sampled. A photograph of a plume taken with a shutter speed of a fraction of a second will tend to show a long, narrow, sinuous plume, in effect a plume realization or instantaneous plume. However, if a time exposure of several minutes is made, the resulting plume will tend to appear broader, smoother, and more symmetric about its mean axis (see Gifford [93] and Nappo [200] for examples of plume photographs taken over different exposure times).

This behaviour is illustrated schematically in Fig. 5a: as cross-plume concentration measurements are made for longer sampling times, the resulting concentration distribution becomes broader and

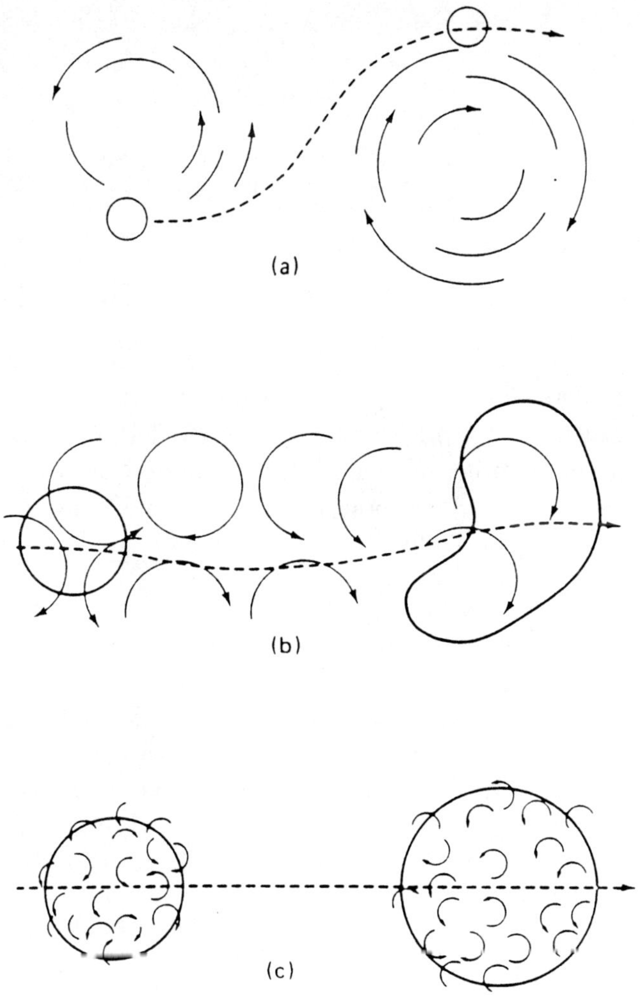

Figure 4: Idealized dispersion patterns of a pollutant 'cloud' for three dominant scales of motion: (a) uniform field of eddies large in size compared to size of pollutant cloud; (b) uniform field of eddies comparable in size to pollutant cloud; (c) uniform field of eddies small in size compared to size of pollutant cloud (from Slade [197]).

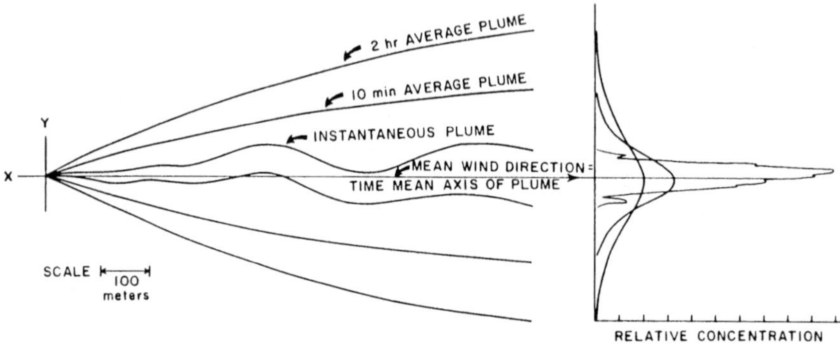

Figure 5: Effects of averaging time on plume shape and intensity: (a) outlines of a pollutant plume viewed from above at an instant in time, over a 10-minute period, and over a 2-hour period; (b) corresponding cross-plume relative concentration (from Slade [201]).

smoother and peak concentration values become smaller (Fig. 5b). *Observing time* or *sampling time* τ is thus an important quantity when studying continuous releases. This dependence of plume concentration on sampling time is the result of averaging over flow fluctuations or eddies with time scales shorter than the sampling period. Since eddy time scales are normally directly proportional to spatial scale, increasing the sampling time incorporates a greater and greater proportion of all eddy contributions and hence captures a greater range of plume displacements. The role of sampling time is discussed in more detail in Section 6.3.

Atmospheric motions larger than the plume dimensions tend to transport the plume intact, a process known as *meander*, while smaller motions tend to act at the plume boundary, resulting in further mixing, growth, and dilution. For a plume, *all* scales of motion contribute initially to mixing, but as the plume grows, the smaller scales contribute less and less to dilution of the plume, instead acting merely to oscillate fluid elements back and forth (see Section 6.1.1). On a scale of days, synoptic-scale flow fluctuations can also change the direction of plume transport, a process referred to as 'synoptic swinging' by Smith [202].

Under stable conditions, plume meander can dominate the reduction of mean concentrations due to increased sampling time: a simple analogy is to consider the effect on the resulting spray pattern of waving a garden hose back and forth versus holding it steady (e.g., Kristensen et al. [203]; Hanna [204]; Etling [205]). Meander is also responsible for most concentration fluctuations in the early stages of *elevated* plume transport (e.g., Hanna [206, 207]; Moore et al. [208]; Fackrell and Robins [209]; Peterson and Lamb [210]).

For statistically-stationary flows, the limiting mean value of time averaging will be the Eulerian ensemble-mean plume, that is, the mean plume resulting from the average of an infinite number of instantaneous, independent plume realizations for identical source characteristics and flow conditions. Recognizing this, Nappo [200] employed another photographic technique to estimate the Eulerian ensemble-mean plume. He overlaid four georeferenced plume snapshots taken at approximately 15-minute intervals from an overflying aircraft to produce a fixed-reference-frame composite plume consisting of nearly independent, instantaneous plume samples or realizations. Both plume growth and widening and plume meander were accounted for by this technique (see also Groll and aufm Kampe [211]).

In a steady flow, a plume can also be approximated as the *sequential* release of many small puffs, which are equivalent to plume sections or segments (e.g., Frenkiel [10, 161]). Plume diffusion can thus be studied by considering the behavior of a sequence of puffs as they pass a given fixed point. The greater the number of puffs considered (assuming a fixed puff release rate), the greater the effective sampling time. Successive plume sections may follow different trajectories as a result of larger-scale flow fluctuations so that, as mentioned before, mean centerline concentrations downwind from a source will depend inversely on sampling time.

In an Eulerian reference frame, the emphasis will be on the positions of the centers of mass of this sequence of *independent* fluid elements relative to a fixed point or axis rather than on the separation of such a set of elements relative to their common center of mass and to each other. Plume diffusion viewed from an Eulerian perspective is thus frequently referred to as *one-particle* or *absolute* diffusion since it can be studied by considering an ensemble of independent marked-fluid-parcel trajectories (e.g., Batchelor [192]; Section 6.1 of this study).

Note that plume diffusion, unlike the three-dimensional growth of a puff, is essentially a two-dimensional process in the cross-wind and vertical directions, and this is a significant difference when considering the dilution of a pollutant release. A puff can be diluted by growth in all of the cross-wind, along-wind, and vertical directions whereas along-wind expansion is not a factor for the dilution of a plume. This difference may be even greater in a shear flow since the effective diffusivity in the shear direction can be greatly enhanced (e.g., Csanady [4, p. 136]), and vertical shear is usually most pronounced in the along-wind direction (i.e., 'speed' shear; see Section 8).

5.3 Relative diffusion vs. absolute diffusion

The distinction between relative diffusion and absolute diffusion is a fairly subtle one. In fact, the study of atmospheric dispersion was well established before Batchelor [192, 195] and Brier [212] discussed this difference and its relation to the choice of observing reference frame. Batchelor and Brier showed that absolute diffusion is properly studied in an Eulerian reference frame while relative diffusion must be studied in a Lagrangian reference frame.

Earlier in this section, a puff release was used to illustrate the phenomenon of relative diffusion while a plume release was used to illustrate the phenomenon of absolute diffusion. It is also possible, however, to calculate the absolute diffusion of a puff and the relative diffusion of a plume; neither type of release is uniquely associated with one type of diffusion or the other (e.g., Csanady [4, Sec. 4.1]).

5.3.1 Absolute diffusion of a puff

To study the absolute diffusion of a puff, the observer's frame of reference is chosen to be fixed relative to the release point, contrary to the usual practice of following the puff. A composite or ensemble-average puff for a particular travel time can then be constructed by superposing a set of independent puff realizations corresponding to that travel time. But rather than superposing puff centers, superposition is carried out by lining up, or registering, the source location and other background reference points. The concentration field of the resulting Eulerian ensemble-average puff then corresponds to the probability density function describing the likelihood that a particle released from the instantaneous source will be found at each point in space after traveling for the specified travel time (Batchelor [192]; Batchelor and Townsend [155]; Csanady [4]). This Eulerian puff concentration field for absolute diffusion will be much larger in spatial scale and will have a lower peak value than the Lagrangian center-of-mass-based puff concentration field for relative diffusion since the scatter of the *positions* of the puff centers of mass is considered as well as the growth of each individual puff about its center of mass (e.g., Garrett [3]). For the Lagrangian average, only the *shape* and *size* of the cloud relative to its center of mass is considered (Batchelor [195]). In essence, relative diffusion results only from stirring and mixing whereas absolute diffusion also includes the effects of whole-body advection (Durbin [213]).

It should be noted that *neither* of these ensemble-average puffs is likely to match a single puff realization very well (e.g., Garrett [3]). An ensemble-average puff concentration field will be much smoother, larger, and more symmetric than that of any puff realization. A Lagrangian ensemble-average puff may also have what Chatwin and Sullivan [214, 215] have termed a 'core-bulk' structure, with a high-concentration core enveloped by a low-concentration outer layer (see Fig. 6). No puff realization is likely to have a comparable structure. Chatwin and Sullivan [216] suggested that two advantages of the Lagrangian ensemble average over the Eulerian ensemble average are that (a) the Lagrangian average involves less 'smearing', that is, the Lagrangian ensemble puff will be closer in size to a 'real' puff, and (b) the effects of flow unsteadiness on ensemble averaging are minimized. Gifford et al. [8] and Maryon and Buckland [9] have noted too that for complicated flow fields or for travel times of more than a few days, a single measure such as σ_y may not be sufficient to

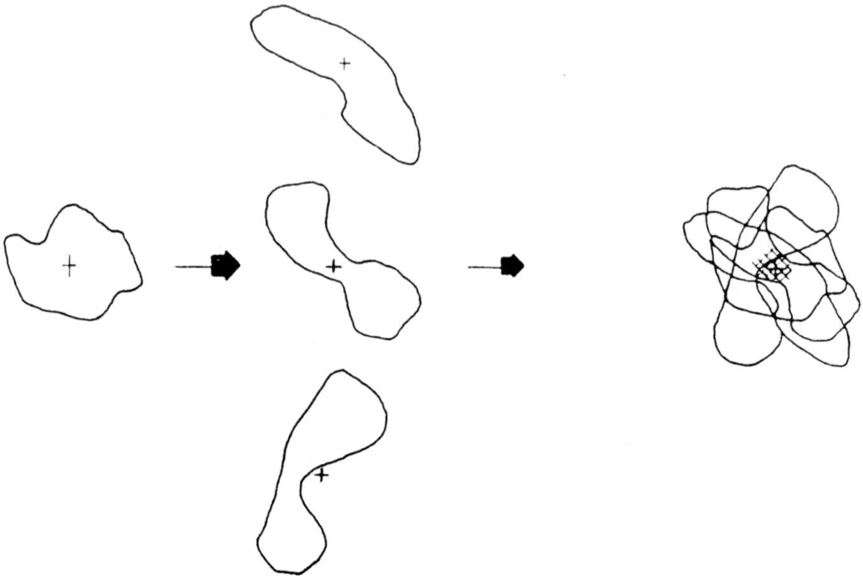

Figure 6: Sketch of core-bulk structure resulting from the superposition of the centers of mass (indicated by 'plus' signs) of four puff realizations. The small cross-hatched area marks the core region (from Chatwin and Sullivan [214]).

describe the shape of a strongly distorted puff or cloud.

Note that the absolute diffusion of a puff has considerable relevance for mesoscale tracer experiments, in which observations of the movement of a tracer cloud across a fixed (i.e., Eulerian) observing network are available. As already noted, however, in Section 4.6, only one mesoscale realization is usually available for each set of flow conditions, precluding the calculation of sample ensemble averages.

5.3.2 Relative diffusion of a plume

To study the relative diffusion of a plume, it is necessary (a) to examine *instantaneous* plumes *and* (b) to perform ensemble averaging *relative* to the centerlines of the instantaneous plumes, that is, relative to a moving reference frame. Nappo [200] measured plume relative diffusion by superposing four plume snapshots relative to their centerlines (i.e., a Lagrangian superposition in addition to the Eulerian superposition already discussed in Section 5.2) and then measuring plume widths. Other plume measurements made with short sampling times (e.g., aircraft, boat, or van traverses over, under, or through a plume) will also measure near-instantaneous plume widths and hence will describe relative diffusion rather than absolute diffusion (e.g., Murthy and Csanady [217]; Gifford [74]; Chatwin and Sullivan [218]; Carras and Williams [52, 53]; McElroy et al. [219]; Groll and aufm Kampe [211]; Groll et al. [220]). Figure 3, for example, is based on near-instantaneous plume traverses and hence describes plume relative diffusion. If a plume cross-section can be assumed to be composed entirely of marked fluid released at the same time, then cross-plume widths at different distances will correspond to puff widths at different travel times. However, this assumption will not hold in shear flows, where marked fluid at different locations in the plume cross-section is likely to have been released at different times (e.g., Uthe et al. [221]; Carras and Williams [52]).

Knowledge of plume relative diffusion is important when studying concentration fluctuations. One example is odor studies, where near-instantaneous concentrations are important (e.g., Högström [222]; Csanady [4]; Fackrell and Robins [209]). Plume meander, on the other hand, will affect odor intermittency or frequency, i.e., whether the plume is present at a particular location or not. Interestingly, one effect of relative diffusion is to reduce concentration fluctuation variance $\overline{c'^2}$ by increasing plume width and hence reducing (a) the number of zero concentrations due to meander and (b) the peak-to-mean concentration ratios (Durbin [213]).

The role of sampling time in plume measurements should now be clearer. For one thing, sampling time provides an *operational* definition of turbulence, separating eddy scales with shorter lifetimes that contribute to turbulent diffusion from longer-lived flow scales that contribute to the 'mean' flow. Sampling over a long period at a fixed point in a statistically-stationary flow field is one way to estimate a local ensemble average. In the same way, observing a plume with long-exposure photographs or other extended-sampling-period techniques will permit estimates to be made of absolute plume diffusion and the Eulerian ensemble-average plume over an area or volume. Plume σ_y s based on time-averaged measurements will thus be larger than instantaneous σ_y s because meander effects will also contribute (e.g., Moore et al. [208]).

Sampling time does not play a comparable role in puff observations since time averages in a moving reference frame will have no comparable connection to ensemble averages (e.g., Csanady [4, Sec. 4.1]). Another way, however, to estimate an ensemble average is to average over a number of independent measurements or realizations. In this case, near-instantaneous plume measurements can be used to estimate *either* Eulerian or Lagrangian ensemble averages depending on the choice of reference frame used for averaging (e.g., Nappo [200]). See Section 6.3 for a further discussion of sampling time.

5.4 Travel time

A third dispersion time scale, in addition to release time and sampling time, is *travel time*, the length of time that the tracer is carried by the flow. For MAD, this quantity will vary between several hours and several days. In general, the longer the travel time, the greater the impact that flow inhomogeneities, both spatial and temporal, will have on dispersion because of the greater likelihood that fluid elements will encounter such fluctuations (e.g., Fig. 2). As discussed in the next section, travel time acts as a finite averaging time in absolute diffusion, and partly determines cloud size, and hence the maximum eddy space scale that can affect puff growth, in relative diffusion. Theory also suggests that relative and absolute diffusion become less and less distinguishable and peak-to-mean ratios approach unity at longer travel times (see Section 6.2.2). This is an important consideration for mesoscale travel times since (a) most atmospheric dispersion models do not represent relative diffusion very well and (b) many mesoscale diffusion measurements are near-instantaneous and thus describe relative diffusion rather than absolute diffusion. See Section 6.3 for a further discussion of travel time.

6 Statistical theory of turbulent diffusion and MAD

The statistical theory of turbulent diffusion has provided a valuable theoretical framework for studying short-range relative and absolute diffusion (e.g., Monin and Yaglom [12]; Hinze [6]; Pasquill and Smith [14]). Consideration of the large-time asymptotic limits of some of the equations developed to describe short-range turbulent diffusion can also provide some insight into MAD. And as discussed in this section, this theoretical description also helps to clarify the roles and significance of the various dispersion time scales for MAD and the close similarity between relative and absolute diffusion on the mesoscale.

6.1 Taylor's theorem and one-particle diffusion

Taylor [96], in an early and seminal paper, developed a statistical description of absolute or one-particle diffusion. Assuming a *statistically homogeneous* and *stationary* turbulent flow, Taylor considered the motion of infinitesimal particles leaving the origin at $t = 0$ and being tossed back and forth by turbulent velocity fluctuations. The particles are assumed to assume the fluid velocity completely and to be passive dynamically. If a particle has velocity component u in the x-direction, then after a period of time T, it will have reached point X. If the mean-square displacement of a large number of particles released in succession (or of many realizations of the release of a single particle) from the origin and travelling for time T is denoted by the quantity $\overline{X^2}$, then

$$\overline{X^2}(T) = 2\overline{u'^2} \int_0^T \int_0^t R_L(\xi)\,\mathrm{d}\xi\,\mathrm{d}t \quad . \tag{9}$$

Taylor commented that eqn (9) is "rather remarkable" since it reduces the problem of absolute diffusion to consideration of a single quantity, the (one-particle, two-point) Lagrangian velocity autocovariance $\overline{u'(t)\,u'(t+\xi)}$ (cf. eqn (2)). Taylor then showed that eqn (9) approaches the expression

$$\overline{X^2} = \overline{u'^2} T^2 \tag{10}$$

asymptotically for small travel time T and the expression

$$\overline{X^2} = 2\overline{u'^2} T_L T \tag{11}$$

for large travel time, where T_L is the Lagrangian integral time scale [see eqn (3)].

Note that eqn (11) has the well-known parabolic time dependence of diffusion in homogeneous turbulence (e.g., Batchelor [192]). Note also that release time and sampling time have both been assumed to be much larger than travel time T since absolute diffusion is being considered (e.g., Hay and Pasquill [223]). And note that the flow is assumed to be statistically homogeneous but not necessarily *isotropic*. This requirement explains why Taylor's equation is usually applied to horizontal diffusion and less often to vertical diffusion, but, as pointed out in Section 3.2, horizontal diffusion is the main concern for MAD and so Taylor's equation is still useful. The more complicated but more realistic case of shear flows, in which horizontal diffusion can be enhanced by the interaction of differential horizontal advection and vertical diffusion, is discussed in Section 8.

6.1.1 Spectral representation

Eqn (9) can be rewritten in a spectral or frequency-domain form as (e.g., Hinze [6]; Pasquill and Smith [14])

$$\overline{X^2} = \overline{u'^2} T^2 \int_0^\infty F_L(n) \, \frac{\sin^2 \pi n T}{(\pi n T)^2} \, \mathrm{d}n \quad , \tag{12}$$

where n is the cyclic frequency (in Hertz or cycles per second) and $F_L(n)$, the normalized energy spectrum function, is the cosine transform of $R_L(\xi)$, which is a real and symmetric function:

$$F_L(n) = 4 \int_0^\infty R_L(\xi) \cos 2\pi n \xi \, \mathrm{d}\xi \quad . \tag{13}$$

Eqn (12) thus relates particle dispersion to the KE frequency spectrum of a fluid particle (e.g., Batchelor [192]).

This spectral form of Taylor's equation provides another way of looking at the asymptotic limits given by eqns (10) and (11). The spectral filter or weighting function $\sin^2\pi nT/(\pi nT)^2$ depends on the travel time T. When T is small, the filter function will be very broad since the first zero lies at $n = 1/T$. The energy spectrum integral in eqn (12) can then be approximated as $\int_0^\infty F_L(n)\,dn$ so that eqn (12) will have the small-time limit given by eqn (10) since $\int_0^\infty F_L(n)\,dn = 1$. For long travel times, on the other hand, the spectral filter function becomes very narrow so that only the contributions of very low frequencies are considered. For the large-time limit the energy spectrum integral can be approximated as $F_L(0)\int_0^\infty \sin^2\pi nT/(\pi nT)^2\,dn$, which has the value $2T_L/T$ (e.g., Monin and Yaglom [224, p. 528]), thus yielding eqn (11). Physically, these asymptotic limits suggest that eddies of all frequencies initially contribute to particle dispersion exactly as they do to the turbulent kinctic energy. For longer travel times, however, contributions from higher-frequency eddies diminish systematically. This is because these smaller eddies will merely *oscillate* a particle about its mean position while advection by lower-frequency eddies will still displace particles in a *sustained* fashion on that time scale (e.g., Pasquill and Smith [14]).

Eqn (12) can be viewed in another way too. Let u'_T represent the mean particle fluctuation velocity over travel time T. Then

$$X(T) = u'_T T \quad , \tag{14}$$

$$\overline{X^2}(T) = \overline{u'^2_T}\,T^2 \quad . \tag{15}$$

Eqn (15) has the same form as the small-time asymptotic limit given by eqn (10) but holds for all times. However, $\overline{u'^2_T}$ will be smaller than the actual turbulence intensity $\overline{u'^2}$ since the travel time T can be thought of as a finite *averaging* time that smooths out the shortest time scales contributing to $\overline{u'^2}$; that is, u'_T is an average or smoothed velocity rather than an instantaneous velocity. It can then be shown that averaging in time over travel time T is equivalent to applying a 'low-pass' spectral filter function in the frequency domain so that (Kahn [225]; Pasquill [226])

$$\overline{u_T'^2} = \overline{u'^2} \int_0^\infty F_L(n) \, \frac{\sin^2 \pi n T}{(\pi n T)^2} \, \mathrm{d}n \quad ; \tag{16}$$

substitution of eqn (16) into eqn (15) then yields eqn (12).

6.2 Two-particle diffusion

Relative diffusion can also be described by a theoretical model very similar to Taylor's theorem (eqn (9)). Consider two particles moving simultaneously through a statistically stationary and homogeneous turbulent flow. Let $X_1(t)$ and $X_2(t)$ be the x-coordinates of the two particles at time t and let $X_s(t)$ be the distance separating the two particles in the x-direction, where the initial particle separation is $X_s(0)$. If we define δu to be the relative velocity $u_1' - u_2'$ in the x-direction, then the mean-square separation $\overline{X_s^2}$ varies in time as (Batchelor [195])

$$\overline{X_s^2}(T) = \overline{X_s^2}(0) + 2 \int_0^T \int_0^t \overline{\delta u(t) \, \delta u(t - \xi)} \, \mathrm{d}\xi \, \mathrm{d}t \quad , \tag{17}$$

an equation that, with the exception of the initial separation term, is formally identical to eqn (9). Note, however, that the integrand in eqn (17) involves the correlation between velocities of *two* different fluid particles at *two* different times (e.g., Batchelor [153]; Smith and Hay [198]). This is in contrast to eqn (9), in which the integrand depends only upon the correlation between the velocities of a *single* particle at two different times.

6.2.1 Relative diffusion regimes

It is possible to use dimensional arguments (e.g., Batchelor [195]; Batchelor and Townsend [155]) to obtain the time dependencies of $\overline{X_s^2}(T)$ at small and intermediate travel times T. For small T, the mean-square particle separation depends upon T^2 just as in eqn (10). The intermediate regime, sometimes called the 'accelerated' regime, is of considerable interest since the time dependence is proportional to T^3. This growth rate is considerably faster than the T^2 dependence in the initial stages of absolute diffusion or the T dependence of molecular diffusion or of absolute diffusion at large travel times (see eqn (11)). A physical explanation is that eddies contributing to the relative diffusion of two particles tend to be of the same scale as the particle separation. As the separation distance increases, larger and larger eddies contribute to particle separation – in effect, the variance of the relative velocity also increases with time so long as the particle separation is still within the inertial range, that is, considerably smaller than the scale of the wavelengths of the energy-bearing eddies.

In a paper discussing the observational evidence for the accelerated regime, Gifford [74] argued, based on a number of different tropospheric relative diffusion data sets, that the intermediate regime ends somewhere between 1000 and 3000 s. Gifford also acknowledged, however, that Corrsin [227], Saffman [228], and Smith [229] had derived $\overline{X_s^2}(T) \propto T^3$ laws for horizontal relative diffusion in *shear* flows that would apply at time scales beyond the three-dimensional inertial range (cf. Section 8).

6.2.2 Large-travel-time limit

From the viewpoint of MAD, however, the most interesting relative-diffusion regime is that of large travel times. At very large T, the pair of particles should be so far apart that their velocities are completely uncorrelated. In this case,

$$\overline{\delta u(t)\, \delta u(t-\xi)} = 2\overline{u(t)\, u(t-\xi)} \quad , \tag{18}$$

where u is the absolute u-velocity component of *either* particle, and

$$\overline{X_s^2} = \overline{X_s^2}(0) + 4 \int_0^T \int_0^t \overline{u(t)\, u(t-\xi)}\, \mathrm{d}\xi\, \mathrm{d}t \quad , \tag{19}$$

so that by eqns (2) and (9), relative diffusion has the same long-time-limit behaviour as absolute diffusion.

At what time or space scales might we expect eqn (19) to hold? Based on Gifford's [74] analysis of available tropospheric relative-diffusion measurements, the accelerated regime does not hold past 3000 s. Fig. 7 shows a plot of both time-averaged (over a one-hour period) and instantaneous σ_y s vs. travel time for an oil fog plume about 5 km long generated not long after sunrise (June 17, 1977, roughly 0700–0830 MDT). The instantaneous plume σ_y s can be seen to to obey a T^3 law at longer times and to approach the average-plume or absolute-diffusion values as time increases. Carras and Williams [100] suggested a value of eight hours based on a large number of CBL plume measurements in Australia. It is worth noting in this regard that Heffter [65] did not distinguish between relative and absolute diffusion in his study of the variation of horizontal diffusivity for travel times of more than one hour.

Gifford has argued (Gifford [75, 104, 76]; Barr and Gifford [106]; Gifford et al. [8]) that the accelerated diffusion regime "... exists for 10 to 15 hours or more of downwind travel from the source, i.e., to length scales of up to several hundred kilometers, and that the dominant time scale, T_L, is on the order of 10^4 s" (Gifford [104, p. 165]). It is still not clear whether this is in fact completely true, particularly in view of the competing contributions from vertical shear (Section 8). From the point of view of a one-particle Lagrangian dispersion model, however, extending the time over which the velocities of two particles are correlated will not affect the results obtained provided that the scale of motion moving the two particles in concert can be resolved and included in the mean wind field used by the model (see end of Section 8). That is, a one-particle dispersion model is capable of treating grid-scale relative diffusion but not subgrid-scale relative diffusion.

As discussed in the next section, relative and absolute diffusion will also become indistinguishable as the travel time T approaches the sampling time τ. Gifford [104] has shown too that the sampling time τ can reduce the value of T_L estimated from experimental diffusion data if the sampling period is less than a few hours.

6.3 Release, travel, sampling, and averaging times

The release, travel, and sampling times t_r, T, and τ together determine the character of the dispersion problem being considered. Hanna [231] observed that these three dispersion time scales each define spectral 'windows' that filter the turbulence KE spectrum and connect observed diffusion with the turbulent flow properties. They

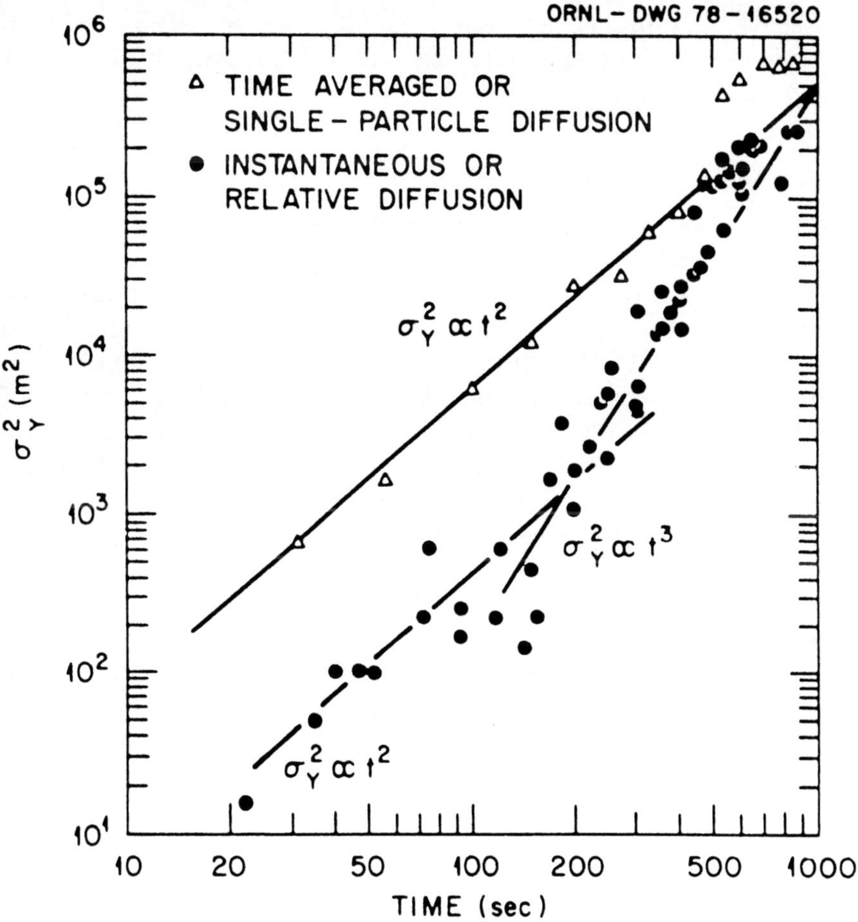

Figure 7: Plot of σ_y vs. travel time for both instantaneous and time-averaged plumes (from Nappo [230]).

also determine the type of diffusion: "If travel time is much greater than release time or if sampling time is much less than travel time, then puff diffusion is appropriate. If sampling time and release time are much greater than travel time, then continuous plume diffusion is important" (Hanna, [231, p. 276]). For mesoscale tracer experiments, for example, travel time is usually considerably larger than both release time and sampling time so that relative diffusion is the appropriate analysis framework.

6.3.1 Absolute diffusion case

In deriving eqn (9), sampling time and release time are assumed to be much larger than travel time so that absolute diffusion is being described (e.g., Hay and Pasquill [223]). In fact, the sampling time and, correspondingly, the release time (otherwise, the plume would not be 'attached' to the source but would rather correspond to an elongated puff), are both assumed to be large enough for all scales of turbulence to be accounted for in $\overline{u'^2}$ and $R(\xi)$. In the real world, however, estimates of plume or absolute diffusion based on particles released serially from a fixed point over a certain finite sampling time will not sample much of the low-frequency end of the energy spectrum. The complementary exclusion of the high-frequency end of the energy spectrum that results from the equivalence of travel time and averaging time has already been described by eqn (16). Rewriting eqn (12) for a finite sampling time or sampling period τ then yields the equation (e.g., Ogura [232]; Pasquill and Smith [14]):

$$\overline{X^2} = \overline{u'^2}\,T^2 \int_0^\infty F_L(n) \left[1 - \frac{\sin^2 \pi n \tau}{(\pi n \tau)^2}\right] \frac{\sin^2 \pi n T}{(\pi n T)^2}\, \mathrm{d}n \quad . \tag{20}$$

Thus, spectral contributions corresponding to time scales less than the travel time T or greater than the sampling time τ are diminished or completely filtered out by one of the two sine terms in the integrand. The actual reasons for this 'band-pass' filtering are quite different, however: high-frequency contributions to $\overline{u'^2}$ are removed because they are ineffective for particle dispersion (that is, net particle displacement), but low-frequency contributions are excluded from $\overline{u'^2}$ because the sampling time is too short to allow them to contribute to $\overline{X^2}$.

Note that eqn (20) can be approximated as (Kahn [225]; Hino [233]; Wollenweber and Panofsky [234])

$$\overline{X^2} \approx \overline{u'^2}\, T^2 \int_{1/2\tau}^{1/2T} F_L(n)\, dn \quad . \tag{21}$$

This equation shows more clearly than eqn (20) how particle diffusion or mean-square particle displacement for a finite travel time T and a finite observation or sampling interval τ is determined by a *finite* spectral interval. To quote Ogura [232, p. 152] on the role of sampling time in this equation: "... only eddies whose frequencies are larger than $2\pi/\tau$ contribute appreciably to the mean rate of diffusion averaged over the period of τ. In other words, for the turbulent diffusion observed during the period τ, large eddies whose frequencies are less than $2\pi/\tau$ should be regarded as a mean flow, not as turbulent components, and consequently they contribute nothing to the dispersion of particles. Their role is no more than that of determining the center line of dispersed plumes."

How should eqn (21) be interpreted if $\tau < T$? The simplest answer is that eqn (21) is not valid for this condition since it is based on Taylor's equation (eqn (9)), which requires that $T << \tau$. This answer is not very satisfying for MAD, however, since $\tau < T$ often occurs in MAD observations. Ogura [232] argued that this case is inadmissible for describing absolute diffusion because diffusion would then be occurring relative to a local, possibly moving, origin rather than relative to the fixed source (consider the extreme example of $\tau = 0$). Pasquill and Smith [14] agreed with Ogura, arguing that having $\tau < T$ introduces an *effective* sampling time larger than τ. This is important because the sampling period τ *defines* the mean wind direction and hence the reference frame for absolute diffusion. A larger effective sampling time would incorporate the influence of lower-frequency eddies and hence might yield a different mean wind direction. These arguments again suggest that the appropriate analysis framework for MAD is a relative-diffusion one.

6.3.2 Relative diffusion case

The relationship between the three characteristic dispersion time scales is somewhat different for the diffusion of a cloud or cluster of particles. In this case, the finite size of the cloud or cluster acts as an effective *spatial* band-pass filter. Spatial spectral scales larger than the scale of the cloud cannot contribute to its growth but rather can only move it as a whole (e.g., Fig. 4a). The smallest scales are also ineffective in contributing to cloud growth, since they can only oscillate particles locally and cannot maintain sustained displacement. Only the eddy spectral scales comparable in size to the cloud size contribute significantly to its growth and dispersion. Moreover, the characteristics of this spatial 'band-pass' filter change in time as the puff grows (Smith and Hay [198]). The combined effects of the turbulence intensity (and KE spectral slope) of the flow and the travel time T determine the cloud or cluster size and hence the effective spatial filter bandwidth. Skupniewicz [235] referred to this filter as a "sliding spectral window".

Sampling time for cloud or cluster diffusion, on the other hand, is *instantaneous* and perhaps is best thought of from the Lagrangian perspective. That is, the measurement is taken from the point of view of an observer moving with the cloud center rather than from that of an observer at a fixed point in space. However, if we compare puff diffusion to plume diffusion in spectral terms, we do better to think of the *travel* time as the effective sampling time for puff diffusion since it, like sampling time in the plume diffusion case, determines the maximum eddy time scale that can affect puff growth (though only indirectly through its influence on cloud size). Another way to think of this is to note that as T increases, the growing cloud will sample a larger spatial region than did the initial cloud, whose size was determined by the release time t_r. In fact, Ogura [232, p. 155] stated that "... it is impossible in relative diffusion to distinguish between the (travel) time and the (sampling) time T".

Smith and Hay [198] considered the problem of cloud growth and proposed a β approximation for the relative-velocity Lagrangian and Eulerian time correlograms analogous to the one proposed by Hay and Pasquill [223] for absolute diffusion. By using this assumption and by applying an ensemble average over all realizations of the velocity field and a spatial average over the cloud volume to separate relative diffusion from the mean transport, they arrived at the following expression for the rate of change of cloud size with time in terms of the filtered three-dimensional KE spectrum $E(k)$:

$$\frac{\mathrm{d}}{\mathrm{d}t}\sigma^2 = \frac{\pi\beta}{6U}\int_0^\infty E(k)\,\frac{1-e^{-\sigma^2 k^2}}{k}\,\mathrm{d}k \quad , \tag{22}$$

where σ is the concentration standard deviation about the cloud center of mass, $\beta = T_L/T_E$, and U is the mean transport wind speed. The exponential term multiplying the energy spectral density function in the integrand of eqn (22) has a mid-range maximum and approaches zero at both ends of the integration range. It thus acts as a band-pass filter. Moreover, as σ increases with T, the maximum of the filter will slide over towards smaller KE wavenumbers (longer wavelengths).

Returning briefly to absolute diffusion, absolute diffusion results from the contributions of both meander and relative diffusion, but the contribution of meander decreases monotonically with travel time T. Physically, this results from the dependence of the size of the puff or plume on T. As shown by eqn (22), as a puff (or a plume section) grows in size, larger and larger eddies become effective in deforming and spreading the puff. On the other hand, since meander results from the whole-body advection of the puff, correspondingly fewer eddies remain that are larger than the puff and hence capable of advecting it as a whole. For a fixed sampling time τ, the portion of the energy spectrum that can contribute to meander thus shrinks as T increases and approaches the value of τ. As a consequence, the value of meander approaches a constant value (e.g., Lee and Stone [236]; Pasquill and Smith [14]).

7 Mesoscale spectra

Equations (20) and (22) relate mean-square horizontal particle displacement to the KE frequency spectrum for absolute and relative diffusion, respectively. Taylor's transformation based on Taylor's "frozen-turbulence" hypothesis can then be used to convert frequency spectra to wavenumber spectra (e.g., Taylor [237]; Brown and Robinson [238]; Lilly and Petersen [239]; Gage and Nastrom [137]). This conversion is useful because most available observed mesoscale energy spectra are based on spatial rather than temporal measurements.

Figure 8 shows a composite horizontal KE wavenumber spectrum covering the horizontal spatial range from 5 to over 10,000 km. The Global Atmospheric Sampling Program (GASP) data referred to in this figure are a set of observed horizontal wind, temperature, water vapour, ozone, and carbon monoxide measurements taken in the upper troposphere during the period 1975-79 on over 6900 flights by long-haul commercial aircraft, including 311 flights more than 10,000 km in length (Nastrom et al. [240]). The GASP data set was the first data set that permitted the calculation of 'climatological' mesoscale *spatial* energy spectra. (Satellite-borne instruments now offer another large-scale observing tool; for example, Tjemkes and Visser [241] have calculated comparable horizontal wavenumber spectra of temperature, water vapour, and cloud liquid water based on satellite sounder data.)

Two important spectral regimes are shown in Fig. 8. In the 5000 km to 500 km wavelength range (left side of the figure), the slope of the spectrum is about -3 whereas at wavelengths less than 500 km the slope is about -5/3. As discussed in Section 4, a k^{-3} KE spectral slope had been predicted earlier by theoretical models for two-dimensional and geostrophic turbulence for a downscale, enstrophy-cascading inertial range at atmospheric scales smaller than the energy-injecting wavenumbers associated with baroclinic instability (e.g., Kraichnan [117, 118]; Leith [242, 119]; Charney [126]; Lilly [122]). Earlier observed large-scale atmospheric KE spectra had also provided supporting evidence for this prediction (e.g., Chen and Wiin-Nielsen [128]; Balsley and Carter [243]; Boer and Shepherd [130]).

The existence of a mesoscale $k^{-5/3}$ KE spectral-slope region at spatial scales less than 500–800 km had also been inferred earlier by Golitsyn [244] and Gage [245] on the basis of other measurements. Gage [245] and Gage and Nastrom [137, 246] have argued that this spectral region corresponds to a mesoscale *reverse*- or *upscale*-energy-cascade inertial range, and Lilly [141] hypothesized that a fraction of the three-dimensional turbulent energy created at small scales by convective and shearing instabilities 'leaks' upscale to form quasi-two-dimensional 'stratified turbulence' or 'vortical modes' in a process analogous to the 'wake collapse' phenomenon in a stratified environ-

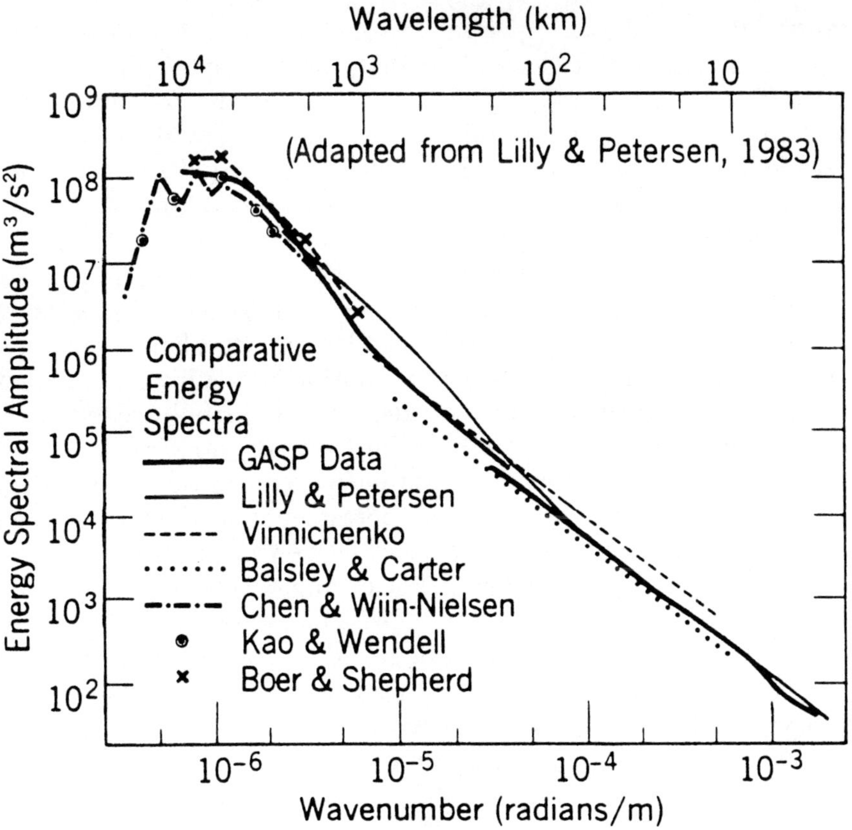

Figure 8: Composite horizontal kinetic-energy spectrum comparing GASP-based spectra to earlier results adapted from Lilly and Petersen [239] (from Nastrom and Gage [131]).

ment (see also Lin and Pao [247]; Riley et al. [248]; McWilliams [136]; Hopfinger [249]; Müller et al. [142]; Etling [205]).

Gifford [125] and Gifford et al. [8] represented the tropospheric turbulence KE spectrum shown in Fig. 8 using a simple energy-spectrum model and similarity-theory arguments (and argued that similarity theory would hold regardless of the *direction* of the energy transfer in the different spectral regions). They then determined that characteristic time scales bounding the enstrophy-cascade spectral region are $2\pi f^{-1}$, and f^{-1}, implying that the transition from the large-scale baroclinic-instability wavenumbers to the enstrophy-cascade spectral range is on the order of a day and the transition from the k^{-3} to the $k^{-5/3}$ spectral region is about three hours (cf. Section 6.2.2).

The mesoscale $k^{-5/3}$ KE spectral-slope regime is significant for a number of reasons. Firstly, there is much more KE present at higher mesoscale wavenumbers (or smaller wavelengths) than can be explained by the decay of geostrophic turbulence (Lilly [141]). Secondly, it reveals much about mesoscale wave and turbulence dynamics (e.g., Gage and Nastrom [250]). Thirdly, it provides a description of mesoscale variability and measurement representativeness (Lilly [251, 252]). And fourthly, it has important implications for numerical weather prediction from the point of view of predictability and subgrid-scale parameterizations. For example, there is no sign of a broad spectral 'mesoscale gap' (e.g., Nastrom and Gage [131]). Nastrom et al. [253] also found that calculated horizontal-variance power spectra of three trace constituents – ozone, water vapour, and carbon monoxide – all followed a $k^{-5/3}$ power law at wavelengths of less than 500–800 km as well (cf. Tjemkes and Visser [241]).

7.1 Implications for MAD

Figure 8 also has a number of implications for MAD. Firstly, the spectrum shown is an ensemble-mean spectrum based on the average of many samples or realizations. For example, in his conceptual model of the $k^{-5/3}$ reverse-cascade mesoscale KE spectrum, Lilly [141] had to address the issue that the injection of turbulent energy by decaying convective elements is *sporadic* in time and space. Other authors have suggested breaking internal waves and shear-induced turbulence as additional sources of mesoscale energy (e.g., Gage and Nastrom [137]), but these mechanisms are also intermittent in time and space. Thus, the appearance of a smooth, continuous spectrum does not mean that KE associated with a seamless downscale or upscale spectral cascade will necessarily be present in all wavenumbers in all atmospheric flow realizations. In fact, this is unlikely to be the case. Otherwise, it is difficult to explain the four cases discussed by Moore et al. [254], in which pairs of 300-hPa constant-level balloons released simultaneously displayed exceedingly small separation after long travel times. In the most extreme case, two balloons were recovered within 2 km of each other after travelling approximately 1800 km from Utah to Wisconsin.

Secondly, MAD is qualitatively different in the enstrophy-cascade region from the reverse-energy-cascade region. As discussed in Section 4.1.1, pollutant clouds can be deformed, distorted, and convolved in the former range but not diluted; relative diffusion or dilution can only be supported in the reverse-energy-cascade region and at higher wavenumbers/smaller spatial scales (e.g., Gifford [125]; Gifford et al. [8]; Maryon and Buckland [9]).

Thirdly, since eddy time and space scales are usually correlated (Section 5.2), then averaging times will have to be much larger than the eddy characteristic time scales of mesoscale eddies (hours to days) in order to estimate mesoscale ensemble averages (e.g., Mellor [109]). The implication for MAD realizations is that mesoscale fluctuations have to be represented individually and explicitly rather than statistically (cf. Section 4.6).

And fourthly, as already noted in Section 4.6, in initializing mesoscale meteorological models, it may not be possible to include much mesoscale energy in the models's initial flow fields due to the spatial coarseness of the synoptic observing network even though such energy may be present in the real atmosphere.

8 Shear-enhanced dispersion

The presence of velocity gradients in the mean flow complicates the treatment of flow dynamics and dispersion, whatever the scale. For example, many flow instabilities occur at least partly due to flow shear, and velocity gradients introduce differential advection into the dispersion problem. Flow shears may enhance turbulent diffusion by increasing turbulence intensities. In addition, the presence of velocity gradients provides another mechanism for dispersion through the *interaction* of differential advection and turbulent diffusion. In the PBL, for example, vertical shear of the horizontal flow may result in pollutant at different levels being advected at different speeds (speed shear) or in different directions (direction shear). Vertical diffusion will then mix the pollutant through a larger volume than if no shear were present.

For the case of a shear flow, Smith [229] obtained a more complicated version of Taylor's theorem (eqn (9)), which consisted of three terms: one term was identical to the right-hand side of eqn (9), the second term described the direct effect of wind shear independent of horizontal turbulence, and the third term incorporated the correlation of the horizontal and vertical velocity components. One advantage of Smith's statistical treatment of wind shear is that it provides a method with which to estimate the travel distance at which shear effects begin to outweigh simple diffusional spread. For example, Pasquill and Smith [14, p. 177] suggested that shear enhancement might be expected to become important at travel distances on the order of 10 kilometers (see also Gifford [74]).

8.1 Simultaneous vs. delayed shear enhancement

The interaction of turbulent mixing with differential advection can occur at the *same* time or at a *later* time. Differential advection with *simultaneous* mixing has been recognized for many years as a contributing factor to dispersion in shear flows (e.g., Eckart [2]; Taylor [255, 256, 257]; Aris [258]; Saffman [228]; Högström [259]; Tyldesley and Wallington [260]; Csanady [261, 94]; Maul [262]; Taylor [263]; Millán [188]; Lamb et al. [31]), but it appears that Pasquill [264] was the first to mention the combination of differential advection with *subsequent* or *delayed* vertical mixing as a mechanism in MAD.

Pasquill discussed the differences in the contribution of vertical shear to dispersion between night time and daytime in his paper. In the daytime for unstable conditions, vertical turbulent mixing acts to reduce vertical wind shear and hence also reduces shear-enhanced dispersion. However, vertical shear will still be present due to surface friction, baroclinicity, and other sources (e.g., Kahl and Samson

[265]), and its interaction with vertical diffusion results in simultaneous mixing and shear-enhanced lateral diffusion. At night under stable conditions, much larger vertical shears are possible but vertical turbulence levels are lower, resulting in potential or delayed shear dispersion. That is, differential advection of the pollutant is followed later by vertical mixing and dilution, either through turbulent bursts due to Richardson-number breakdown (e.g., Narasimha and Kailas [187]; Nappo [266]) or during the growth of the daytime convective boundary layer after sunrise. This mechanism will also apply to PBL transitions from stable to unstable conditions due to advection as in the case of advection from cold land to warm water in winter or from cold water to warm land in summer.

8.2 The diurnal cycle and the low-level jet

Clearly the diurnal heating/cooling cycle will provide a regular source of vertical wind shear on MAD timescales, including the frequent formation of a nocturnal low-level jet (LLJ). The nocturnal LLJ has been shown to be important for mesoscale transport of pollutants (e.g., Husar et al. [267]; Carhart et al. [102]; Corsmeier et al. [268]), but it will also play an important role in MAD through Pasquill's delayed mixing mechanism.

At night, the formation of a nocturnal surface inversion due to longwave radiational cooling decouples the nocturnal boundary layer from the remainder of the well-mixed daytime PBL above (termed the residual layer by Stull [269]). The residual layer then begins to sweep through an inertial oscillation, resulting in the formation of a nocturnal low-level jet. The increased stratification reduces turbulent mixing, but vertical variations in horizontal advection due to the direction and speed shear caused by the inertial oscillation may then distort a pollutant cloud.

For example, Uthe et al. [221] presented four nighttime lidar scans of a shear-distorted elevated power plant plume about 40 km downwind of its source. Fig. 9 shows a schematic representation of their observations. (Note that they also had to assume that the stack plume rise varies somewhat due to changing meteorological conditions in order to have the pollutant distributed through a finite vertical

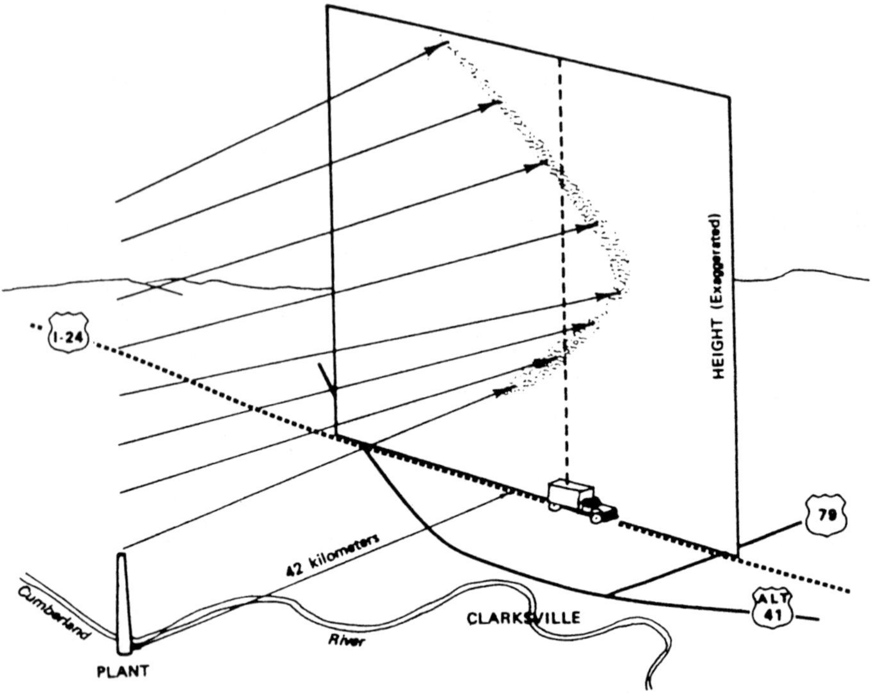

Figure 9: Schematic representation of the effects of directional wind shear and varying plume rise on an elevated power plant plume under stable conditions (from Uthe et al. [221].

depth.) When morning arrives and the daytime PBL begins to grow anew, the shear-distorted cloud again mixes in the vertical, resulting in a wider, ground-level plume 'footprint'.

Carras and Williams [52] showed significant shear contributions to the lateral diffusion of the Mt Isa smelter plume using a simple empirical approach. They calculated a set of constant-height forward trajectories starting from Mt Isa for eight levels within the daytime PBL based on a time sequence of observed three- or six-hourly winds at four meteorological stations located across northern Australia. They then selected one trajectory for each level from a set of sequential releases so that the *arrival* time at the downwind distance of interest was the same even though the eight trajectories selected could have different *release* times. The predicted plume width then corresponded to the cross-wind width of the envelope of trajectories. All levels could be assumed to contribute to the predicted near-surface concentrations since most observations were made at midday when a well-mixed PBL would be expected. Predicted plume widths using this approach agreed with measured values to an accuracy of about $\pm 30\%$.

A number of numerical studies of MAD have also suggested that delayed mixing can play an important role in lateral mesoscale diffusion in the PBL in conjunction with the development of the LLJ. For example, Moran et al. [270] and McNider et al. [103] used a mesoscale meteorological model coupled to a mesoscale Lagrangian particle dispersion model (LPDM) to simulate the diffusion of the Mt Isa plume in Australia. Motivated by these studies, McNider et al. [107] and Singh et al. [271] formulated an analytical model for the contribution of the nocturnal LLJ to MAD.

Moran and Pielke used a mesoscale LPDM coupled with a different mesoscale meteorological model to examine the role of the LLJ in the 1980 Great Plains tracer experiment (Moran [78]; Moran and Pielke [39, 180]) and in CAPTEX (Moran [78]; Moran and Pielke [179]). Carhart et al. [102] compared the predictions of seven other numerical MAD models with observations from the 1980 Great Plains tracer experiment: none of these models represented the occurrence of the nocturnal LLJ during the experiment well and none of their predictions compared very well with observed tracer concentrations and other cloud characteristics such as width and arrival time. Other numerical simulations of one or more of the seven CAPTEX tracer releases include Draxler [87], Lee [272], Kao and Yamada [273], Brost et al. [274], Godowitch [275], and Shi et al. [276]. Based on their simulations of three CAPTEX tracer releases with a displacement-based LPDM driven by diagnosed wind and temperature fields, Shi et al. [276, p. 3701] concluded that "... the dominant dispersion mechanism over long distances is the combination of vertical mixing and vertical wind shears".

9 Summary

Mesoscale atmospheric dispersion has been defined in this chapter as the transport and diffusion of air pollutants or other atmospheric constituents over horizontal distances of 2 to 2000 km and time periods of 1 to 48 hours. Conventional wisdom used to hold that for pollutant dispersion over mesoscale distances, sufficient dilution will have taken place for there to be little practical need to know pollutant concentrations at these distances. More recently, however, it has been recognized that (a) if pollutant sources are strong, numerous, or long-lived enough or (b) if the pollutants themselves are sufficiently toxic or active in some way, then ambient pollutant concentrations or long-term dosages may still be significant even after mesoscale dispersion and dilution.

The multi-country radionuclide contamination experienced during the 1986 Chernobyl disaster in the former U.S.S.R. is one recent example. Other anthropogenic pollutants or their precursors are responsible for regional-scale acidic deposition, oxidant episodes, and visibility degradation even after transport and diffusion over mesoscale distances. Heavy metals in aerosol form and synthetic organics such as pesticides can be transported to, and then accumulate in, sensitive regions. In all of these cases even remote areas such as the Arctic, which are far from the sources of such pollutants, can be affected. Natural aerosols produced by volcanic eruptions, forest fires, or sandstorms, and biological organisms such as pollens, seeds, bacteria, spores, viruses, and insects, can also affect regions far from their source.

Observations of MAD have been obtained from balloon releases, from formal mesoscale tracer experiments, and from so-called 'sources of opportunity' such as large point sources, urban plumes, major fires, volcanic eruptions, and both accidental and intentional radionuclide releases. Formal tracer experiments have the advantages of control over (a) the magnitude and timing of the release(s) of the selected tracer(s) and (b) the design and implementation of the sampling network and the sampling protocol.

Mesoscale atmospheric dispersion was shown in this chapter to be an even more complicated phenomenon than short-range dispersion since additional atmospheric processes may play a role and some simplifying assumptions made in studying short-range dispersion are not applicable. For example, over mesoscale time and space scales, the mean wind field can seldom be considered steady or horizontally homogeneous. In fact, it is very important to specify the operational definitions of turbulence and diffusion – that is, the type of averaging and the scale of averaging being employed – when distinguishing between mean and fluctuating quantities on these scales since neither the ergodic hypothesis nor the Reynolds conditions are likely to hold. Possible averages include space, time, space-time, and ensemble averages. Also, MAD 'realizations' such as mesoscale tracer experiments are often time-varying and several days long so that it is seldom possible to construct sample ensemble averages; individual realizations are studied instead.

Other differences between mesoscale and shorter-range dispersion were noted. Horizontal diffusion is usually more important than vertical diffusion in MAD as compared to short-range dispersion due to the limited vertical extent of the PBL and the free troposphere. Vertical shear often plays a much more important role in MAD, as does vertical transport due to synoptic-scale or mesoscale circulations or to cloud venting. Mesoscale concentration fields also tend to be more complex, with multiple maxima.

The nature of atmospheric turbulence also depends on the time and space scales of concern. Atmospheric turbulence occurs across a wide range of flow scales: in addition to the familiar three-dimensional, PBL inertial-range turbulence, there is also stratified turbulence, geostrophic turbulence, and two-dimensional turbulence, but all of these turbulent flows have in common irregular, aperiodic randomness, complex nonlinear dynamics, a continuous spectrum of eddies, and statistically distinct average values. Turbulence is also local, non-propagating, incoherent, and strongly dissipative. One important property of small-scale and mesoscale turbulence is its ability to disperse fluid particles initially grouped together. Geostrophic and two-dimensional turbulence can also separate particles through deformation and folding but cannot reduce pollutant concentrations through dilution. Stratified fluids can also support other flow modes, including IGW and vortical modes, and random fields of these flow modes have some properties in common with turbulence, but not the strong capability to disperse fluid particles that are initially grouped together.

Types of pollutant releases and types of diffusion were also reviewed in this chapter. Pollutant releases range in duration from instantaneous *puff*-type releases to continuous *plume*-type releases. There are also two basic forms of turbulent diffusion: (a) diffusion of a cloud about its center of mass, and (b) diffusion from a fixed, continuously emitting source. The former, which is known as relative diffusion or two-particle diffusion, requires a Lagrangian reference frame moving with the center of mass of the fluid element, while the latter, which is known as absolute diffusion or one-particle diffusion, requires a fixed or Eulerian reference frame. While puff releases are generally associated with relative diffusion, and plume releases are most often associated with absolute diffusion, what actually distinguishes these two basic types of turbulent diffusion from each other is the analysis reference frame employed, since it is also possible to consider the absolute diffusion of puffs and the relative diffusion of plumes.

Puff growth and dilution is determined by the relative size of the puff and the dominant motions of the flow field in which it is embedded. Flow elements or eddies that are large compared to the puff advect it as a whole, eddies comparable in size to the puff deform and mix it rapidly, and eddies much smaller than the puff contribute to slow growth and dilution at its boundaries. For a plume, the whole-body advection by flow elements larger than the plume is called meander while smaller scales contribute to relative diffusion and dilution. The absolute diffusion of a plume is the sum of meander and relative diffusion. A plume can be approximated as a sequential release of puffs, but puff growth cannot be represented as the superposition of the growth of puff component elements since the rate of growth of a puff depends directly on the size of the puff. Also, puffs can grow in three dimensions, but plumes can grow only in two dimensions.

Characteristic time scales for MAD include the diurnal period and daylight period, the inertial period, release duration, travel time, and sampling time. Characteristic space scales include the physiographic or synoptic forcing length scale, the Rossby radius of deformation, the source size, the cloud size, and the travel distance.

The release duration, travel time, and sampling time determine the type of diffusion being considered. If either release time or sampling time is much smaller than travel time, then a puff-diffusion framework is appropriate. If both sampling time and release time are much greater than travel time, then continuous plume diffusion is being considered. For plume diffusion, sampling time provides an operational partitioning between turbulence and the mean flow: eddies with shorter lifetimes than the sampling time contribute to turbulent diffusion, whereas eddies with longer lifetimes contribute to the mean flow. In effect the sampling time defines a high-pass spectral filter. Plume widths vary directly with sampling time while maximum centerline concentrations vary indirectly with sampling time. Travel time, on the other hand, acts as an effective averaging time for plume diffusion that smooths out the contributions of smaller eddy time scales; it thus defines a low-pass spectral filter. For puff diffusion, the situation is quite different: sampling time is instantaneous and travel time acts as the effective sampling time, since it determines (indirectly through its relationship to cloud size) the maximum eddy time scale that can affect puff growth .

The distinction between absolute and relative diffusion is less important for MAD than for shorter-range dispersion. For travel times much larger than the Lagrangian integral time scale T_L, the mean-square displacement due to plume meander approaches a constant value so that absolute and relative diffusion have the same time dependence. Absolute and relative plume diffusion also become indistinguishable as travel time approaches sampling time, but in this case it is because the portion of the turbulent KE spectrum contributing to plume meander approaches zero rather than a non-zero asymptotic limit.

Ensemble-average mesoscale KE wavenumber spectra have been shown to have both a downscale enstrophy-cascade k^{-3} inertial range and an upscale reverse-energy-cascade $k^{-5/3}$ inertial range. The latter should be able to support an accelerated relative-diffusion regime whereas the k^{-3} inertial range cannot contribute directly to relative diffusion. It is unlikely, though, that an individual mesoscale flow field will possess a smooth, continuous KE spectrum since the energy required to maintain the upscale $k^{-5/3}$ inertial range will be injected sporadically in both time and space.

Finally, as described in Section 8, horizontal diffusion can be enhanced by vertical shear through either the simultaneous or delayed interaction of horizontal differential advection and vertical diffusion over one or more diurnal periods. Pasquill [264] appears to have been the first to mention the combination of differential advection with *delayed* vertical mixing as a mechanism in MAD. This turbulent-dispersion mechanism has been shown to be important in several MAD case studies, particularly in the presence of a nocturnal low-level jet. In general, delayed shear enhancement is likely to occur (a) for warm-season overland PBL dispersion over one or more diurnal cycles, (b) for transport from stable to unstable conditions such as cold-air outbreaks, and (c) for intermittent turbulent mixing associated with Kelvin-Helmholtz instability or moist convection.

Acknowledgements

Much of the material in this chapter has been drawn from my doctoral dissertation [78], and I would again like to thank my doctoral advisor, Prof. Roger A. Pielke, Sr. of Colorado State University, for his guidance and support during my doctoral studies at that institution. Dr. Paul Makar of the Meteorological Service of Canada and Prof. Grant Edwards of the University of Guelph provided helpful comments on draft versions of this chapter. I am also obliged to the following organizations and companies for their kind permission to reproduce previously published illustrations: the Swedish Geophysical Society, publisher of *Tellus*, for Fig. 1; Elsevier Science Ltd., publishers of *Atmospheric Environment*, for Figs. 2 and 3; The Institute of Mathematics and its Applications, Southend-on-Sea, England, for Fig. 6; the American Meteorological Society for Fig. 7; and the Air and Waste Management Association, publisher of the *Journal of the Air Pollution Control Association* [now the *Journal of the Air and Waste Management Association*], for Fig. 9.

References

[1] R. A. Pielke, R. W. Arritt, M. Segal, M. D. Moran and R. T. McNider, Mesoscale numerical modeling of pollutant transport in complex terrain. *Boundary-Layer Meteor.*, **41**, pp. 59–74, 1987.

[2] C. Eckart, An analysis of the stirring and mixing processes in incompressible fluids. *J. Mar. Res.*, **7**, pp. 265–275, 1948.

[3] C. Garrett, On the initial streakiness of a dispersing tracer in two- and three-dimensional turbulence. *Dyn. Atmos. Oceans*, **7**, pp. 265–277, 1983.

[4] G. T. Csanady, *Turbulent Diffusion in the Environment.* D. Reidel Publishing Co., Dordrecht, 1973.

[5] P. Welander, Studies on the general development of motion in a two-dimensional ideal fluid. *Tellus*, **7**, pp. 141–156, 1955.

[6] J. O. Hinze, *Turbulence, 2nd Edition.* McGraw-Hill Book Co., New York, 1975.

[7] S.-K. Kao, Basic characteristics of global scale diffusion in the troposphere. *Adv. Geophys.*, **18B**, pp. 15–32, 1974.

[8] F. A. Gifford, S. Barr, R. C. Malone and E. J. Mroz, Tropospheric relative diffusion to hemispheric scales. *Atmos. Environ.*, **22**, pp. 1871–1879, 1988.

[9] R.H. Maryon and A.T. Buckland, Tropospheric dispersion: the first ten days after a puff release. *Quart. J. Roy. Meteor. Soc.*, **121**, pp. 1799–1833, 1995.

[10] F. N. Frenkiel, Application of the statistical theory of turbulent diffusion to micrometeorology. *J. Meteor.*, **9**, pp. 252–259, 1952.

[11] G. K. Batchelor, *The Theory of Homogeneous Turbulence.* Cambridge University Press, 1953.

[12] A. S. Monin and A. M. Yaglom, *Statistical Fluid Mechanics, Vol. 1.* MIT Press, Cambridge, Massachusetts, 1971.

[13] H. Tennekes and J. L. Lumley, *A First Course in Turbulence.* MIT Press, Cambridge, Massachussetts, 1972.

[14] F. Pasquill and F. B. Smith, *Atmospheric Diffusion, Third Edition.* John Wiley & Sons, New York, 1983.

[15] L. F. Richardson, *Weather Prediction by Numerical Process.* Cambridge University Press, 1922. (Reprinted by Dover Publications, New York, 1965).

[16] C. Persson, H. Rodhe and L.-E. De Geer, The Chernobyl accident — a meteorological analysis of how radionuclides reached and were deposited in Sweden. *Ambio*, **16**, pp. 20–31, 1987.

[17] F. B. Smith and M. J. Clark, Wet and dry deposition of Chernobyl releases. *Nature*, **322**, pp. 245–249, 1988.

[18] J. Crabtree, The travel and diffusion of the radioactive material emitted during the Windscale accident. *Quart. J. Roy. Meteor. Soc.*, **85**, pp. 362–370, 1959.

[19] H. W. Feely, H. Seitz, R. J. Lagomarsino and P. E. Biscaye, Transport and fallout of stratospheric radioactive debris. *Tellus*, **18**, pp. 316–327, 1966.

[20] R. J. List, L. P. Salter and K. Telegadas, Radioactive debris as a tracer for investigating stratospheric motions. *Tellus*, **18**, pp. 345–353, 1966.

[21] E. R. Reiter, *Atmospheric Transport Processes. Part 4: Radioactive Tracers.* Technical Information Center, U. S. Dept. of Energy, Washington, D. C., 1978. (Available as TID-27114 from National Technical Information Service, U.S. Department of Commerce, Springfield, Virginia 22161).

[22] W. A. Lyons, Evidence of transport of hazy air masses from satellite imagery. *Ann. N. Y. Acad. Sci.*, **338**, pp. 418–433, 1980.

[23] NRC, *Acid Deposition: Atmospheric Processes in Eastern North America.* National Research Council, National Academy Press, Washington, D. C., 1983.

[24] J.A. Pudykiewicz, A. Kallaur and P.K. Smolarkiewicz, Semi-Lagrangian modelling of tropospheric ozone. *Tellus*, **49B**, pp. 231–248, 1997.

[25] I.J. Ackermann, H. Hass, M. Memmesheimer, A. Ebel, F.S. Binkowski and U. Shankar, Modal aerosol dynamics model for Europe: Development and first applications. *Atmos. Environ.*, **32**, pp. 2981–2999, 1998.

[26] R. S. Stolarski, The Antarctic ozone hole. *Sci. Amer.*, **258**, pp. January, 30–36, 1988.

[27] N. B. A. Trivett, L. A. Barrie, J. W. Bottenheim, J.-P. Blanchet, G. den Hartog, R. M. Hoff and R. E. Mickle, An experimental investigation of Arctic haze at Alert, N. W. T., March 1985. *Atmosphere–Ocean*, **26**, pp. 341–376, 1988.

[28] W. Maenhaut, P. Cornille, J. M. Pacyna and V. Vitols, Trace element composition and origin of the atmospheric aerosol in the Norwegian Arctic. *Atmos. Environ.*, **23**, pp. 2551–2569, 1989.

[29] L.A. Barrie. Occurrence and trends of pollution in the Arctic troposphere, In E.W. Wolff and R.C. Bales, editors, *Chemical Exchange Between the Atmosphere and Polar Snow*, volume NATO ASI Series, Vol. I43, pages 93–129. Springer-Verlag, Berlin, 1996.

[30] E. Aguado, Effect of advected pollutants on solar radiation attenuation: Mojave Desert, California. *Atmos. Environ.*, **24B**, pp. 153–157, 1990.

[31] B. K. Lamb, X.-F. Zhang, R. E. Eskridge, R. Benner, H. Westberg, J. Mitchell, X.-C. Zou and J. Li, Elevated plume transport and dispersion: 20–150 km downwind of Beijing, P.R.C. *Atmos. Environ.*, **24A**, pp. 859–870, 1990.

[32] M. Trainer, B.A Ridley, M.P. Buhr, G. Kok, J. Walega, G. Hübler, D.D. Parrish and F.C. Fehsenfeld, Regional ozone and urban plumes in the southeastern United States: Birmingham, a case study. *J. Geophys. Res.*, **100**, pp. 18823–18834, 1995.

[33] J. Crabtree and M. Kitchen, The long-range travel and dispersion of the plume from the Mount St Helens volcano. *Atmos. Environ.*, **18**, pp. 1073–1079, 1984.

[34] Y. S. Chung, Air pollution detection by satellites: the transport and deposition of air pollutants over oceans. *Atmos. Environ.*, **20**, pp. 617–630, 1986.

[35] G. A. D'Almeida, A model for Saharan dust transport. *J. Climate Appl. Meteor.*, **25**, pp. 903–916, 1986.

[36] M. Segal, J. Weaver and J. F. W. Purdom, Some effects of the Yellowstone fire smoke plume on northeast Colorado at the end of summer 1988. *Mon. Wea. Rev.*, **117**, pp. 2278–2284, 1989.

[37] R. A. Ferrare, R. S. Fraser and Y. J. Kaufman, Satellite measurements of large-scale air pollution: measurements of forest fire smoke. *J. Geophys. Res.*, **95**, pp. 9911–9925, 1990.

[38] D. E. Pedgley, *Windborne Pests and Diseases.* Ellis-Horwood Ltd, Chichester, 1982.

[39] M. D. Moran and R. A. Pielke, Evaluation of a mesoscale atmospheric dispersion modeling system with observations from the 1980 Great Plains mesoscale tracer experiment. Part I: Datasets and meteorological simulations. *J. Appl. Meteor.*, **35**, pp. 281–307, 1996.

[40] J. K. Angell, D. H. Pack, L. Machta, C. R. Dickson and W. H. Hoecker, Three-dimensional air trajectories determined from tetroon flights in the planetary boundary layer of the Los Angeles Basin. *J. Appl. Meteor.*, **11**, pp. 451–471, 1972.

[41] D. H. Pack, G. J. Ferber, J. L. Heffter, K. Telegadas, J. K. Angell, W. H. Hoecker and L. Machta, Meteorology of long-range transport. *Atmos. Environ.*, **12**, pp. 425–444, 1978.

[42] J. F. Clarke, T. L. Clark, J. K. S. Ching, P. L. Haagenson, R. B. Husar and D. E. Patterson, Assessment of model simulation of long-distance transport. *Atmos. Environ.*, **17**, pp. 2449–2462, 1983.

[43] J. K. Angell, On the use of tetroons for the estimation of atmospheric dispersion on the mesoscale. *Mon. Wea. Rev.*, **90**, pp. 263–270, 1962.

[44] P. Thomas and S. Vogt, Mesoscale atmospheric dispersion experiments using tracer and tetroons. *Atmos. Environ.*, **24A**, pp. 1271–1284, 1990.

[45] J. K. Angell and D. H. Pack, Atmospheric lateral diffusion estimates from tetroons. *J. Appl. Meteor.*, **4**, pp. 418–425, 1965.

[46] S.-K. Kao and A. A. al-Gain, Large-scale dispersion of clusters of particles in the atmosphere. *J. Atmos. Sci.*, **25**, pp. 214–221, 1968.

[47] J. Er-El and R. L. Peskin, Relative diffusion of constant-level balloons in the Southern Hemisphere. *J. Atmos. Sci.*, **38**, pp. 2264–2274, 1981.

[48] K. R. Peterson, Continuous point source behavior out to 160 miles. *J. Appl. Meteor.*, **7**, pp. 217–226, 1968.

[49] D. Randerson, J. G. Garcia and V. S. Whitehead, Photogrammetric and photometric investigation of a smoke plume viewed from space. *J. Appl. Meteor.*, **10**, pp. 1122–1131, 1971.

[50] M. M. Millán and Y. S. Chung, Detection of a plume 400 km from the source. *Atmos. Environ.*, **11**, pp. 939–944, 1977.

[51] N. V. Gillani, Project MISST: mesoscale plume modeling of the dispersion, transformation and ground removal of SO_2. *Atmos. Environ.*, **12**, pp. 569–588, 1978.

[52] J. N. Carras and D. J. Williams, The long-range dispersion of a plume from an isolated point source. *Atmos. Environ.*, **15**, pp. 2205–2217, 1981.

[53] J. N. Carras and D. J. Williams, Measurements of relative σ_y up to 1800 km from a single source. *Atmos. Environ.*, **22**, pp. 1061–1069, 1988.

[54] R. M. Hoff and A. J. Gallant, The use of an available SO_2 tracer during the 1983 CAPTEX experiment. *Atmos. Environ.*, **19**, pp. 1573–1575, 1985.

[55] R. M. Brown, L. A. Cohen and M. E. Smith, Diffusion measurements in the 10–100 km range. *J. Appl. Meteor.*, **11**, pp. 323–334, 1972.

[56] P. L. Haagenson and A. L. Morris, Forecasting the behavior of the St. Louis, Missouri, pollutant plume. *J. Appl. Meteor.*, **13**, pp. 901–909, 1974.

[57] H. G. Bastable, D. P. Rogers and D. E. Schorran, Tracers of opportunity and pollutant transport in southern California. *Atmos. Environ.*, **24B**, pp. 137–151, 1990.

[58] R. W. Davies, Large-scale diffusion from an oil fire. *Adv. Geophys.*, **6**, pp. 413–415, 1959.

[59] D. Randerson, A study of air pollution sources as viewed by earth satellites. *J. Air Pollu. Control Assoc.*, **18**, pp. 249–253, 1968.

[60] D. Randerson, Temporal changes in horizontal diffusion parameters of a single nuclear debris cloud. *J. Appl. Meteor.*, **11**, pp. 670–673, 1972.

[61] D. Randerson. Quantitative analysis of atmospheric pollution phenomena, In *Skylab Explores the Earth, NASA SP-380*, pages 381–405. Scientific and Technical Information Office, National Aeronautics and Space Administration, Washington, D.C., 1977.

[62] S. S. Limaye, V. E. Suomi, C. Velden and G. Tripoli, Satellite observations of smoke from oil fires in Kuwait. *Science*, **252**, pp. 1536–1539, 1991.

[63] Y. S. Chung, A. Gallant, F. Fanaki and M. Millán, On the observations of Mount St Helens volcanic emissions. *Atmosphere–Ocean*, **19**, pp. 172–178, 1981.

[64] D. A. Wheeler, Atmospheric dispersal and deposition of radioactive material from Chernobyl. *Atmos. Environ.*, **22**, pp. 853–863, 1988.

[65] J. L. Heffter, The variation of horizontal diffusion parameters with time for travel periods of one hour or longer. *J. Appl. Meteor.*, **4**, pp. 153–156, 1965.

[66] W. F. Dabberdt and R. N. Dietz. Gaseous tracer technology and applications, In *Probing the Atmospheric Boundary Layer*, pages 103–128. American Meteorological Society, Boston, 1986.

[67] G. J. Ferber, J. L. Heffter, R. R. Draxler, R. J. Lagomarsino, F. L. Thomas, R. N. Dietz and C. M. Benkovitz, Cross-Appalachian Tracer Experiment (CAPTEX '83) final report. NOAA Tech. Memo. ERL ARL-142, Air Resources Laboratory, National Oceanic and Atmospheric Administration, Silver Spring, Maryland, January 1986.

[68] R. J. Lagomarsino, T. J. Weber and N. Latner, The Across North America Tracer Experiment (ANATEX). Volume IV:

Quality assurance of ground-level samplers and quality assessment of EML analysis of surface samples. Report No. EML-523, Environmental Measurements Laboratory, U. S. Department of Energy, New York, New York, February 1991.

[69] G. J. Ferber, K. Telegadas, J. L. Heffter, C. R. Dickson, R. N. Dietz and P. W. Krey, Demonstration of a long-range atmospheric tracer system using perfluorocarbons. NOAA Tech. Memo. ERL ARL-101, Air Resources Laboratory, National Oceanic and Atmospheric Administration, Silver Spring, Maryland, April 1981.

[70] S.-E. Gryning, The ø experiment — a Nordic mesoscale dispersion experiment over a land-water-land area. *Bull. Amer. Meteor. Soc.*, **66**, pp. 1403–1407, 1985.

[71] R. R. Draxler and J. L. Heffter, editors, *Across North America Tracer Experiment (ANATEX). Volume I: Description, ground-level sampling at primary sites, and meteorology*, Silver Spring, Maryland, 83 pp, 1989. NOAA Tech. Memo. ERL ARL-167, January, Air Resources Laboratory, National Oceanic and Atmospheric Administration.

[72] P. L. Haagenson, K. Gao and Y.-H. Kuo, Evaluation of meteorological analyses, simulations, and long-range transport calculations using ANATEX surface tracer data. *J. Appl. Meteor.*, **29**, pp. 1268–1283, 1990.

[73] E. Bauer, Dispersion of tracers in the atmosphere and ocean: survey and comparison of experimental data. *J. Geophys. Res.*, **79**, pp. 789–795, 1974.

[74] F. A. Gifford, Tropospheric relative diffusion observations. *J. Appl. Meteor.*, **16**, pp. 311–313, 1977.

[75] F. A. Gifford, Atmospheric diffusion in the mesoscale range: the evidence of recent plume width observations. In *Proc. 6th AMS Symp. on Turbulence and Diffusion*, pages 300–304, March 22-25, American Meteorological Society, Boston, 1983.

[76] F. A. Gifford. Atmospheric diffusion in the range 20 to 2000 kilometers, In C. De Wispelaere, F. A. Schiermeier and N. V. Gillani, editors, *Air Pollution Modeling and Its Application V*, pages 247–252. Plenum Press, New York, 1986.

[77] R. R. Draxler. Diffusion and transport experiments, In D. Randerson, editor, *Atmospheric Science and Power Production*, chapter 9, pages 367–422. Report No. DOE/TIC-27601, Office of Scientific and Technical Information, U.S. Department of Energy, Washington, D.C., 1984.

[78] M. D. Moran, Numerical modelling of mesoscale atmospheric dispersion. Atmospheric Science Paper 513, Department of Atmospheric Science, Colorado State University, Fort Collins, Colorado, 1992. (Available from Prof. R.A. Pielke, Department of Atmospheric Science, Colorado State University, Fort Collins, CO 80523).

[79] F. Pasquill, Limitations and prospects in the estimation of dispersion of pollution on a regional scale. *Adv. Geophys.*, **18B**, pp. 1–13, 1974.

[80] S. R. Hanna, G. A. Briggs and R. P. Hosker, Jr., *Handbook on Atmospheric Diffusion.* DOE/TIC-11223, U.S. Department of Energy, Washington, D.C., 1982. (Available from the National Technical Information Service, U.S. Department of Commerce, Springfield, Virginia 22161).

[81] L. L. Wendell, Mesoscale wind fields and transport estimates determined from a network of wind towers. *Mon. Wea. Rev.*, **100**, pp. 565–578, 1972.

[82] B. W. Atkinson, *Meso-scale Atmospheric Circulations.* Academic Press, New York, 1981.

[83] R. R. Dickerson, G. J. Huffman, W. T. Luke, L. J. Nunnermacker, K. E. Pickering, A. C. D. Leslie, C. G. Lindsey, W. G. N.

Slinn, T. J. Kelly, P. H. Daum, A. C. Delany, J. P. Greenberg, P. R. Zimmerman, J. F. Boatman, J. D. Ray and D. H. Stedman, Thunderstorms: An important mechanism in the transport of air pollutants. *Science*, **235**, pp. 460–465, 1987.

[84] J. K. S. Ching, S. T. Shipley and E. V. Browell, Evidence for cloud venting of mixed layer ozone and aerosols. *Atmos. Environ.*, **22**, pp. 225–242, 1988.

[85] B. E. Martner and R. A. Kropfli, TRACIR: A radar technique for observing the exchange of air between clouds and their environment. *Atmos. Environ.*, **23**, pp. 2715–2721, 1989.

[86] F. M. Vukovich and J. K. S. Ching, A semi-empirical approach to estimate vertical transport by nonprecipitating convective clouds on a regional scale. *Atmos. Environ.*, **24A**, pp. 2153–2168, 1990.

[87] R. R. Draxler, Sensitivity of a trajectory model to the spatial and temporal resolution of the meteorological data during CAPTEX. *J. Climate Appl. Meteor.*, **26**, pp. 1577–1588, 1987.

[88] P. L. Haagenson, Y.-H. Kuo, M. Skumanich and N. L. Seaman, Tracer verification of trajectory models. *J. Climate Appl. Meteor.*, **26**, pp. 410–426, 1987.

[89] D. Martin, C. Mithieux and B. Strauss, On the use of the synoptic vertical wind component in a transport trajectory model. *Atmos. Environ.*, **21**, pp. 45–52, 1987.

[90] M. Segal, R. A. Pielke, R. W. Arritt, M. D. Moran, C.-H. Yu and D. Henderson, Application of a mesoscale atmospheric dispersion modeling system to the estimation of SO_2 concentrations from major elevated sources in southern Florida. *Atmos. Environ.*, **22**, pp. 1319–1334, 1988.

[91] M. Uliasz, Development of the mesoscale dispersion modeling system using personal computers. Part II: Numerical simulations. *Z. Meteor.*, **40**, pp. 285–298, 1990.

[92] A.M. Thompson, K.E. Pickering, R.R. Dickerson, W.G. Ellis, Jr., D.J. Jacob, J.R. Scala, W.-K. Tao, D.P. McNamara and J. Simpson, Convective transport over the central United States and its role in regional CO and ozone budgets. *J. Geophys. Res.*, **99**, pp. 18703–18711, 1994.

[93] F. A. Gifford. An outline of theories of diffusion in the lower layers of the atmosphere, In D. Slade, editor, *Meteorology and Atomic Energy 1968*, chapter 3, pages 65–116. U.S. Atomic Energy Commission, Washington, D.C., 1968. (Available as Publication TID-24190 from the National Technical Information Service, U.S. Department of Commerce, Springfield, Virginia 22161).

[94] G. T. Csanady, Crosswind shear effects on atmospheric diffusion. *Atmos. Environ.*, **6**, pp. 221–232, 1972.

[95] R. R. Draxler, Overview and preliminary results from the Across North America Tracer Experiment (ANATEX). In *Proc. Sixth AMS/APCA Conf. on Applications of Air Pollution Meteorology*, pages 82–85, Jan. 30–Feb. 3, Anaheim, California, American Meteorological Society, Boston, 1989.

[96] G. I. Taylor, Diffusion by continuous movements. *Proc. London Math. Soc. Ser. 2*, **20**, pp. 196–212, 1921.

[97] S. R. Hanna, Lagrangian and Eulerian time-scale relations in the daytime boundary layer. *J. Appl. Meteor.*, **20**, pp. 242–249, 1981.

[98] F. A. Gifford, The time-scale of atmospheric diffusion considered in relation to the universal diffusion function, f_1. *Atmos. Environ.*, **21**, pp. 1315–1320, 1987.

[99] H. A. Panofsky, Discussion of "A Laboratory Study of Dispersion from an Elevated Source within a Modeled Convective Planetary Boundary Layer". *Atmos. Environ.*, **12**, pp. 2036, 1978.

[100] J. N. Carras and D. J. Williams, Experimental studies of plume dispersion in convective conditions — II. Horizontal plume dispersion. *Atmos. Environ.*, **20**, pp. 2307–2314, 1986.

[101] J. L. Heffter, Air Resources Laboratory Atmospheric Transport and Dispersion model (ARL-ATAD). NOAA Tech. Memo. ERL ARL-81, Air Resources Laboratory, National Oceanic and Atmospheric Administration, Silver Spring, Maryland, February 1980.

[102] R. A. Carhart, A. J. Policastro, M. Wastag and L. Coke, Evaluation of eight short-term long-range transport models using field data. *Atmos. Environ.*, **23**, pp. 85–105, 1989.

[103] R. T. McNider, M. D. Moran and R. A. Pielke, Influence of diurnal and inertial boundary-layer oscillations in long-range dispersion. *Atmos. Environ.*, **22**, pp. 2445–2462, 1988.

[104] F. A. Gifford, The random force theory: applications to meso- and large-scale atmospheric diffusion. *Boundary-Layer Meteor.*, **30**, pp. 159–175, 1984.

[105] F. A. Gifford, Does atmospheric turbulence have an outer scale? *Agric. Forest Meteor.*, **47**, pp. 155–161, 1989.

[106] S. Barr and F. A. Gifford, The random force theory applied to regional scale tropospheric diffusion. *Atmos. Environ.*, **21**, pp. 1737–1741, 1987.

[107] R. T. McNider, M. P. Singh and J. T. Lin, Diurnal wind-structure variations and dispersion of pollutants in the boundary layer. *Atmos. Environ.*, **27A**, pp. 2199–2214, 1993.

[108] C. M. Sheih, On lateral dispersion coefficients as functions of averaging time. *J. Appl. Meteor.*, **19**, pp. 557–561, 1980.

[109] G. L. Mellor, Ensemble average turbulence closure. *Adv. Geophys.*, **28B**, pp. 345–358, 1985.

[110] J. W. Hutchings, Turbulence theory applied to large-scale atmospheric phenomena. *J. Meteor.*, **12**, pp. 263–271, 1955.

[111] B. Saltzman, Equations governing the energetics of the larger scales of atmospheric turbulence in the domain of wave number. *J. Meteor.*, **14**, pp. 513–523, 1957.

[112] A. Defant, Die Bestimmung der Turbulenzgrössen der atmosphärischen Zirkulation aussertropischer Breiten. *Sitz.-Ber. Akad. Wiss. Wien*, **130**, pp. 383–403, 1921.

[113] L. Onsager, Statistical hydrodynamics. *Nuovo Cimento (Suppl.)*, **6**, pp. 279–287, 1949.

[114] T. D. Lee, Difference between turbulence in a two-dimensional fluid and in a three-dimensional fluid. *J. Appl. Phys.*, **22**, pp. 524, 1951.

[115] Y. Ogura, The structure of two-dimensionally isotropic turbulence. *J. Meteor. Soc. Japan*, **30**, pp. 59–64, 1952.

[116] R. Fjørtoft, On the changes in the spectral distribution of kinetic energy for twodimensional, nondivergent flow. *Tellus*, **5**, pp. 225–230, 1953.

[117] R. H. Kraichnan, Inertial ranges in two-dimensional turbulence. *Phys. Fluids*, **10**, pp. 1417–1423, 1967.

[118] R. H. Kraichnan, Inertial-range transfer in two- and three-dimensional turbulence. *J. Fluid Mech.*, **47**, pp. 525–535, 1971.

[119] C. E. Leith, Atmospheric predictability and two-dimensional turbulence. *J. Atmos. Sci.*, **28**, pp. 145–161, 1971.

[120] J.-T. Lin, Relative dispersion in the enstrophy-cascading inertial range of homogeneous two-dimensional turbulence. *J. Atmos. Sci.*, **29**, pp. 394–396, 1972.

[121] R. H. Kraichnan and D. Montgomery, Two-dimensional turbulence. *Rep. Prog. Phys.*, **43**, pp. 547–619, 1980.

[122] D. K. Lilly, Two-dimensional turbulence generated by energy at two scales. *J. Atmos. Sci.*, **46**, pp. 2026–2030, 1989.

[123] H. Tennekes. A comparative pathology of atmospheric turbulence in two and three dimensions, In M. Ghil, R. Benzi and G. Parisi, editors, *Turbulence and Predictability in Geophysical Fluid Dynamics and Climate Dynamics*, pages 45–70. North-Holland Physics Publishing, Amsterdam, 1985.

[124] G. K. Batchelor, Computation of the energy spectrum in homogeneous two-dimensional turbulence. *Phys. Fluids Suppl. II*,, pages 233–239, 1969.

[125] F. A. Gifford, A similarity theory of the tropospheric turbulence energy spectrum. *J. Atmos. Sci.*, **45**, pp. 1370–1379, 1988.

[126] J. G. Charney, Geostrophic turbulence. *J. Atmos. Sci.*, **28**, pp. 1087–1095, 1971.

[127] F. Baer, Hemispherical spectral statistics of APE. *J. Atmos. Sci.*, **74**, pp. 932–941, 1974.

[128] T.-C. Chen and A. Wiin-Nielsen, On nonlinear cascades of atmospheric energy and enstrophy in a two-dimensional spectral index. *Tellus*, **30**, pp. 313–322, 1978.

[129] S. J. Lambert, A diagnostic study of global energy and enstrophy fluxes and spectra. *Tellus*, **33**, pp. 411–414, 1981.

[130] G. J. Boer and T. G. Shepherd, Large-scale two-dimensional turbulence in the atmosphere. *J. Atmos. Sci.*, **40**, pp. 164–184, 1983.

[131] G. D. Nastrom and K. S. Gage, A climatology of atmospheric wavenumber spectra of wind and temperature observed by commercial aircraft. *J. Atmos. Sci.*, **42**, pp. 950–960, 1985.

[132] O. Morita and M. Uryu, Geostrophic turbulence in a rotating annulus of fluid. *J. Atmos. Sci.*, **46**, pp. 2349–2355, 1989.

[133] P. B. Rhines. Geostrophic turbulence, In *Annual Review of Fluid Mechanics, Vol. 11*, pages 401–441. Annual Reviews, Inc., Palo Alto, California, 1979.

[134] J. R. Herring, Statistical theory of quasi-geostrophic turbulence. *J. Atmos. Sci.*, **37**, pp. 969–977, 1980.

[135] G. Holloway and S. S. Kristmannsson, Stirring and transport of tracer fields by geostrophic turbulence. *J. Fluid Mech.*, **141**, pp. 27–50, 1984.

[136] J. C. McWilliams, A uniformly valid model spanning the regimes of geostrophic and isotropic, stratified turbulence: balanced turbulence. *J. Atmos. Sci.*, **42**, pp. 1773–1774, 1985.

[137] K. S. Gage and G. D. Nastrom, Theoretical interpretation of atmospheric wavenumber spectra of wind and temperature observed by commercial aircraft during GASP. *J. Atmos. Sci.*, **43**, pp. 729–739, 1986.

[138] C. O. Hines, Gravity waves in the atmosphere. *Nature*, **239**, pp. 73–78, 1972.

[139] E. M. Dewan, On the nature of atmospheric waves and turbulence. *Radio Sci.*, **20**, pp. 1301–1307, 1985.

[140] W. H. Hooke. Gravity waves, In P. S. Ray, editor, *Mesoscale Meteorology and Forecasting*, chapter 12, pages 272–288. American Meteorological Society, Boston, 1986.

[141] D. K. Lilly, Stratified turbulence and the mesoscale variability of the atmosphere. *J. Atmos. Sci.*, **40**, pp. 749–761, 1983.

[142] P. Müller, R.-C. Lien and R. Williams, Estimates of potential vorticity at small scales in the ocean. *J. Phys. Ocean.*, **18**, pp. 401–416, 1988.

[143] M.-P. Lelong and J. J. Riley, Internal wave–vortical mode interactions in strongly stratified flows. *J. Fluid Mech.*, **232**, pp. 1–19, 1991.

[144] N. E. Busch, Waves and turbulence. *Radio Sci.*, **4**, pp. 1377–1379, 1969.

[145] R. W. Stewart, Turbulence and waves in a stratified atmosphere. *Radio Sci.*, **4**, pp. 1269–1278, 1969.

[146] D. N. Axford, Spectral analysis of an aircraft observation of gravity waves. *Quart. J. Roy. Meteor. Soc.*, **97**, pp. 313–321, 1971.

[147] J. I. Metcalf, Gravity waves in a low-level inversion. *J. Atmos. Sci.*, **32**, pp. 351–361, 1975.

[148] S. SethuRaman, The observed generation and breaking of atmospheric internal gravity waves over the ocean. *Boundary-Layer Meteor.*, **12**, pp. 331–349, 1977.

[149] J. C. R. Hunt, J. C. Kaimal and J. E. Gaynor, Some observations of turbulence structure in stable layers. *Quart. J. Roy. Meteor. Soc.*, **111**, pp. 793–815, 1985.

[150] J. J. Finnigan, Kinetic energy transfer between internal gravity waves and turbulence. *J. Atmos. Sci.*, **45**, pp. 486–505, 1988.

[151] J. L. Lumley, The spectrum of nearly inertial turbulence in a stably stratified fluid. *J. Atmos. Sci.*, **21**, pp. 99–102, 1964.

[152] A. Court, Undulence and turbulence. *Quart. J. Roy. Meteor. Soc.*, **91**, pp. 234–235, 1965.

[153] G. K. Batchelor, The application of the similarity theory of turbulence to atmospheric diffusion. *Quart. J. Roy. Meteor. Soc.*, **76**, pp. 133–146, 1950.

[154] S. Corrsin, Outline of some topics in homogeneous turbulent flow. *J. Geophys. Res.*, **64**, pp. 2134–2150, 1959.

[155] G. K. Batchelor and A. A. Townsend. Turbulent diffusion, In G. K. Batchelor and R. M. Davies, editors, *Surveys in Mechanics*, pages 352–399. Cambridge University Press, Cambridge, 1956.

[156] H. Stommel, Horizontal diffusion due to oceanic turbulence. *J. Mar. Res.*, **8**, pp. 199–225, 1949.

[157] L. F. Richardson, Transforms for the eddy-diffusion of clusters. *Proc. Roy. Soc. London, Series A*, **214**, pp. 1–20, 1952.

[158] P. D. Thompson, Prognostic equations for the mean motions of simple fluid systems and their relation to the theory of large scale atmospheric turbulence. *Tellus*, **6**, pp. 150–164, 1954.

[159] R. G. Lamb, Air pollution models as descriptors of cause-effect relationships. *Atmos. Environ.*, **18**, pp. 591–606, 1984.

[160] O. Reynolds, On the dynamical theory of incompressible viscous fluids and the determination of the criterion. *Phil. Trans. Roy. Soc. London, Ser. A*, **186**, pp. 123–164, 1895.

[161] F. N. Frenkiel, Turbulent diffusion: mean concentration distribution in a flow field of homogeneous turbulence. *Adv. Appl. Mech.*, **3**, pp. 61–107, 1953.

[162] J. C. Wyngaard. Boundary-layer modeling, In F. T. M. Nieuwstadt and H. van Dop, editors, *Atmospheric Turbulence and Air Pollution Modelling*, chapter 3, pages 69–106. D. Reidel Publishing Company, Boston, 1982.

[163] W. R. Cotton and R. A. Anthes, *Storm and Cloud Dynamics*. Academic Press, Inc., San Diego, 1989.

[164] J. Kampé de Fériet, Averaging processes and Reynolds equations in atmospheric turbulence. *J. Meteor.*, **8**, pp. 358–361, 1951.

[165] J. Kampé de Fériet, Introduction to the statistical theory of turbulence: Part I. *J. Soc. Indust. Appl. Math.*, **2**, pp. 1–9, 1954.

[166] J. Kampé de Fériet, Turbulent atmospheric diffusion: the first twenty-five years, 1920–1945. *Adv. Geophys.*, **18A**, pp. 1–23, 1974.

[167] R. A. Pielke, *Mesoscale Meteorological Modeling.* Academic Press, Orlando, Florida, 1984.

[168] L. Mahrt, Grid-averaged surface fluxes. *Mon. Wea. Rev.*, **115**, pp. 1550–1560, 1987.

[169] W. R. Cotton. Averaging and the parameterization of physical processes in mesoscale models, In P. S. Ray, editor, *Mesoscale Meteorology and Forecasting*, chapter 26, pages 614–635. American Meteorological Society, Boston, 1986.

[170] W. S. Lewellen, Parameterization of subgrid-scale fluxes and estimation of dispersion. In *Modeling the Urban Boundary Layer*, pages 381–409, American Meteorological Society, Boston, 1987.

[171] R. G. Lamb. Diffusion in the convective boundary layer, In F. T. M. Nieuwstadt and H. van Dop, editors, *Atmospheric Turbulence and Air Pollution Modelling*, pages 159–229. D. Reidel Publishing Co., Boston, 1982.

[172] H. A. Panofsky and J. A. Dutton, *Atmospheric Turbulence.* J. Wiley & Sons, New York, 1984.

[173] C.-H. Moeng and J. C. Wyngaard, Evaluation of turbulent transport and dissipation closures in second-order modeling. *J. Atmos. Sci.*, **46**, pp. 2311–2330, 1989.

[174] M. G. Hadfield, W. R. Cotton and R. A. Pielke, Large-eddy simulations of thermally forced circulations in the convective boundary layer. Part I: A small-scale circulation with zero wind. *Boundary-Layer Meteor.*, **57**, pp. 79–114, 1991.

[175] R. L. Walko, W. R. Cotton and R. A. Pielke, Large-eddy simulations of the effects of hilly terrain on the convective boundary layer. *Boundary-Layer Meteor.*, **58**, pp. 133–150, 1992.

[176] J. C. Wyngaard. On surface-layer turbulence, In D. A. Haugen, editor, *Workshop on Micrometeorology*, chapter 3, pages 101–149. American Meteorological Society, Boston, 1973.

[177] A. J. Yule. Phase scrambling effects and turbulence data analysis, In L. J. S. Bradbury, F. Durst, B. E. Launder, F. W. Schmidt and J. H. Whitelaw, editors, *Turbulent Shear Flows 2*, pages 263–281. Springer-Verlag, New York, 1980.

[178] J. W. Deardorff, The use of subgrid transport equations in a three-dimensional model of atmospheric turbulence. *J. Fluids Engineering*, **95**, pp. 429–438, 1973.

[179] M. D. Moran and R. A. Pielke, Delayed shear dispersion in mesoscale atmospheric dispersion. In *Proc. 8th Joint AMS/AWMA Conf. on Applications of Air Pollution Meteorology*, pages 96–103, January 24–27, Nashville, Tennessee, American Meteorological Society, Boston, 1994.

[180] M. D. Moran and R. A. Pielke, Evaluation of a mesoscale atmospheric dispersion modeling system with observations from the 1980 Great Plains mesoscale tracer experiment. Part II: Dispersion simulations. *J. Appl. Meteor.*, **35**, pp. 308–329, 1996.

[181] A. Andrén, Evaluation of a turbulence closure scheme suitable for air-pollution applications. *J. Appl. Meteor.*, **29**, pp. 224–239, 1990.

[182] A. Andrén, A meso-scale plume dispersion model: preliminary evaluation in a heterogeneous area. *Atmos. Environ.*, **24A**, pp. 883–896, 1990.

[183] R. G. Lamb and S. K. Hati, The representation of atmospheric motions in models of regional-scale air pollution. *J. Climate Appl. Meteor.*, **26**, pp. 837–846, 1987.

[184] R. I. Sykes, The variance in time-averaged samples from an intermittent plume. *Atmos. Environ.*, **18**, pp. 121–123, 1984.

[185] R. L. Elsberry. Observation and analysis of tropical cyclones, In *A Global View of Tropical Cyclones*, chapter 1, pages 1–12. University of Chicago Press, Chicago, 1989. (Available from Prof.

R. L. Elsberry, Naval Postgraduate School, Monterey, California 93943).

[186] P. Ruscher and L. Mahrt, Coherent structures in the very stable atmospheric boundary layer. *Boundary-Layer Meteor.*, **47**, pp. 41–54, 1989.

[187] R. Narasimha and S. V. Kailas, Turbulent bursts in the atmosphere. *Atmos. Environ.*, **24A**, pp. 1635–1645, 1990.

[188] M. M. Millán. The regional transport of tall stack plumes, In S. Sandroni, editor, *Regional and Long-range Transport of Air Pollution*, pages 249–280. Elsevier Science Publishers B.V., Amsterdam, 1987.

[189] F. T. M. Nieuwstadt and R. A. Brost, The decay of convective turbulence. *J. Atmos. Sci.*, **43**, pp. 532–546, 1986.

[190] G. L. Mellor and T. Yamada, A hierarchy of turbulence closure models for planetary boundary layers. *J. Atmos. Sci.*, **31**, pp. 1791–1806, 1974.

[191] H. M. Helfand and J. C. Labraga, Design of a nonsingular level 2.5 second-order closure model for the prediction of atmospheric turbulence. *J. Atmos. Sci.*, **45**, pp. 113–132, 1988.

[192] G. K. Batchelor, Diffusion in a field of homogeneous turbulence. *Austral. J. Sci. Res.*, **A2**, pp. 437–450, 1949.

[193] K. Herterich and K. Hasselmann, The horizontal diffusion of tracers by surface waves. *J. Phys. Ocean.*, **12**, pp. 704–711, 1982.

[194] B. G. Sanderson and A. Okubo, Diffusion by internal waves. *J. Geophys. Res.*, **93C**, pp. 3570–3582, 1988.

[195] G. K. Batchelor, Diffusion in a field of homogeneous turbulence. II: The relative motion of particles. *Proc. Camb. Phil. Soc.*, **48**, pp. 345–362, 1952.

[196] L. F. Richardson, Atmospheric diffusion shown on a distance-neighbour graph. *Proc. Roy. Soc. London*, **A110**, pp. 709–737, 1926.

[197] D. H. Slade, editor, *Meteorology and Atomic Energy 1968.* U.S. Atomic Energy Commission, Washington, D.C., 1968. (Available as Publication TID-24190 from the National Technical Information Service, U.S. Department of Commerce, Springfield, Virginia 22161).

[198] F. B. Smith and J. S. Hay, The expansion of clusters of particles in the atmosphere. *Quart. J. Roy. Meteor. Soc.*, **87**, pp. 82–101, 1961.

[199] T. Mikkelsen, S. E. Larsen and H. L. Pécseli. Spectral parameterization of large-scale atmospheric diffusion, In H. van Dop, editor, *Air Pollution Modeling and Its Application VI*, pages 579–591. Plenum Press, New York, 1988.

[200] C. J. Nappo, Atmospheric turbulence and diffusion estimates derived from observations of a smoke plume. *Atmos. Environ.*, **15**, pp. 541–547, 1981.

[201] D. H. Slade, editor, *Meteorology and Atomic Energy 1968.* U.S. Atomic Energy Commission, Washington, D.C., 1968. (Available as Publication TID-24190 from the National Technical Information Service, U.S. Department of Commerce, Springfield, Virginia 22161).

[202] F. B. Smith, The character and importance of plume lateral spread affecting the concentration downwind of isolated sources of hazardous airborne material. In *Proc. WMO Symp. on Long-Range Transport of Pollutants and Its Relation to General Circulation Including Stratospheric/Tropospheric Exchange Processes*, pages 241–251, WMO Publ. No. 538, October 1–5, Sofia, Bulgaria, World Meteorological Organization, Geneva, 1979.

[203] L. Kristensen, N. O. Jensen and E. L. Petersen, Lateral dispersion of pollutants in a very stable atmosphere — the effect of meandering. *Atmos. Environ.*, **15**, pp. 837–844, 1981.

[204] S. R. Hanna, Spectra of concentration fluctuations: the two time scales of a meandering plume. *Atmos. Environ.*, **20**, pp. 1131–1137, 1986.

[205] D. Etling, On plume meandering under stable stratification. *Atmos. Environ.*, **24A**, pp. 1979–1985, 1990.

[206] S. R. Hanna, Concentration fluctuations in a smoke plume. *Atmos. Environ.*, **18**, pp. 1091–1106, 1984.

[207] S. R. Hanna, Lateral dispersion from tall stacks. *J. Climate Appl. Meteor.*, **25**, pp. 1426–1433, 1986.

[208] G. E. Moore, L. B. Milich and M. K. Liu, Plume behaviors observed using lidar and SF_6 tracer at a flat and hilly site. *Atmos. Environ.*, **22**, pp. 1673–1688, 1988.

[209] J. E. Fackrell and A. G. Robins, The effects of source size on concentration fluctuations in plumes. *Boundary-Layer Meteor.*, **22**, pp. 335–350, 1990.

[210] H. Peterson and B. Lamb, Comparison of results from a meandering-plume model with measured atmospheric tracer concentration fluctuations. *J. Appl. Meteor.*, **31**, pp. 553–564, 1992.

[211] A. Groll and W. aufm Kampe, Single particle and relative diffusion over sea areas. In *Proc. Sixth AMS Symp. on Turbulence and Diffusion*, pages 5–8, March 22–25, Boston, American Meteorological Society, Boston, 1983.

[212] G. W. Brier, The statistical theory of turbulence and the problem of diffusion in the atmosphere. *J. Meteor.*, **7**, pp. 283–290, 1950.

[213] P. A. Durbin, A stochastic model of two-particle dispersion and concentration fluctuations in homogeneous turbulence. *J. Fluid Mech.*, **100**, pp. 279–302, 1980.

[214] P. C. Chatwin and P. J. Sullivan. The basic structure of clouds of diffusing contaminant, In C. J. Harris, editor, *Mathematical Modelling of Turbulent Diffusion in the Environment*, pages 3–32. Academic Press, New York, 1979.

[215] P. C. Chatwin and P. J. Sullivan. The core-bulk structure associated with diffusing clouds, In L. J. S. Bradbury, F. Durst, B. E. Launder, F. W. Schmidt and J. H. Whitelaw, editors, *Turbulent Shear Flows 2*, pages 379–389. Springer-Verlag, New York, 1980.

[216] P. C. Chatwin and P. J. Sullivan, The relative diffusion of a cloud of passive contaminant in incompressible turbulent flow. *J. Fluid Mech.*, **91**, pp. 337–355, 1979.

[217] C. R. Murthy and G. T. Csanady, Experimental studies of relative diffusion in Lake Huron. *J. Phys. Ocean.*, **1**, pp. 17–24, 1971.

[218] P. C. Chatwin and P. J. Sullivan, Measurements of concentration fluctuations in relative turbulent diffusion. *J. Fluid Mech.*, **94**, pp. 83–101, 1979.

[219] J. L. McElroy, J. A. Eckert and C. J. Hager, Airborne downlooking lidar measurements during STATE 78. *Atmos. Environ.*, **15**, pp. 2223–2230, 1981.

[220] A. Groll, W. aufm Kampe and H. Weber. Downwind hazard distances for pollutants over land and sea, In C. De Wispelaere, editor, *Air Pollution Modeling and Its Application IV*, pages 585–600. Plenum Press, New York, 1985.

[221] E. E. Uthe, F. L. Ludwig and F. Pooler, Jr., Lidar observations of the diurnal behavior of the Cumberland power plant plume. *J. Air Pollu. Control Assoc.*, **30**, pp. 889–893, 1980.

[222] U. Högström, A method for detecting odour frequencies from a point source. *Atmos. Environ.*, **6**, pp. 103–121, 1972.

[223] J. S. Hay and F. Pasquill, Diffusion from a continuous source in relation to the spectrum and scale of turbulence. *Adv. Geophys.*, **6**, pp. 345–365, 1959.

[224] A. S. Monin and A. M. Yaglom, *Statistical Fluid Mechanics, Vol. 2.* MIT Press, Cambridge, Massachusetts, 1975.

[225] A. B. Kahn, A generalization of average-correlation methods of spectrum analysis. *J. Meteor.*, **14**, pp. 9–17, 1957.

[226] F. Pasquill. The dispersion of materials in the atmospheric boundary layer — the basis for generalization, In D. A. Haugen, editor, *Lectures on Air Pollution and Environmental Impact Analyses*, pages 1–34. American Meteorological Society, Boston, 1975.

[227] S. Corrsin, Remarks on turbulent heat transfer. In *Proc. First Iowa Symp. on Thermodynamics*, pages 5–30, University of Iowa, Iowa City, Iowa, 1953.

[228] P. G. Saffman, The effect of wind shear on horizontal spread from an instantaneous ground source. *Quart. J. Roy. Meteor. Soc.*, **88**, pp. 382–393, 1962.

[229] F. B. Smith, The role of wind shear in horizontal diffusion of ambient particles. *Quart. J. Roy. Meteor. Soc.*, **91**, pp. 318–329, 1965.

[230] C. J. Nappo, Relative and single particle diffusion estimates determined from smoke plume photographs. In *Proc. Fourth AMS Symp. on Turbulence, Diffusion, and Air Pollution*, pages 46–47, Jan. 15–18, Reno, NV, American Meteorological Society, Boston, 1979.

[231] S. R. Hanna. Applications in air pollution modeling, In F. T. M. Nieuwstadt and H. van Dop, editors, *Atmospheric Turbulence*

and Air Pollution Modelling, pages 275–310. D. Reidel Publishing Co., Dordrecht, 1982.

[232] Y. Ogura, Diffusion from a continuous source in relation to a finite observation interval. *Adv. Geophys.*, **6**, pp. 149–159, 1959.

[233] M. Hino, Maximum ground-level concentration and sampling time. *Atmos. Environ.*, **2**, pp. 149–165, 1968.

[234] G. C. Wollenweber and H. A. Panofsky, The effect of sampling time on velocity variances. In *Proc. 8th AMS Symp. on Turbulence and Diffusion*, pages 116–118, April 25–29, San Diego, American Meteorological Society, Boston, 1988.

[235] C. E. Skupniewicz, Measurements of overwater diffusion: the separation of relative diffusion and meander. *J. Climate Appl. Meteor.*, **26**, pp. 949–958, 1987.

[236] J. T. Lee and G. L. Stone, The use of Eulerian initial conditions in a Lagrangian model of turbulent diffusion. *Atmos. Environ.*, **17**, pp. 2477–2481, 1983.

[237] G. I. Taylor, The spectrum of turbulence. *Proc. Roy. Soc. London*, **A164**, pp. 476–490, 1938.

[238] P. S. Brown, Jr. and G. D. Robinson, The variance spectrum of tropospheric winds over Eastern Europe. *J. Atmos. Sci.*, **36**, pp. 270–286, 1979.

[239] D. K. Lilly and E. L. Petersen, Aircraft measurement of atmospheric kinetic energy spectra. *Tellus*, **35A**, pp. 379–382, 1983.

[240] G. D. Nastrom, K. S. Gage and W. H. Jasperson, Kinetic energy spectrum of large- and mesoscale atmospheric processes. *Nature*, **310**, pp. 36–38, 1984.

[241] S.A. Tjemkes and M. Visser, Horizontal variability of temperature, specific humidity, and cloud liquid water as derived from spaceborne observations. *J. Geophys. Res.*, **99**, pp. 23089–23105, 1994.

[242] C. E. Leith, Diffusion approximation for two-dimensional turbulence. *Phys. Fluids*, **11**, pp. 671–673, 1968.

[243] B. B. Balsley and D. A. Carter, The spectrum of atmospheric velocity fluctuations at 8 km and 86 km. *Geophys. Res. Lett.*, **9**, pp. 465–468, 1982.

[244] G. S. Golitsyn, Introduction to the dynamics of planetary atmospheres. NASA Tech. Translation TT F-15,627, National Aeronautics and Space Administration, Washington, D. C., 1974. (Russian title: "Vvdeniye v Dinamiku Planetnykh Atmosfer").

[245] K. S. Gage, Evidence for a $k^{-5/3}$ law inertial range in mesoscale two-dimensional turbulence. *J. Atmos. Sci.*, **36**, pp. 1950–1954, 1979.

[246] K. S. Gage and G. D. Nastrom, Further discussion of the dynamical processes that contribute to the spectrum of mesoscale atmospheric motions. In *8th AMS Symp. on Turbulence and Diffusion*, pages 217–220, April 25-29, San Diego, American Meteorological Society, Boston, 1988.

[247] J.-T. Lin and Y.-H. Pao. Wakes in stratified fluids, In *Annual Review of Fluid Mechanics, Vol. 11*, pages 317–338. Annual Reviews, Inc., Palo Alto, California, 1979.

[248] J. J. Riley, R. W. Metcalfe and M. A. Weissman. Direct numerical simulations of homogeneous turbulence in density-stratified fluids, In B. J. West, editor, *Nonlinear Properties of Internal Waves*, pages 79–112. American Institute of Physics, New York, 1981. (La Jolla Institute, 1981, AIP Conf. Proc. No. 106).

[249] E. J. Hopfinger, Turbulence in stratified fluids: a review. *J. Geophys. Res.*, **92**, pp. 5287–5303, 1987.

[250] K. S. Gage and G. D. Nastrom, Evidence for coexisting spectra of stratified turbulence and internal waves in mesoscale atmospheric velocity fields. In *Proc. 7th AMS Symp. on Turbulence and Diffusion*, pages 176–179, Nov. 12-15, Boulder, Colorado, American Meteorological Society, Boston, 1985.

[251] D. K. Lilly. Mesoscale variability of the atmosphere, In D. K. Lilly and T. Gal-Chen, editors, *Mesoscale Meteorology — Theories, Observations and Models*, pages 13–24. D. Reidel Publishing Company, Boston, 1983.

[252] D. K. Lilly. The atmospheric noise spectrum – waves or stratified turbulence?, In P. Müller and R. Pujalet, editors, *Internal Gravity Waves and Small-Scale Turbulence*, pages 285–294. Hawaii Institute of Geophysics, University of Hawaii, Honolulu, Hawaii, 1984. (Proc. ‘Aha Huliko‘a Hawaiian Winter Workshop, University of Hawaii at Manoa, Jan. 17-20, 1984).

[253] G. D. Nastrom, W. H. Jasperson and K. S. Gage, Horizontal spectra of atmospheric tracers measured during the Global Atmospheric Sampling Program. *J. Geophys. Res.*, **91**, pp. 13201–13209, 1986.

[254] C. B. Moore, J. R. Smith and A. Gaalswyk, On the use of constant-level balloons to measure horizontal motions in the atmosphere. *J. Meteor.*, **11**, pp. 167–172, 1954.

[255] G. I. Taylor, Dispersion of soluble matter in solvent flowing slowly through a tube. *Proc. Roy. Soc. London A*, **219**, pp. 186–203, 1953.

[256] G. I. Taylor, The dispersion of matter in turbulent flow through a pipe. *Proc. Roy. Soc. London A*, **223**, pp. 446–467, 1954.

[257] G. I. Taylor, Conditions under which dispersion of a solute in a stream of solvent can be used to measure molecular diffusion. *Proc. Roy. Soc. London A*, **225**, pp. 473–477, 1954.

[258] R. Aris, On the dispersion of a solute in a fluid flowing through a tube. *Proc. Roy. Soc. Lond.*, **A235**, pp. 67–77, 1956.

[259] U. Högström, An experimental study on atmospheric diffusion. *Tellus*, **16**, pp. 205–251, 1964.

[260] J. B. Tyldesley and C. F. Wallington, The effect of wind shear and vertical diffusion on horizontal dispersion. *Quart. J. Roy. Meteor. Soc.*, **91**, pp. 158–174, 1965.

[261] G. T. Csanady, Diffusion in an Ekman layer. *J. Atmos. Sci.*, **26**, pp. 414–426, 1969.

[262] P. R. Maul, The effect of the turning of the wind vector with height on the ground level trajectory of a diffusing cloud. *Atmos. Environ.*, **12**, pp. 1045–1050, 1978.

[263] A. D. Taylor, Puff growth in an Ekman layer. *J. Atmos. Sci.*, **39**, pp. 837–850, 1982.

[264] F. Pasquill, The influence of the turning of wind with height on crosswind diffusion. *Proc. Royal Soc. London A*, **265**, pp. 173–181, 1969.

[265] J. D. Kahl and P. J. Samson, Shear effects on wind interpolation accuracy. *J. Appl. Meteor.*, **27**, pp. 1299–1301, 1988.

[266] C.J. Nappo, Sporadic breakdowns of stability in the PBL over simple and complex terrain. *Boundary-Layer Meteor.*, **54**, pp. 69–87, 1991.

[267] R.B. Husar, D.E. Patterson, J.D. Husar, N.V. Gillani and W.E. Wilson, Jr., Sulfur budget of a power plant plume. *Atmos. Environ.*, **12**, pp. 549–568, 1978.

[268] U. Corsmeier, N. Kalthoff, O. Kolle, M. Kotzian and F. Fiedler, Ozone concentration jump in the stable nocturnal boundary layer during a LLJ-event. *Atmos. Environ.*, **31**, pp. 1977–1989, 1997.

[269] R. B. Stull, *An Introduction to Boundary Layer Meteorology*. Kluwer Academic Publishers, Dordrecht, 1988.

[270] M. D. Moran, R. T. McNider and R. A. Pielke, Diurnal influences on mesoscale atmospheric dispersion. In *Proc. Third AMS Conference on Mesoscale Processes*, pages 220–221, August 21-26, Vancouver, B.C., American Meteorological Society, Boston, 1987.

[271] M. P. Singh, R. T. McNider and J. T. Lin, An analytical study of diurnal wind-structure variations in the boundary layer and the low-level nocturnal jet. *Boundary-Layer Meteor.*, **63**, pp. 397–423, 1993.

[272] I. Y. Lee, Numerical simulations of cross-Appalachian transport and diffusion. *Boundary-Layer Meteor.*, **39**, pp. 53–66, 1987.

[273] C.-Y. J. Kao and T. Yamada, Use of the CAPTEX data for evaluations of a long-range transport numerical model with a four-dimensional data assimilation technique. *Mon. Wea. Rev.*, **116**, pp. 293–306, 1988.

[274] R. A. Brost, P. L. Haagenson and Y.-H. Kuo, Eulerian simulation of tracer distribution during CAPTEX. *J. Appl. Meteor.*, **27**, pp. 579–593, 1988.

[275] J. M. Godowitch, Evaluation and sensitivity analyses results of the MESOPUFF II model with CAPTEX measurements. Report EPA/600/3-89/056, Atmospheric Sciences Modeling Division, Atmospheric Research and Exposure Assessment Laboratory, U.S. Environmental Protection Agency, Research Triangle Park, NC 27711, April 1989.

[276] B.-Q. Shi, J. D. Kahl, Z. D. Christidis and P. J. Samson, Simulation of the three-dimensional distribution of tracer during the Cross-Appalachian Tracer Experiment. *J. Geophys. Res.*, **95**, pp. 3693–3703, 1990.

Chapter 3

Atmospheric boundary layers and turbulence

S. Pal Arya
Department of Marine, Earth and Atmospheric Sciences
North Carolina State University
Raleigh, NC 27695-8208, USA

Abstract

In this chapter, a broad overview of theoretical, numerical, and experimental studies of the atmospheric planetary boundary layer (PBL) and its mean flow and turbulence structure is presented. The short introduction includes some definitions and concepts, different types of atmospheric boundary layers, and their particular characteristics that influence mesoscale transport and dispersion. Following the commonly used conservation laws and equations, the various PBL theories and models based on the mean conservation equations are discussed. The emphasis is on the first-order and turbulence kinetic energy models. Similarity scaling and empirical relations based on the same are more thoroughly reviewed. Finally, the rationale and basis of parameterizing the PBL height, subgrid scale fluxes and variances in mesoscale models are given.

1 Introduction

1.1 Definitions

Any description of mesoscale atmospheric dispersion must include transport by mean winds and diffusion or mixing by turbulence in the bottom layer of the atmosphere, variously called the atmospheric boundary layer (ABL) or the planetary boundary layer (PBL). The PBL forms as a consequence of the interaction of the atmosphere with the underlying land or water surface on relatively short time scales of the order of a day or less. Because of the various dynamical instability mechanisms (e.g., viscous, shear, and buoyancy) operating there, the PBL flow is always turbulent, at least intermittently, if not continuously. Turbulence characterizes the apparently chaotic nature of many fluid flows, which is manifested in the form of highly irregular, almost random, temporal and spatial fluctuations in velocity, temperature, and scalar concentrations around their mean values.

Turbulence is strong, rigorous and more or less continuous throughout the PBL under very windy and near-neutral stability conditions. It is essentially generated by surface friction and wind shear and, hence, is referred to as shear-generated or mechanical turbulence. Vigorous turbulence can also be generated by convection over a heated surface and is generally referred to as convective turbulence. More often, over a heated surface, both the shear and convective mechanisms of turbulence generation operate simultaneously, especially in the lowest 10% of the PBL, called the surface layer. Wind shear is strongest near the surface and its influence is largely confined to the surface layer. Convective turbulence dominates in the overlying, deep mixed layer, while much weaker shear-generated turbulence occurs in the transition layer near the top of the PBL. Such a PBL is generally referred to as the convective boundary layer (CBL) whose three-layer structure is schematically shown elsewhere (see e.g., Arya [1]). In contrast to this, the atmospheric boundary layer over a cold surface (e.g., any land surface during a clear night) is called the stable boundary layer (SBL), because it is stably stratified due to increasing temperature or potential temperature with height (also referred to as an inversion condition). Turbulence in the SBL is essentially shear generated, but is much weakened and sometimes entirely suppressed in the upper part by the action of buoyancy. Above the shallow surface layer, turbulence in the SBL is highly intermittent (episodic), patchy, and frequently mixed with internal gravity waves. In terms of the boundary layer height or thickness (h), the SBL is much thinner (typically, h ~ 100 m) than the CBL (typically h ~ 1000 m).

The atmospheric boundary layer during the transitional near-neutral stability regime between unstable and stable conditions is also called the near-neutral boundary layer (NNBL). The theoretician's ideal of the strictly neutral boundary layer is so rare in the atmosphere that systematic observations of the same do not exist. The surface is either slightly warmer or cooler than the air above. The morning and evening transition periods, when the sensible heat exchange between the surface and the atmosphere vanishes momentarily as it changes sign from negative (downward flux) to positive (upward flux) and vice-versa, are often marked by rapid changes in the PBL height, mean winds, and turbulence. Thus, the transitional NNBL is rapidly evolving under changing surface energy fluxes and is essentially nonstationary. A more stationary NNBL is formed under very windy and cloudy conditions over land and more frequently over open sea and ocean surfaces.

The above classification of the PBL according to three broad stability regimes is somewhat artificial and for convenience only. More specific and quantitative PBL stability parameters, such as the bulk Richardson number and the ratio h/L of the PBL height to the Obukhov length, are found to vary continuously over wide ranges. One of these can be used to represent more systematically the important effects of stability on the mean flow and turbulence structure of the PBL as it varies from extremely unstable or convective to extremely stable conditions in the atmosphere. Still, in describing the typical observed structure of the PBL under convective, near-neutral and stable conditions, we will often use the broader, qualitative stability characterizations and also the corresponding designations of the atmospheric boundary layer as CBL, NNBL and SBL. Since most of the PBL theories and models are more generic in their formulation and concepts, however, these will be reviewed without regard to the PBL stability. But, their lack of validity and other limitations under certain stability conditions will be discussed. In describing the PBL theories and models, it is more important to specify the presence or lack of certain simplifying conditions, such as stationarity and horizontal homogeneity of the PBL.

1.2 Horizontally-homogeneous PBL

The atmospheric boundary layer formed under and driven by large-scale flows (e.g., geostrophic or gradient winds) over a reasonably flat and uniform surface can essentially be considered horizontally-homogeneous, especially over horizontal dimensions of up to 100 km, i.e., smaller than the typical resolution of the synoptic observing network. Such an idealized PBL may evolve gradually with time in response to the similar evolution of the large-scale flow with time. Even when the large-scale flow is steady over several days, the PBL evolves diurnally as a result of the diurnal heating and cooling of the surface. Diurnal changes in the PBL are observed to become small and, sometimes, even insignificant over open ocean and sea surfaces and also extensive snow- and ice-covered surfaces over the polar regions during winter. Over a short period of the order of an hour, however, even a diurnally evolving PBL over a bare and dry land surface can be considered to be quasi-stationary when one considers its short-term averaged properties. This idealized (horizontally-homogeneous and quasi-stationary) PBL has been the focus of many experiments as well as simpler theories and models that will be discussed in more detail in later sections. Here we will briefly mention only some of the features of the PBL that influence mesoscale pollutant transport and dispersion.

The distribution of mean winds (vector $\overline{\underset{\sim}{V}}$, or components $\overline{u}, \overline{\mathrm{v}}, \text{and } \overline{\mathrm{w}}$) with height (z) in the PBL is very important, because it essentially determines the speeds and directions at which pollutants will be transported at different heights. In general, both the mean wind speed $\left(\overline{\mathrm{V}}\right)$ and wind direction $\left(\overline{\mathrm{D}}\right)$ vary (increase) with height, more so in the lower part of the PBL. In simpler short-range dispersion theories and models (see e.g., Pasquill and Smith [2]; Arya [3]), only the average or effective transport speed and direction across the plume or the PBL depth is required as an input to the model. In more sophisticated theories and models, average wind speed and direction shears across the pollutant plume or puff are also considered. Wind shears and their effects on dispersion are found to be generally small or insignificant under very unstable and convective conditions, but strong enough to be considered under stably stratified conditions. Wind speed shear enhances the vertical dispersion, while the wind direction shear leads to increased lateral (cross-wind) dispersion with increasing distance from the source (Pasquill and Smith [2]; Arya [3]). Both these effects are important in describing mesoscale dispersion of pollutants in the troposphere.

Another important feature of the PBL influencing pollutant dispersion is turbulence, which is best quantified by the standard deviations of velocity fluctuations $(\sigma_u, \sigma_v, \text{and}\, \sigma_w)$ in different directions. The ensemble-averaged dimensions of a diffusing puff or plume are found to increase with time or distance from the source in proportion to σ_u, σ_v, and σ_w. Their normalized values, called turbulence intensities $i_u = \sigma_u/\overline{V}, i_v = \sigma_v/\overline{V}$, etc., where $\overline{V}$ is the mean wind speed, can be used to specify puff or plume diffusion parameters. Turbulence intensities are strongly dependent on stability and also vary with height in the PBL. By definition, turbulence disappears at the top of the PBL which represents a fairly sharp interface between the turbulent PBL and the nonturbulent free atmosphere.

The average height or depth of the PBL (h) is another important parameter affecting dispersion, especially in the vertical. It limits the depth of vertical mixing and, hence, the vertical dispersion parameter, particularly in the daytime unstable or convective boundary layer which is often capped by a strong inversion. The capping inversion may also be a limiting factor in estimating plume rise for determining the effective release height. The PBL depth or mixing height is an important length scale which determines the size of the largest energy-containing eddies, average wind speed and direction shears in the PBL, as well as the overall stability of the PBL. The PBL depth shows strong diurnal variations in response to the diurnal heating and cooling of the surface. It increases throughout the day, becomes maximum (typically, 1 km) in the late afternoon, and then rapidly collapses to its typical nighttime value of the order of 100 m. Actually, at night mixing height rarely stays constant, but usually oscillates around its mean value in response to periodic turbulence generation episodes. The maximum mixing height during the afternoon also shows large seasonal and spatial variations (Holzworth [4]; Arya [3]).

1.3 Nonhomogeneous boundary layers

The idealised horizontally-homogeneous PBL may be expected to develop over extensive flat and uniform water surfaces, such as open seas, oceans and large lakes, as well as over a variety of land surfaces, such as deserts, grasslands, and forests, altogether covering more than 80% of the earth's surface. However, over most of the developed land areas, which are marked by residential, industrial, commercial, agricultural and recreational use, the atmospheric boundary layer is nonhomogeneous and frequently in a state of transition from one land use type to another. Surface inhomogeneities affecting the PBL flow include boundaries between land and water surfaces (coastlines), urban and rural areas (urban heat and roughness islands), different types of vegetation, and hills and valleys. In going over these land use and terrain inhomogeneities, the PBL flow encounters sudden or gradual changes in surface roughness, temperature, wetness or elevation. Quite often, changes in several surface properties occur simultaneously. The mesoscale atmospheric circulation systems, such as land-sea breezes, urban heat islands, mountain and valley winds, and others associated with complex topography, are likely to have nonhomogeneous boundary layers in them. Boundary layers associated with moving mesoscale weather systems, such as fronts, wind storms and thunderstorms, are also necessarily nonhomogeneous and nonstationary, but these are not usually of much concern in mesoscale atmospheric dispersion.

Mean flow and turbulence characteristics in the developing (nonhomogeneous) atmospheric boundary layer following a sudden change in the surface roughness and/or temperature are strongly influenced by the underlying surface up to a certain height and by the upwind surface above that height. The modified layer near the surface grows with downwind distance from the location of surface discontinuity. Such a modified layer growing within the larger PBL is called the internal boundary layer (IBL). Internal boundary layers commonly form and grow downwind of discontinuities in surface roughness, temperature, wetness and elevation. Over a patchwork of different land uses, for example, one can visualize multiple IBLs forming and growing in succession. In the upper part of the PBL, however, the individual effects of several different upwind surfaces may not be discernible, as it represents the integrated influence of the various surfaces. Most of the studies of the internal boundary layer have been focused on either a step change in the surface roughness (e.g., rough to smooth and vice versa) or in thermal properties (e.g., land to water surface and vice versa). These indicate that the IBL height or thickness (h_i) grows as $h_i \propto x^p$, where the exponent $p \cong 0.8$ for the roughness change under neutral conditions, but may vary between 0.5 and 1.0 for changes in thermal properties and stability (Bradley [5]; Stunder and SethuRaman [6]. More detailed reviews of nonhomogeneous atmospheric boundary layers are given elsewhere (Oke [7], Chs. 7 and 8; Arya [8], Ch. 11, Stull [9], Ch. 14).

2 PBL theories and models

2.1 Conservation laws and equations

Most PBL theories and models are based on the fundamental laws of conservation of mass, energy and momentum, which are expressed in the form of partial differential equations containing velocity, pressure, temperature or density, and concentrations of water vapor and trace gases, if needed. Using the simplifying assumptions or approximations, these conservation equations can be expressed as follows (see Arya [3], Ch. 2):

$$\frac{\partial u}{\partial x} + \frac{\partial v}{\partial y} + \frac{\partial w}{\partial z} = 0 , \tag{1}$$

$$\frac{Du}{Dt} = -\frac{1}{\rho}\frac{\partial p}{\partial x} + f\,v + \nu \nabla^2 u , \tag{2}$$

$$\frac{Dv}{Dt} = -\frac{1}{\rho}\frac{\partial p}{\partial y} - f\,u + \nu \nabla^2 v , \tag{3}$$

$$\frac{Dw}{Dt} = -\frac{1}{\rho}\frac{\partial p}{\partial z} - g + \nu \nabla^2 w , \tag{4}$$

$$\frac{D\theta}{Dt} = \kappa \nabla^2 \theta + S_H , \tag{5}$$

$$\frac{Dq}{Dt} = D_w \nabla^2 q + S_W . \tag{6}$$

Here, u, v and w are the velocity components in the x, y and z directions, using a Cartesian coordinate system, ρ is density, p is pressure, θ is potential temperature, and q is specific humidity. All these variables are instantaneous and for a particular point (x, y, z) in space. The molecular properties included in the above equations are the kinematic viscosity, ν, the thermal diffusivity, κ, and the diffusivity of water vapor, D_W, while S_H and S_W represent any sources (or sinks) of heat and water vapor which could result in the rate of warming and moistening, respectively, of an elementary air parcel. The operators D/Dt and ∇^2 have the usual meanings of total or substantial derivative and Laplacian, respectively, and the Coriolis parameter depends on the earth's angular rotational speed Ω and the latitude ϕ as

$$f = 2\,\Omega \sin \phi. \tag{7}$$

The significance of the various terms in the equations of motion can be explained. The left-hand sides represent the inertia terms which include local as well as advective accelerations. These are balanced by pressure-gradient and Coriolis or gravitational accelerations and also molecular diffusion terms. The terms in the equations for θ and q have similar interpretations, except for the absence of pressure gradient, gravity and Coriolis effects. Mathematically, all conservation equations are of similar form, as are the well-known advection-diffusion equations of trace species (see Arya [3], Ch. 2). These form a closed set which can be solved, in principle, analytically or numerically. In actual practice, however, eqns. (1)-(6) can be solved only for nonturbulent (inviscid) flows.

The flow in the atmospheric boundary layer is always turbulent. The laminar Ekman layer, as a possible analytical solution to the equations of motion over a flat, homogeneous and smooth surface (see e.g., Arya [8], Ch. 7), does not exist in the real atmospheric environment, because it is found to become dynamically unstable at a fairly low value of Reynolds number. For a turbulent flow, only the direct numerical simulation (DNS) attempts fully to resolve all the scales of motion, ranging from the largest energy-containing eddies to the smallest dissipating eddies. But, at present, the DNS technique can be applied only to low Reynolds number flows with the ratio between the largest and smallest scales limited to a few hundred. In the PBL, this ratio is typically 10^6, and generally exceeds 10^4, even for the extremely stable, shallow boundary layer. In the not too distant future, the shallow SBL might become amenable to this most sophisticated direct numerical simulation approach, as computer speed and power continue to grow exponentially.

The simplest PBL models, including some analytical ones, have been traditionally based on the following equations for the mean variables, which are obtained by averaging the equations (1)-(6) for the instantaneous variables after expressing them as sums of their mean and fluctuating parts (i.e., following the so-called Reynolds decomposition):

$$\frac{\partial \bar{u}}{\partial x} + \frac{\partial \bar{v}}{\partial y} + \frac{\partial \bar{w}}{\partial z} = 0, \qquad (8)$$

$$\frac{D\bar{u}}{Dt} = -\frac{1}{\rho_o}\frac{\partial \bar{p}}{\partial x} + f\,\bar{v} - \left(\frac{\partial \overline{u'^2}}{\partial x} + \frac{\partial \overline{u'v'}}{\partial y} + \frac{\partial \overline{u'w'}}{\partial z} \right), \qquad (9)$$

$$\frac{D\bar{v}}{Dt} = -\frac{1}{\rho_o}\frac{\partial \bar{p}}{\partial y} - f\,\bar{u} - \left(\frac{\partial \overline{v'u'}}{\partial x} + \frac{\partial \overline{v'^2}}{\partial y} + \frac{\partial \overline{v'w'}}{\partial z} \right), \qquad (10)$$

$$\frac{D\bar{w}}{Dt} = -\frac{1}{\rho_o}\frac{\partial \bar{p}}{\partial z} - g - \left(\frac{\partial \overline{w'u'}}{\partial x} + \frac{\partial \overline{w'v'}}{\partial y} + \frac{\partial \overline{w'^2}}{\partial z} \right), \qquad (11)$$

$$\frac{D\bar{\theta}}{Dt} = \bar{S}_H - \left(\frac{\partial \overline{\theta'u'}}{\partial x} + \frac{\partial \overline{\theta'v'}}{\partial y} + \frac{\partial \overline{\theta'w'}}{\partial z} \right), \qquad (12)$$

$$\frac{D\bar{q}}{Dt} = \bar{S}_W - \left(\frac{\partial \overline{q'u'}}{\partial x} + \frac{\partial \overline{q'v'}}{\partial y} + \frac{\partial \overline{q'w'}}{\partial z} \right). \tag{13}$$

Here an overbar denotes an ensemble average, prime denotes a turbulent fluctuation, and the usual Reynolds averaging rules or conditions have been used to obtain the mean equations (see e.g., Arya [8], [3]). Note that the mean molecular diffusion terms have been omitted from eqns. (9)-(13), because these are generally found to be negligible outside of very thin molecular sublayers which can possibly form adjacent to smooth surfaces.

When the above mean equations are compared with the corresponding instantaneous equations, one finds that most of the terms, except for the turbulent transport terms in parentheses, are similar and have the same physical significance in the two sets of equations. An important consequence of averaging the instantaneous equations is the appearance of many new (unknown) variables, which are represented by averages of products of two fluctuating variables about their respective means. These are variously called turbulent transports, fluxes or covariances, and are the most important quantities in any turbulent shear flow such as the PBL. Their presence in the Reynolds-averaged equations constitutes the fundamental closure problem (i.e., there are many more unknowns than the number of equations) of turbulence. These equations are otherwise much simpler than the original instantaneous equations, as they deal with only mean quantities which are slowly and smoothly varying functions of time and space, as compared to the rapidly and randomly varying instantaneous variables.

Equations (9)-(13) can be much simplified for the idealised horizontally-homogeneous boundary layer, in the absence of any active condensation or precipitation processes, as

$$\frac{\partial \bar{u}}{\partial t} = f\left(\bar{v} - v_g\right) - \frac{\partial \overline{u'w'}}{\partial z}, \tag{14}$$

$$\frac{\partial \bar{v}}{\partial t} = -f\left(\bar{u} - u_g\right) - \frac{\partial \overline{v'w'}}{\partial z}, \tag{15}$$

$$\frac{\partial \bar{\theta}}{\partial t} = -\frac{\partial \overline{\theta' w'}}{\partial z}, \tag{16}$$

$$\frac{\partial \bar{q}}{\partial t} = -\frac{\partial \overline{q' w'}}{\partial z}, \tag{17}$$

in which horizontal pressure gradients are expressed in terms of the specified geostrophic wind components u_g and v_g. The above equations for the horizontally-homogeneous PBL still contain twice as many unknowns as the number of equations.

The dynamical equations for second moments (variances and covariances) have also been derived. However, they contain third moments and other unknown terms (e.g., molecular dissipations and covariances involving pressure fluctuations) and present the same closure problem of turbulence at a somewhat higher level. Depending on the type of mean flow variables and turbulence moments for which dynamical ensemble-averaged equations are included, different orders of turbulence closure models have been formulated. The simplest are the first-order closure models, which are based on the ensemble-averaged equations for the mean variables given above, with appropriate parameterizations (closure relations) for turbulence variances and covariances.

2.2 First-order closure models

2.2.1 Local closure hypotheses

Assuming a direct analogy between molecular exchanges and turbulent transports (as originally suggested by J. Boussinesq in 1877), turbulent fluxes are expressed in terms of the mean gradients as

$$\overline{u' w'} = -K_m \frac{\partial \bar{u}}{\partial z}; \ \overline{v' w'} = -K_m \frac{\partial \bar{v}}{\partial z}, \tag{18}$$

$$\overline{\theta' w'} = -K_h \frac{\partial \bar{\theta}}{\partial z}; \ \overline{q' w'} = -K_w \frac{\partial \bar{q}}{\partial z}, \tag{19}$$

in which K_m, K_h and K_w are called the eddy diffusivities or exchange coefficients of momentum, heat, and water vapor, respectively. Substituting from eqns. (18)-(19) into eqns. (14)-(17) results in a closed set of equations which can be solved to obtain the mean flow and thermodynamic variables in the PBL as functions of time and height above the surface for an appropriate specification of eddy diffusivities. In the past, many different specifications of eddy diffusivities by different investigators have resulted in a wide variety of the K-theory models of the stationary as well as time-dependent PBLs (see e.g., Holt and Raman [10]; Stull [9]; Sorbjan [11]). For the horizontally-homogeneous PBL, similarity theories provide the most rational and general expressions of eddy diffusivities, especially for neutral and stable stratifications (Businger and Arya [12]; Brost and Wyngaard [13]).

For nonhomogeneous atmospheric boundary layers also, the turbulent transport terms in eqns. (9)-(13) are often reduced to only those containing the vertical gradients, using the so-called boundary-layer approximations for gradually-developing flows, but all the horizontal and vertical advection terms are retained. The flux-gradient relations are further generalized by also including the horizontal gradients, e.g.,

$$\overline{u'w'} = -K_m \left(\frac{\partial \overline{u}}{\partial z} + \frac{\partial \overline{w}}{\partial x} \right). \tag{20}$$

Even more generalized flux-gradient relations utilizing eddy diffusivities as first-order or second-order tensors have been proposed (Kao, [14]). For example, it is common practice to specify the eddy diffusivity of a diffusing tracer as a vector with different components (K_x, K_y and K_z) in different directions.

In general, the gradient-transport relations are not the expressions of any sound physical laws in the same sense as their molecular counterparts are. These are not based on any rigorous theory, but only on an intuitive analogy between molecular and turbulent transfer processes. This analogy is found to be rather weak and qualitative only. Gradient-transport relations are the simplest and somewhat crude parameterizations of turbulent transports, implying down-gradient transfers of momentum, heat, etc. This basic premise of down-gradient transport becomes highly questionable and even invalid under certain conditions, particularly when large-scale convective motions dominate turbulent transports. It is a well known fact that in the convective mixed layer, the vertical momentum, heat and water vapor fluxes are not related to the vanishingly small gradients of mean velocity, potential temperature and specific humidity in that layer. The gradient-transport theory, utilizing a local flux-gradient relationship (closure hypothesis), is, therefore, not valid in the CBL, especially in the mixed layer. In the neutral and stably stratified boundary layers with strong gradients of mean variables, there are no such conceptual difficulties, but eddy diffusivities are specified largely on an ad hoc basis.

A more rational approach to specifying eddy diffusivities is provided by Prandtl's mixing length hypothesis, which can be used to derive the following alternative relations (Arya [3], [8]):

$$K_m = \ell_m^2 \left| \frac{\partial \overline{\underset{\sim}{V}}}{\partial z} \right| ; K_h = \ell_m \ell_h \left| \frac{\partial \overline{\underset{\sim}{V}}}{\partial z} \right| , \tag{21}$$

$$K_m = c\, \ell_m \sigma_w ; \; K_h = c\, \ell_h \sigma_w , \tag{22}$$

where ℓ_m and ℓ_h are the average mixing lengths for the transfer of momentum and heat, respectively, $\overline{\underset{\sim}{V}}$ is the mean vector velocity, and c is an empirical constant. One can specify a mixing length more rationally by identifying it with the characteristic large-eddy length scale which can be determined from the measurement of autocorrelation or spectrum function of vertical velocity fluctuations. Both the mixing length and the large-eddy length scale are found to be proportional to the height above the surface, especially closer to the surface, and approach some constant value which is either proportional to the PBL depth or limited by stability in the upper part of the PBL. Simple interpolation formulas of the form

$$\frac{1}{\ell_m} = \frac{1}{kz} + \frac{1}{\ell_b} + \frac{1}{\ell_o} , \tag{23}$$

in which $k \cong 0.4$ is the von Karman constant, ℓ_b is a buoyancy-limited mixing length, and ℓ_0 is another limiting value which is assumed proportional to h, $u_* / |f|$, or $G/|f|$. Here, u_* is the friction velocity (defined from the surface stress τ_0) and G is the magnitude of the surface geostrophic wind. The buoyancy-limited mixing length term is included only for modeling the SBL.

2.2.2 Nonlocal closure models

Recognizing the limitations of local closure hypotheses, especially under unstable and convective conditions, several nonlocal closure formulations have been proposed for modeling turbulence and diffusion in the CBL (Blackadar [15]; Stull [9], [16]; Pleim and Chang [17]). Nonlocal closures are based on the basic premise that turbulent eddies of varying sizes can transport flow properties, such as momentum, heat and mass, across finite vertical distances, comparable to their size. Thus, the largest eddies in the CBL with the vertical dimension equal to the PBL depth h, can transfer properties all the way from the surface layer to the top of the CBL and vice-versa, irrespective of the vanishingly small gradients in the mixed layer. Consequently, mean velocity, potential temperature and scalar concentrations at any location (height level) are a result of eddy transport and mixing of fluid parcels coming from many cross-stream locations (height levels).

Stull [16] has proposed a transilient turbulence model of nonlocal mixing across finite distances. In this, a vertical column of air is assumed to be divided between a finite number (N) of equal size grid boxes. Turbulent mixing of fluid coming from any box (denoted by index j) into the reference grid box (denoted by index i) can change any conserved passive scalar property s_i (e.g., potential temperature, specific humidity) of the reference box. If c_{ij} represents the fraction of air in box i that came from box j during a time interval Δt, one can express the mean value of s_i at a future time as

$$\overline{S}_i\,(t+\Delta t) = \sum_{j=1}^{N} c_{ij}\,(t, \Delta t)\,\overline{S}_j\,(t)\,, \tag{24}$$

in which the matrix of mixing coefficients c_{ij} is called the transilient matrix. By definition, the conservation of the transferred property requires that

$$\sum_{j=1}^{N} c_{ij} = \sum_{i=1}^{N} c_{ij} = 1. \tag{25}$$

The vertical turbulent flux of scalar at any height k (defined as the top of the grid box k) is given by

$$\overline{w's'}\,(k) = \frac{\Delta z}{\Delta t} \sum_{i=1}^{k} \sum_{j=1}^{N} c_{ij} \left(\overline{S}_i - \overline{S}_j\right), \tag{26}$$

where Δz is the vertical grid spacing and Δt is the time step for the c_{ij} (t, Δt) matrix. The above discrete-grid expression can also be written in a continuous form (Stull [9]).

The main closure problem in the above model is the specification or estimation of the transilient matrix c_{ij}. These mixing coefficients are expected to be different for different flow situations and cannot be generalized. Different approaches have been suggested for estimating the mixing coefficients, but none of them is very rigorous and completely satisfactory (Stull [9], [16]). For example, one approach utilizes the nonlocal (integral) form of the turbulent kinetic energy equation. After parameterizing the unknown terms in terms of the mean wind and potential temperature parameters, mixing potentials are obtained and c_{ij} are, then, expressed in terms of mixing potentials. The more non-zero elements of matrix in the model, the more complex and expensive would be the numerical solution.

Another approach aimed at reducing the computational cost is to specify the transilient matrix according to some simple conceptual model of mixing in the CBL (see e.g., Fiedler and Moeng [18]; Pleim and Chang [17]). In this way, only a fraction of the elements in the transilient matrix need to be non-zero. Fiedler and Moeng [18] used a 20 x 20 matrix which includes non-zero elements for the tridiagonal portion, as in simple eddy diffusion models, as well as in the two columns down the left side of the matrix which represent buoyant plume transport from the lowest two layers in the surface layer to every other layer in the CBL.

The values of these matrix elements were determined so as to fit the model results to those of a large eddy simulation (LES).

A similar, but much simpler, nonlocal integral closure model for the CBL was proposed by Blackadar [15]. Unlike the Fiedler and Moeng [18] model, however, upward and downward mixing in Blackadar's model are symmetric, and local mixing between adjacent layers is not allowed. This is not consistent with the observed asymmetry of the large-eddy motions in the CBL, which are characterized by narrow fast-moving updrafts and thicker slow-moving compensating downdrafts. Consequently, the vertical transport and dispersion in the CBL are also found to be asymmetrical (see e.g., Arya [3], Ch. 8). More recently, Pleim and Chang [17] have proposed an asymmetrical convective model (ACM) of mixing in CBL, which is computationally as simple as Blackadar's model, but incorporates a modified scheme for downward mixing. The upward transport originates in the bottom most layer, which constitutes the surface layer, from which air is allowed to mix with each of the upper layers in the CBL directly. The downward transport and mixing, on the other hand, proceed from each layer to the next lower layer only in a cascading manner. This stimulates rapid upward transport from the surface layer by buoyant plumes and more gradual compensatory subsidence in downdrafts. The governing equations for the ACM are derived from mass continuity using the above asymmetric mixing scheme and the σ-coordinate system, for its intended application in mesoscale meteorological and dispersion models (see Pleim and Chang [17]). The model has been tested and shown to perform much better than some of the local closure schemes including one based on the turbulence kinetic energy equation (Alapaty et al. [19]).

2.2.3 Integral models

Integral or slab models use the vertically integrated or layer-averaged forms of equations of motion and thermodynamic variables (see e.g., Driedonks [20]; Arya and Byun [21]; Stull [19]; Sorbjan [11]; Byun and Arya [22]). These equations for the layer-averaged wind components $\left(\overline{u}_m \text{ and } \overline{v}_m\right)$, potential temperature $\left(\overline{\theta}_m\right)$, and so on, also contain the surface fluxes of momentum, heat, etc., and those at the top of the mixed layer or the PBL. Parameterizations of the surface fluxes (e.g., through the bulk transfer relations) and the entrainment fluxes at the top of the mixed layer (e.g., through the usual entrainment relations) provide the necessary closure to the set of equations which can be numerically solved. In addition, the rate equations for the mixed layer depth or the PBL height, transition-layer thickness, and changes or jumps in the mean variable across the transition layer are also carried in some integral models. Depending on the sophistication with which the transition or entrainment layer at the top of the mixed layer is represented and the type of closure used, the various zero-order and first-order slab models have been proposed for the horizontally-homogeneous CBL (Driedonks [20]; Arya and Byun [21]; Stull [9]). Simpler zero-order jump models may be adequate for predicting the mixed-layer depth and potential temperature, but more sophisticated integral models are required for better simulation of the thickness of, and entrainment processes occurring in, the transition layer.

The integral modeling approach has also been extended and usefully applied to nonhomogeneous atmospheric boundary layers developing over varying land use and mesoscale topography (Keyser and Anthes [23]; Deardorff et al. [24]). For example, an integral model for the time-dependent daytime unstable urban boundary layer over varying land use and topography was formulated and tested against the selected data from the Regional Air Pollution Study (RAPS) over and around St. Louis, U.S.A. (Byun [25]; Byun and Arya [22]). For reasons of simplicity and limited computer resources, the dynamics of the overlying free atmosphere and its interaction with the mixed layer were not included in the model.

2.3 Turbulence kinetic energy models

In the first-order local closure models, one has to specify eddy diffusivities or mixing lengths as functions of the characteristic length and velocity scales of turbulence. Any generalized explicit specification of the same, based on PBL similarity theories and empirical turbulence data, can be justified only for an idealized, horizontally-homogeneous PBL. In more complex boundary-layer and mesoscale flows, eddy diffusivities are expected to vary with time and space, in response to the variations of turbulence structure. One can still relate eddy diffusivities to velocity and length scales of turbulence, but the latter must be computed from their dynamical equations. The most appropriate dynamical equation for the turbulence velocity scale, which can be taken as the square-root of TKE, is the TKE equation

$$\frac{DE}{Dt} = -\left[\overline{u'w'}\frac{\partial u}{\partial z} + \overline{v'w'}\frac{\partial v}{\partial z}\right] + \frac{g}{T_{vo}}\overline{w'\theta'_v} - \frac{\partial}{\partial z}\left(\overline{w'E'} + \frac{\overline{w'p'}}{\rho_o}\right) - \varepsilon, \quad (27)$$

where E and E′ are the mean and fluctuating turbulence kinetic energies per unit mass

$$E = \frac{1}{2}\left(\overline{u'^2} + \overline{v'^2} + \overline{w'^2}\right), \quad (28)$$

$$E' = \frac{1}{2}\left(u'^2 + v'^2 + w'^2\right). \quad (29)$$

$\overline{w'\theta'_v}$ is the virtual heat flux and ε is the rate of energy dissipation. The various terms on the right-hand side of eqn. (27) represent shear production, buoyancy production or destruction, turbulent transport, and the rate of dissipation of the TKE.

All the TKE models carry eqn. (27), in addition to the equations for mean variables, in which turbulent fluxes or covariances are parameterized using the usual first-order closure relations. Thus, the turbulent transport term is parameterized as

$$\overline{w'E'} + \frac{\overline{w'p'}}{\rho_o} = -K_E\frac{\partial E}{\partial z}, \quad (30)$$

where K_E is the TKE diffusivity which is assumed equal or proportional to K_m. In the parameterized form eqn. (27) can be written as

$$\frac{DE}{Dt} = K_m \left[\left(\frac{\partial \bar{u}}{\partial z} \right)^2 + \left(\frac{\partial \bar{v}}{\partial z} \right)^2 \right] - \frac{g}{T_{vo}} K_h \frac{\partial \bar{\theta}_v}{\partial z} + \frac{\partial}{\partial z}\left(K_E \frac{\partial E}{\partial z} \right) - \varepsilon, \tag{31}$$

in which the ratios of eddy diffusivities, viz., $\sigma_h = K_m / K_h$ and $\sigma_E = K_m / K_E$, are generally specified as constants (e.g., σ_h = 0.9 and σ_E = 1 are the widely-used standard values). The closure is still not complete without the specification of K_m and ε. Different approaches or models have been used for this purpose.

2.3.1 The parameterized length-scale model

The simplest parameterizations of eddy viscosity and the rate of energy dissipation are based on the TKE and appropriate length scales; the most commonly used parametric relations are:

$$K_m = C \ell E^{1/2}, \tag{32}$$

$$\varepsilon = C_\varepsilon E^{3/2} / \ell_\varepsilon. \tag{33}$$

Here, ℓ is the large-eddy length scale, ℓ_ε is the dissipation length scale, and C and C_ε are empirical constants whose values may depend on how ℓ and ℓ_ε are specified. In most of the TKE models it is assumed that $\ell_\varepsilon = \ell = \ell_m$, where the mixing length is usually specified as in eqn. (23), while a few investigators have used different expressions for ℓ and ℓ_ε (Therry and Lacarrere [26]).

Substituting from the well known similarity relations for the various parameters in the neutral surface layer (e.g., $K_m = kzu_*, \varepsilon = u_*^3 / kz$, $\ell = \ell_\varepsilon = kz$, and $E \cong 5.5\ u_*^2$), one can estimate $C \cong 0.43$ and $C_\varepsilon = C^3 \cong 0.08$. Somewhat different values of these constants might be more appropriate when modeling the entire PBL under different stability conditions (Detering and Etling [27]; Lacser and Arya [28]).

A wide variety of length-scale parameterizations have been proposed in the literature (Holt and Raman [10]). Sensitivity tests, including a comparative study of several different parameterizations have shown that, while mean wind and temperature profiles may not be very sensitive to different formulations, the TKE and the rate of energy dissipation are quite sensitive to the same (Lacser and Arya [28]).

2.3.2 The E-ε model

In most computational fluid dynamics (CFD) models of turbulent flows, the dynamical equation for the rate of energy dissipation is carried in addition to the TKE equation. In these so-called k-ε or E-ε models, the eddy viscosity is usually expressed (parameterized) in terms of E and ε as

$$K_m = C_k E^2/\varepsilon, \tag{34}$$

where C_k is an empirical constant that may actually depend on the type of flow. The derivation of the dynamical equation for ε is quite involved (see e.g., Lumley [29]); it contains many new unknown terms that are considerably simplified and parameterized. The highly parameterized form used in the standard E-ε model is given as

$$\frac{D\varepsilon}{Dt} = C_1 \frac{\varepsilon}{E}\left[-\overline{u'w'}\frac{\partial \bar{u}}{\partial z} - \overline{v'w'}\frac{\partial \bar{v}}{\partial z} + \frac{g}{T_{vo}}\overline{\theta'_v w'}\right] - C_2 \frac{\varepsilon^2}{E} + \frac{\partial}{\partial z}\left(K_\varepsilon \frac{\partial \varepsilon}{\partial z}\right), \tag{35}$$

where C_1 and C_2 are empirical constants, and K_ε is the diffusivity of dissipation rate which is usually linked to that of momentum as $K_\varepsilon = K_m/\sigma_\varepsilon$, where σ_ε is another diffusivity ratio.

The various terms on the right-hand side of eqn. (35) represent production, destruction and turbulent transport of the dissipation rate respectively. In their parameterized forms, represented in eqn. (35), the production and destruction terms have been simply assumed to be proportional to the production and dissipation of the TKE. The various constants (C_k, C_1, C_2, etc.) in the E-ε model, often used in commercial CFD codes, have been estimated from the laboratory experimental data. Their "standard" values ($C_k = 0.09$, $C_1 = 1.44$, $C_2 = 1.92$, etc.) may not be appropriate for the stratified, rotating atmospheric PBL (see e.g., Detering and Etling [27]). But sufficiently accurate measurements of the various terms in E- and ε- equations are not available for a direct empirical determination of these constants for the PBL. Different values have been suggested by different investigators on the basis of comparisons of model-predicted and observed winds and turbulence (see e.g., Holt and Raman [10]).

2.3.3 The ε-ℓ model

In this model the eddy viscosity is determined by eqn. (32), with both E and ℓ computed from their dynamical equations. Thus, instead of the ε-equation, a prognostic equation for the length scale ℓ (more appropriately, the product $E\ell$) is included in the model. Mellor and Yamada [30] proposed such a model as a compromise between their level 2 and level 3 models. The parameterized form of the prognostic equation for $E\ell$ looks similar to that for the dissipation rate ε (Mellor and Yamada [31]); both have the same limitations and large uncertainties in the estimates of the various closure constants.

A comparison of different TKE-closure schemes with limited field experimental data by Holt and Raman [10] reveals that mean profiles of wind, potential temperature and specific humidity show little sensitivity to the type of closure scheme. Thus, in the determination of K_m, using a diagnostic formulation of ℓ or a prognostic determination of ε makes very little difference. However, the E-ε model with the modified Detering and Etling [27] scheme performs best in the simulation of turbulence structure, because the energy dissipation rate is calculated from its prognostic equation rather than simply parameterized by eqn. (33) with a specified length scale.

2.4 Higher-order closure models

The gradient transport theory with local closure relations has some serious flaws and becomes invalid in convective conditions when turbulent transports are not necessarily down the mean gradients. Most of the TKE-closure models also utilize the local gradient-transport concept and relations, and fail to account for significant counter-gradient transports by large convective eddies. For this reason, second- and higher-order closure models of turbulence and diffusion in the PBL have been proposed by many investigators (see e.g., Donaldson [32]; Mellor and Yamada [30]; Wyngaard et al. [33]; Andre et al. [34]; Lumley [29]; Sun [35]). Second-order closure models are based on the dynamical equations for turbulent transports or fluxes (Stull [9]; Sorbjan [11]). These equations contain the unknown third moments, correlations involving pressure and velocity or other scalar fluctuations, and the molecular dissipation terms, all of which have to be parameterized to close the set of equations. Thus, the closure problem shifts to the higher order second-moment equations.

The second-moment equations contain much more physics of turbulent transports than the TKE equation. Therefore, it has been argued that the second-order closure approach should be more general and accurate than the first-order or one and a half-order (TKE) closure approach. Similarly, a third-order closure model utilizing the dynamical equations for all the first, second and third moments might be considered even more general, because it contains more physics of turbulence. Unfortunately, the governing equations also become more complex and numerous, and their closure becomes much more difficult and ad-hoc in the absence of experimental data on higher moments of turbulence (see e.g., Andre et al. [34]; Lumley [36]). Here we mention only some of the commonly used closure approaches to parameterize the turbulent transport (third moments), molecular dissipation and pressure terms in the second-moment equations.

The third-moment terms, representing turbulent transports of second moments, are very important to the dynamics of buoyancy-driven unstable and convective boundary layers. They are found to be rather insignificant in stably stratified flows, such as the SBL. The simplest parameterization of turbulent transport terms is based on the gradient-transport concept, whereby third moments are expressed as products of the spatial gradients of second moments and some form of turbulent diffusivity. The diffusivity is expressed in terms of large-eddy length and velocity scales (e.g., $K \sim E^{1/2}\ \ell$). Alternatively, the turbulent diffusion time scale defined by the ratio E/ε can be used (e.g., $K \sim E^2/\varepsilon$). However, this type of turbulent-transport modeling has been found to be deficient and inconsistent with the observed turbulence structure in the CBL. A more complex, but general, functional-expansion approach (Lumley [29], [36]) can remove some of the limitations of the simpler approach, but it also introduces many more terms and empirical constants in the parametric relations.

An accurate representation or parameterization of molecular dissipation terms in second-moment equations, especially the variance equations, is important, because these terms determine the rate at which turbulence variances and the TKE are destroyed. In some models, ε also determines the turbulence length and time scales that are used in the parameterization of turbulent transport and pressure terms. In the simplest approach, dissipations are expressed in terms of the appropriate length and velocity scales, as in eqn. (33). Alternatively, dynamical equations are carried for the rates of dissipation of TKE and variances of scalar fluctuations. Such equations are very complex and only the highly simplified and parameterized forms are used in second-order closure models (Wyngaard et al. [33]; Lumley [36]). The molecular dissipation terms in the dynamical equations for the covariances or turbulent fluxes are usually neglected, using Kolmogorov's local isotropy hypothesis for large Reynolds-number flows.

The satisfactory parameterization of the covariances between pressure fluctuations and spatial gradients of velocity, temperature, and other scalar fluctuations, which are also called return-to-isotropy terms, is probably the most difficult and challenging problem in higher-order closure modeling. These are the primary destruction terms in the covariance equations; they are also responsible for the distribution of turbulence kinetic energy among three components. In the simplest parameterizations, pressure terms are assumed to be proportional to the second moments (variances and covariances) they tend to destroy. But this approach does not properly account for the interactions of turbulence with mean flow shear and buoyancy. More refined parameterizations can approximately account for the above interactions (Launder et al. [37]; Lumley [36]), but contain many more terms and empirical constants.

The applications of the second-order closure approach to dispersion modeling are mostly limited to small-scale or local air pollution problems involving individual sources (Lewellen and Teske [38];

Yamada [39]; Sun [35]). Mellor and Yamada [31] have also proposed a simpler version (level 2 or 2.5) of their model for applications to three-dimensional mesoscale atmospheric dispersion. More sophisticated and complex higher-order closure approaches, as well as large-eddy simulation, may not be computationally and practically feasible for use in mesoscale or regional dispersion and air-quality models for large domains such as Europe or the U.S.A.

3 Similarity scaling and empirical relations

In this section, we discuss some observed aspects of mean flow and turbulence structure of idealized PBLs over flat and homogeneous surfaces. The observed wind, temperature and specific humidity profiles in such PBLs are best represented in an appropriate similarity theory framework. A similarity theory, based on dimensional analysis, provides the most appropriate scales for normalizing the variables into dimensionless groups or similarity parameters, organizing experimental data in the most efficient manner, and empirically deriving the similarity relations or predictions. Most generalized parameterizations of the PBL are usually based on some form of similarity scaling.

3.1 Surface-layer similarity scaling

In the lower 5-10% of the PBL, called the surface layer, the mean wind direction and the vertical fluxes of momentum, heat and water vapor remain nearly constant with height and can be represented by their surface values. For the stratified surface layer, Monin and Obukhov [40] proposed a similarity hypothesis that mean gradients and turbulence statistics depend only on the height z, the surface stress or momentum flux τ_o/ρ, the virtual surface heat flux $H_{vo}/\rho c_p$, and the buoyancy variable g/T_{vo}, where g is the acceleration due to gravity and T_{vo} is the virtual temperature of the reference atmosphere near the surface. The virtual heat flux can be expressed in terms of the sensible heat flux (H_o) and water vapor flux (E_o) as

$$H_{vo} = H_o + 0.61\, c_p\, T_o\, E_o\,. \tag{36}$$

The Monin-Obukhov (M-O) scaling parameters based on surface fluxes are:

$$u_* = (\tau_o / \rho)^{1/2}; \ \theta_* = -H_o / \rho c_p u_* \\ q_* = -E_o / \rho u_*; L = -\rho c_p T_{vo} u_*^3 / k g H_{vo} , \quad (37)$$

where u_* is the friction velocity, θ_* and q_* are the corresponding temperature and specific humidity scales, L is the Obukhov length, and $k \cong 0.40$ is the von Karman constant. Since the height z above the surface is another length scale, its ratio with the Obukhov length, z/L, forms the dimensionless similarity parameter of the M-O similarity theory. This parameter is a measure of buoyancy or stability effects in the surface layer and is commonly referred to as the M-O stability or similarity parameter.

The M-O similarity theory predicts that mean variables, such as the vertical gradients of mean velocity, potential temperature and specific humidity, as well as turbulence statistics (e.g., variances, correlation coefficients, spectrum functions, etc.), when scaled (normalized) by appropriate combinations of z, u_*, θ_* and q_*, must be unique functions of $\zeta = z/L$ only.

3.1.1 Mean profiles

The most widely used M-O similarity relations for the velocity and potential temperature gradients in the surface layer are:

$$\frac{kz}{u_*}\frac{\partial \bar{u}}{\partial z} = \phi_m(\zeta); \ \frac{kz}{\theta_*}\frac{\partial \bar{\theta}}{\partial z} = \phi_h(\zeta), \quad (38)$$

in which the similarity functions $\phi_m(\zeta)$ and $\phi_h(\zeta)$ have been evaluated empirically from micrometeorological experiments conducted over flat and uniform sites (Businger et al. [41]; Dyer [42]; Hicks [43]; Hogstrom [44], [45]; Garratt and Hicks [46]). Although there are still some differences in the empirical estimates of the above similarity functions by different investigators, the most widely used Businger-Dyer relations are (Arya [3], [8]):

$$\phi_h = \phi_m^2 = (1 - 15\zeta)^{-1/2}, \text{ for } -5 < \zeta < 0, \quad (39)$$

$$\phi_h = \phi_m = 1 + 5\zeta, \text{ for } 1 > \zeta \geq 0 . \tag{40}$$

These imply a simple and explicit relationship between the M-O similarity parameter (ζ) and the gradient Richardson number (Ri), which can be used to estimate the former from measurements of the latter.

The vertical profiles of velocity and potential temperature can be obtained by integrating eqns. (39) and (40) with respect to z, i.e.,

$$\overline{u}/u_* = 2.5\left[ln\left(z/z_o\right) - \psi_m\left(\zeta\right)\right], \tag{41}$$

$$\left(\overline{\theta} - \overline{\theta}_o\right)/\theta_* = 2.5\left[ln\left(z/z_o\right) - \psi_h\left(\zeta\right)\right], \tag{42}$$

in which z_o is the roughness length or parameter, $\overline{\theta}_o$ is the potential temperature at $z = z_o$, and $\psi_m(\zeta)$ and $\psi_h(\zeta)$ are uniquely related to $\phi_m(\zeta)$ and $\phi_h(\zeta)$, respectively (see e.g., Panofsky and Dutton [47]; Arya [3], [8]). The above profiles may be considered as simple modifications of the log-law in which both the stability-dependent functions increase with increasing instability.

The applicability of the above profile relationships is limited to moderately unstable to moderately stable conditions. On the stable side, the profile functions (ϕ_m and ψ_m) have been shown to deviate from their linear forms when $\zeta \geq 1$ (see e.g., Webb [48]; Hicks [43]). Above the shallow surface layer, however, the mean wind profile tends to become linear under very stable conditions. In very unstable and convective conditions, on the other hand, the Businger-Dyer profile relations may be applicable only up to a height of about 50 m (i.e., less than 5% of the CBL height).

An alternative empirical representation of the mean wind distribution frequently used in air pollution dispersion applications is the power-law wind profile

$$\overline{V}/\overline{V}_r = \left(z/z_r\right)^m , \tag{43}$$

where $\overline{V}$ is the mean wind speed at height z, $\overline{V}_r$ is the wind speed at a reference height z_r, and the exponent m has a value between 0 and 1, depending on the surface roughness and stability (m increases with increasing roughness and stability). The above power-law profile does not have a very sound theoretical basis, but it is found to provide a good fit to observed wind profiles in the lower part of the PBL, even well above the top of the surface layer, over a wide range of roughness and stability conditions. For the lower part of the barotropic CBL, Santoso and Stull [49] have proposed a product of power-law and exponential profile functions which matches the mixed-layer value at the top of the layer. They also give a theoretical justification for their proposed similarity profile for the bottom-fifth of the CBL. The simpler power-law profile may still be preferable because of its applicability to a much wider range of stability conditions (Arya [50]).

3.1.2 Turbulence statistics

Turbulence structure of the atmospheric surface layer has been extensively investigated (see e.g., Panofsky and Dutton [47]; Arya [3], [8]; Stull [9]; Sorbjan [11]; Kaimal and Finnigan [51]). The simplest statistics are the standard deviations of velocity, temperature and other scalar fluctuations. In particular, σ_u, σ_v and σ_w are of great interest in dispersion modeling and parameterization. According to the M-O similarity theory, $\sigma_u / u_*, \sigma_v / u_*$ and σ_w / u_* must be some unique functions of ζ only. Measurements of turbulence in the atmospheric surface layer are not entirely consistent, however, with the above theoretical prediction. In particular, σ_u and σ_v do not follow the M-O similarity scaling under unstable and convective conditions; instead, they follow the mixed-layer similarity scaling to be discussed later. However, the statistics of vertical velocity, temperature, and specific humidity fluctuations are well described by the M-O similarity theory. For example, the following empirical relationships have been suggested by the various representations of observed turbulence data using the M-O similarity scaling:

$$\frac{\sigma_u}{u_*} \simeq 2.5; \frac{\sigma_v}{u_*} \simeq 1.9; \frac{\sigma_w}{u_*} \simeq 1.3, \text{ for } \zeta \geq 0, \tag{44}$$

$$\frac{\sigma_w}{u_*} = 1.3\,(1 - 3\zeta)^{1/3}, \text{ for } \zeta < 0. \tag{45}$$

With increasing stability, turbulence becomes very weak and σ_u/u_*, σ_v/u_*, etc. show rather large scatter with no definite dependence on ζ. The above similarity relations for σ_u and σ_v may not remain valid under extremely stable conditions with weak winds and highly variable wind direction.

Combining the similarity relations for σ_u, σ_v and σ_w into the definition of TKE, one can obtain

$$E = 5.75\, u_*^2, \text{ for } \zeta \geq 0, \tag{46}$$

$$E = 0.36\, W_*^2 + 0.85\, u_*^2\, (1 - 3\zeta)^{2/3}, \text{ for } \zeta < 0, \tag{47}$$

in which horizontal components are assumed to follow the mixed layer similarity relations $\sigma_u = \sigma_v \simeq 0.6\, W_*$, where W_* is the convective velocity to be defined later. The corresponding similarity relations for the rate of energy dissipation, obtained from the measurements of ε, are (Wyngaard and Cote [52]; Hogstrom [45]):

$$\varepsilon = \frac{u_*^3}{kz}(1.24 + 4.3\zeta), \text{ for } \zeta \geq 0, \tag{48}$$

$$\varepsilon = \frac{u_*^3}{kz}\left(1 + 0.5|\zeta|^{2/3}\right)^{3/2}, \text{ for } \zeta < 0. \tag{49}$$

The spectra of horizontal and vertical velocity fluctuations in the surface layer show systematic variations with the M-O similarity parameter, except for the horizontal components under unstable and convective conditions (Kaimal et al. [53]). The frequency or wave number corresponding to the peak in the spectrum is often used to estimate the characteristic large-eddy time or length scale for use in certain dispersion theories and models (Hanna [54]; Arya [3]).

3.2 Mixed-layer similarity scaling

In unstable and convective PBLs, there is a mixed layer above the surface layer in which the mean potential temperature is observed to be nearly uniform (Arya [8]; Stull [9]; Garratt [55]; Kaimal and Finnigan

[51]). Variations in specific humidity and mean velocity are also observed to be small over the depth of the mixed layer. Deardorff [56] proposed a similarity hypothesis that turbulence structure in the convective mixed layer depends only on the virtual heat flux $H_{vo}/\rho c_p$, the buoyancy variable g/T_{vo}, the height above the surface z, and the height z_i of the base of inversion which caps the mixed layer. The appropriate scaling parameters for the mixed-layer similarity are z_i and the convective velocity

$$W_* = \left(\frac{g}{T_{vo}} \frac{H_{vo}}{\rho c_p} z_i \right)^{1/3} . \tag{50}$$

The mixed-layer similarity theory predicts that $\sigma_u / W_*, \sigma_v / W_*$, etc., and other appropriately scaled turbulence statistics must be unique functions of z/z_i only. This is strictly valid only within the mixed layer $\left(0.1\, z_i < z \le z_i\right)$ of the CBL with $-z_i/L > 10$. Actual observations of turbulence in the CBL, however, indicate that there is hardly any dependence of σ_u/W_* and σ_v/W_* on z/z_i and, for all practical purposes,

$$\sigma_u / W_* \simeq \sigma_v / W_* \simeq 0.60 . \tag{51}$$

The above mixed-layer scaling can also be extended to the surface layer under free convective conditions, but the following empirical expression is more appropriate for a wider range of instability varying from slightly unstable to free convective conditions (Panofsky et al. [57]):

$$\sigma_u / u_* \simeq \sigma_v / u_* \simeq \left(12 - 0.5 \frac{z_i}{L} \right)^{1/3} . \tag{52}$$

The standard deviation of vertical velocity fluctuations in the CBL has been observed to show a much stronger dependence on z/z_i. It increases with height, attains a broad maximum in the middle of the mixed layer, and decreases again in the upper part of the CBL (Caughey and Palmer [58]; Sorbjan [11]). For most practical applications, $\sigma_w / W_* \simeq 0.60$ represents an average value for the whole mixed layer;

more detailed empirical expressions of the same in different layers are given by Hanna [54].

Combining the mixed-layer similarity relations for σ_u, σ_v and σ_w into the definition of TKE, one obtains

$$E = 0.38\, W_*^2 + 0.9\left(\frac{z}{z_i}\right)^{2/3}\left(1 - 0.8\,\frac{z}{z_i}\right)^2 W_*^2, \tag{53}$$

or, more simply, for most practical purposes,

$$E \simeq 0.54\, W_*^2 \tag{54}$$

in the convective mixed layer. An empirical expression for rate of energy dissipation in the same is:

$$\varepsilon = \frac{W_*^3}{z_i}\left(0.8 - 0.3\,\frac{z}{z_i}\right). \tag{55}$$

The spectra of horizontal and vertical velocity fluctuations in the CBL are best represented in the same mixed-layer similarity framework, using z_i and W_* as the appropriate length and velocity scales. Observed normalized spectra of horizontal velocity fluctuations are more or less independent of z/z_i, but the spectra of vertical velocity fluctuations show a stronger, systematic dependence on the dimensionless height (see e.g., Kaimal et al. [59]; Kaimal and Finnigan [51]). The large-eddy length scales estimated from the observed spectra are given by (Caughey and Palmer [58])

$$\lambda_{mu} \simeq \lambda_{mv} \simeq 1.5\, z_i, \tag{56}$$

$$\lambda_{mw} \simeq 1.8\, z_i\left[1 - \exp\left(-4\,\frac{z}{z_i}\right) - 0.0003\exp\left(8\,\frac{z}{z_i}\right)\right], \tag{57}$$

with a linear variation of $\lambda_{mw} \simeq 5.9\, z$ near the surface ($z < 0.1\, z_i$). Here λ_{mu}, λ_{mv} and λ_{mw} represent the wavelengths corresponding to the maximum contribution to the variance spectra of u, v and w, respectively.

Higher moments of turbulence in the mixed layer are also found to be scaled by W*. In particular, the normalized third moment of vertical velocity $\overline{w'^3}/W_*^3$, or skewness $S = \overline{w'^3}/\sigma_w^3$, is observed to be significantly positive, indicating much larger positive velocities in convective updrafts as compared to the average negative velocity in downdrafts. Since the average vertical velocity is essentially zero in a horizontally-homogeneous CBL, the horizontal area occupied by downdrafts is much (about 50%) larger than that occupied by updrafts. This is also confirmed by estimated probability density functions of vertical velocity in the CBL (Caughey et al. [60]; Quintarelli [61]).

The mixed-layer similarity scaling may not be valid in the transition layer above the inversion base. Turbulence and diffusion processes in this layer are also influenced by entrainment of free atmospheric air, stronger wind speed and direction shears, and potential temperature gradients observed in the transition layer (see e.g., Stull [9]; Boers [62]; Nelson et al. [63]; Arya [1]).

3.3 PBL similarity theories and scaling

More comprehensive reviews of the various PBL similarity theories and scales based on the same are given elsewhere (Tennekes [64]; Stull [9]; Sorbjan [11]). Here we give only a brief discussion of the similarity scales and relations that might be useful to atmospheric dispersion and air quality modeling. Based on atmospheric stability in the PBL, these can be divided into several categories.

3.3.1 Rossby-number similarity

The Rossby-number similarity theory for the stratified PBL is based on a similarity hypothesis proposed by Kazanski and Monin [65] that the PBL structure is determined by all the variables that were included in the M-O similarity hypothesis plus the Coriolis parameter f. The latter yields an addition length scale $u_*/|f|$ which is a measure of the height of the neutral PBL $(h \simeq 0.3\, u_*/|f|)$. The normalized height of the stratified PBL is predicted to be a unique function of the stability parameter, i.e.,

$$h|f|/u_* = F_h(\mu), \tag{58}$$

where $\mu = u_*|f|/L$ is the relevant stability parameter. The similarity theory also predicts that in a barotropic PBL, the dimensionless ageostrophic wind components $(\bar{u} - u_g)/u_*$ and $(\bar{v} - v_g)/u_*$, as well as turbulence statistics, must be unique functions of the dimensionless height, $z|f|/u_*$, and μ only. An appropriate matching of the PBL similarity profiles with those of the surface layer also yields geostrophic drag or resistance laws (Zilitinkevich et al. [66]; Arya [67]) which can be used for parameterizing the PBL in large-scale atmospheric circulation models. These resistance laws contain the surface Rossby number, $G|f|/z_o$, as an important parameter, which explains the name given to this similarity theory and scaling.

Tennekes [64] has given a formal and elegant derivation of the velocity-defect profile relations for the barotropic, neutral PBL:

$$(\bar{u} - u_g)/u_* = F_u\left(\frac{zf}{u_*}\right), \tag{59}$$

$$(\bar{v} - v_g)/u_* = F_v\left(\frac{zf}{u_*}\right). \tag{60}$$

The corresponding geostrophic drag relations, derived from similarity matching, are:

$$u_g/u_* = \frac{1}{k}\left[ln\left(u_* f/z_o\right) - A\right], \tag{61}$$

$$v_g/u_* = -B/k, \tag{62}$$

which are, then, used to express the geostrophic drag coefficient, $c_g = u_*^2/G^2$, and the cross-isobar angle of near-surface winds $\alpha_o = \tan^{-1}\left(-v_g/u_g\right)$ as functions of the surface Rossby number (Ro). In the above relations, A and B are empirical similarity constants and only the magnitude of f (not its sign) is considered. The normalized variances, covariances, and other turbulence statistics in the neutral PBL are also predicted to be universal functions of zf/u_*. Ideally, the height of such a PBL should not be restricted by a low-level inversion and is given by $h = c\, u_*/f$, where c has been estimated to lie between 0.2 and 0.4.

Mean flow and turbulence data obtained in such a near-neutral (slightly unstable) PBL during the Joint Air-Sea Interaction Experiment (JASIN) has been presented in the above similarity framework by Nicholls [68]. He estimated average values of $A \cong 1.4$, $B \cong 4.2$, and $c \cong 0.2$.

The proposed extensions of the above similarity relations consider the dependence of the various similarity functions including A, B, etc., on the stability parameter μ (see e.g., Zilitinkevich and Chalikov [69]; Clarke and Hess [70]; Arya [67]). This also implies that the dimensionless PBL height must also be a unique function of μ, as in eqn. (58). However, Deardorff [71], [72] and others have pointed out that it was not true under unstable and convective conditions in which the PBL height is not at all related to u_*/f or μ, but is determined by the height of the lowest inversion base. For this reason, the Rossby-number similarity scaling cannot be valid under unstable and convective conditions. Its validity to the stable boundary layer (SBL) is also somewhat limited to slightly and moderately stable conditions during which there is continuous exchange between adjacent layers in the vertical, i.e., turbulence is continuous throughout the SBL. Under very stable conditions, turbulence becomes highly intermittent and sporadic, so that upper layers may become decoupled from the surface layer. Any similarity scaling based on the surface fluxes will not be appropriate.

For the SBL with more or less continuous turbulence, Zilitinkevich [73] derived the following diagnostic relationship for steady state or stationary conditions:

$$h = d\left(u_* L / |f|\right)^{1/2}, \tag{63}$$

in which d is an empirical constant. The above relationship implies a simple form of the similarity function in eqn. (58) as

$$h|f| / u_* = d\, \mu^{-1/2}. \tag{64}$$

These SBL height relations have been tested against experimental data obtained at several flat and approximately homogeneous sites (Garratt [74]). The estimated values of d range from 0.2 to 0.7 with an average value of about 0.4 which is also the value suggested by numerical modeling studies (Brost and Wyngaard [13]; Garratt [74]). It has been shown that d may depend on the time after transition, surface cooling rate, terrain slope and possibly other factors.

The above diagnostic similarity relation may not be valid for an evolving SBL for several hours after the transition time. Prognostic equations for the SBL height have been proposed for such an evolving boundary layer (see e.g., Nieuwstadt and Tennekes [75]). The SBL evolves rather slowly in time and rarely reaches a steady or equilibrium state. This may explain why correlations between the observed and theoretically predicted SBL heights are not usually very strong (Arya [76]; Koracin and Berkowicz [77]).

Geostrophic drag and heat transfer relations in the SBL have been extensively studied and the similarity functions A (μ), B (μ), etc., have been evaluated from experimental data as well as numerical models of the SBL (Arya [67], [78]; Zilitinkevich [79]; Rao and Snodgrass [80]). Experimental estimates of the same show large scatter, reflecting possible effects of baroclinicity and other factors (Sorbjan [11]; Garratt [55]).

3.3.2 Generalized similarity scaling

Recognizing the importance of the PBL height (h) as an independent height scale and the lack of general validity of eqn. (58), Deardorff [71] utilized z/h and h/L as the dimensionless height and stability parameters. Zilitinkevich and Deardorff [81] also proposed that $u_* /|f|$ be replaced by h in the PBL similarity formulation for the entire range of stability conditions. It was later recognized, however, that $u_* /|f|$ cannot be ignored entirely, and $h|f|/u_*$ should also be considered as an additional parameter in any generalized similarity theory. Further consideration of baroclinicity added two more similarity parameters to the expanding list of the same (see e.g., Arya and Wyngaard [82]). Subsequently, the similarity functions A, B, etc., have been evaluated as functions of h/L

(Yamada [83]). These are also characterized by large scatter, possibly due to baroclinicity and other parameters. The relations involving layer-averaged (over the PBL depth) winds, temperature, etc., are found to be less sensitive to the effects of baroclinicity (see e.g., Arya [84]; Garratt et al., [85]). For example, the layer-averaged wind speed in an unstable PBL is given by

$$\overline{V}_m / u_* = \frac{1}{k}\left[ln\frac{h}{z_o} - 0.5\, ln\,|h/L| - 2.3\right]. \qquad (65)$$

For representing the turbulence structure of unstable and convective boundary layers, it is much simpler to use the mixed-layer similarity scaling than the generalized similarity scaling discussed above. Therefore, most turbulence and diffusion data in the CBL are presented using the mixed-layer scaling.

Turbulence data in slightly and moderately stable boundary layers with more or less continuous turbulence are best represented in the generalized similarity framework, i.e., as functions of z/h and h/L. The parameter $h|f|/u_*$ appears to be uniquely related to h/L and may not be an independent parameter affecting turbulence structure. Even the influence of the stability parameter does not appear to be overly significant, when normalized turbulence statistics are represented as functions of z/h. For example, in a steady PBL over a flat terrain, the vertical profiles of TKE and fluxes can be represented as (Lenschow et al. [86]; Smedman [87]; Rao and Nappo [88]):

$$E \simeq 5.75\, u_*^2 \left(1 - \frac{z}{h}\right)^{\alpha_1}, \qquad (66)$$

$$\tau = \rho\, u_*^2 \left(1 - \frac{z}{h}\right)^{\alpha_1}, \qquad (67)$$

$$H = H_0 \left(1 - \frac{z}{h}\right)^{\alpha_2}, \qquad (68)$$

in which the values of exponents α_1 and α_2 may depend on the temporal development or evolution of the SBL, terrain slope, and other factors. For a well developed or steady SBL, $\alpha_1 \simeq 1.75$ and $\alpha_2 \simeq 1.5$ (Lenschow et al. [86]; Smedman [87]). Equations (66)-(68) may be considered as extensions of the surface-layer similarity relation (46) and the assumption of constant fluxes. As an extension of eqn. (48), the rate of energy dissipation may be expressed as

$$\varepsilon = \frac{u_*^3}{kz}\left(1.24 + 4.3\frac{z}{L}\right)\left(1 - 0.85\frac{z}{h}\right)^{1.5} . \tag{69}$$

3.4 Local similarity theory/scaling

Nieuwstadt [89] proposed a local similarity hypothesis that, above the surface layer, mean gradients and turbulence statistics in the SBL follow a local similarity scaling based on the local (not surface) fluxes. Thus, analogous to the Monin-Obukhov similarity scales of eqns. (37), the local scaling parameters are defined as

$$\begin{aligned} u_{*\ell} &= (\tau/\rho)^{1/2};\ \theta_{*\ell} = -\,H/\rho\, c_p\, u_{*\ell} \\ \Lambda &= T_v\, u_{*\ell}^2 / k\, g\, \theta_{*\ell} \end{aligned} , \tag{70}$$

where $u_{*\ell}$ and $\theta_{*\ell}$ are the local friction velocity and temperature scale respectively, and Λ is a local buoyancy length scale which is analogous to the Obukhov length in the surface layer.

The local similarity theory predicts that appropriately scaled (normalized) mean velocity and temperature gradients, as well as turbulence statistics, at any height in the SBL must be some universal functions of the local stability parameter z/Λ. Furthermore, under very stable conditions when upper layers become decoupled from the surface layer and even from adjacent layers, and turbulence is sporadic and becomes independent of z, i.e., under the so-called z-less stratification (Monin and Yaglom [90]; Wyngaard [91]), the locally-scaled gradients and turbulence statistics must become constants, independent of z or z/Λ. Nieuwstadt [89] has shown that this is indeed the case for $z/\Lambda >> 1$.

The practical utility of the local similarity theory predictions to the parameterization of the SBL is somewhat questionable, because in order to predict velocity variances and the TKE one needs to know the local fluxes of momentum and heat, which are even more difficult to measure or estimate. Nevertheless, it provides a convenient framework for representing and comparing mean gradients and turbulence data in the SBL.

4 PBL parameterization in mesoscale models

4.1 Subgrid scale fluxes

The governing equations of motion, thermodynamic energy, specific humidity and other scalars used in mesoscale models contain subgrid scale variances and covariances (fluxes) of subgrid scale perturbations of velocity, temperature, specific humidity, etc. (Pielke [92]). A perturbation is defined as the deviation of a variable from its grid-averaged value where averaging is implied over the grid volume ($\Delta x \Delta y \Delta z$) as well as over the finite time increment (Δt). The subgrid scale perturbation fluxes must be parameterized in terms of grid-averaged variables in order to form a closed set of equations for the latter. An accurate or satisfactory parameterization of subgrid scale fluxes is crucial to the satisfactory performance of any mesoscale model, because the terms involving subgrid scale perturbations are usually of the same order or even larger than the corresponding terms involving grid volume-averaged resolved variables (Pielke [92]).

In developing parameterizations for subgrid scale averaged quantities, it needs to be recognized that the preferred representation is an ensemble average over the grid volume rather than simply the grid volume average. The ensemble average represents the most likely value of the subgrid scale quantity (variance or covariance) whereas the grid volume average represents just one realization (Pielke [92]). Thus, most parameterizations discussed here may represent only the most likely (ensemble-averaged) estimates of subgrid scale quantities. Another distinction that needs to be made, but is rarely recognized or even mentioned, is that between the subgrid scale perturbation fluxes and the corresponding turbulent fluxes in the PBL. The former obviously depend on the grid spacings and may not be equal to the corresponding turbulent fluxes, except for the rare case when grid spacings are equal to the dimensions of the largest turbulent eddies. In large-domain mesoscale models with horizontal grid sizes of tens of kilometers, perturbation fluxes are likely to be larger than turbulent fluxes, because the former also contain the contributions of the subgrid mesoscale perturbations in addition to the contribution of turbulence. In mesoscale models of small domains with horizontal grid sizes of less than 1 km (i.e., the size of most energy-containing turbulent eddies), on the other hand, subgrid scale perturbation fluxes might be smaller than the total turbulent fluxes. Since PBL parameterizations in mesoscale models attempt to represent total ensemble-averaged fluxes due to turbulence only, these are most appropriate for mesoscale models with horizontal grid sizes of the order of 1 km.

4.2 Parameterization of turbulent fluxes

Most mesoscale atmospheric circulation and dispersion models attempt to resolve the PBL by placing several discrete grid levels within the lowest 1 km of the model domain. Ideally, the lowest grid level should lie within the surface layer. Anthes et al. [93] have shown that a detailed multi-level resolution of the PBL is essential in a mesoscale model of flow over complex terrain with differential heating (e.g., across land-water boundaries), because significant vertical gradients of meteorological variables occur within the PBL. Different types of parameterizations are used to represent the surface fluxes and turbulence variances, kinetic energy, fluxes, and scales at other discrete levels in the PBL, as these are usually not resolved by most widely used first-order closure models.

4.2.1 Surface fluxes

The surface fluxes of momentum, heat and moisture are usually parameterized by using the following bulk transfer relations:

$$\begin{aligned} \tau_o &= \rho\, C_D\, \overline{V}_1^2 \\ H_o &= \rho\, c_p\, C_H\, \overline{V}_1 \left(\overline{\theta}_o - \overline{\theta}_1\right), \\ E_o &= \rho\, C_E\, \overline{V}_1 \left(\overline{q}_o - \overline{q}_1\right) \end{aligned} \tag{71}$$

where $\overline{V}_1$, $\overline{\theta}_1$, and $\overline{q}_1$ are the mean wind speed, potential temperature, and specific humidity, respectively, at the first grid level (z_1), $\overline{\theta}_o$ and $\overline{q}_o$ are the values at $z = z_0$, and C_D, C_H and C_E are the bulk transfer coefficients of momentum (C_D is also called the drag coefficient), heat and water vapor, respectively. Using a viscous or molecular sublayer parameterization, $\overline{\theta}_o$ and $\overline{q}_o$ can be related to their values at the ground surface (Pielke [92]).

In some mesoscale models, using only crude parameterization of surface fluxes, constant values of bulk-transfer coefficients are specified with some distinction made between land and water surfaces and also between stable and unstable conditions. In most cases with large variations in surface roughness over the model domain and strong diurnal variations of stability in the surface layer, however, it is inappropriate to treat C_D and C_H (usually $C_E = C_H$) as constants. These could easily be specified as continuous functions of surface roughness and stability, using the Monin-Obukhov similarity theory (Arya [78]; Pielke [92]). When z_1 lies in the surface layer, C_D and C_H can be represented as functions of z_1/z_0 and the bulk Richardson number

$$Ri_b = \frac{g}{T_{vo}} \frac{\left(\overline{\theta}_{v1} - \overline{\theta}_{v0}\right) z_1}{\overline{V}_1^2}. \tag{72}$$

Computations show that C_D and C_H can vary over very wide ranges which depend on ranges of z_o and Ri_b values encountered (see e.g., Arya [78].

4.2.2 Fluxes and variances at interior levels

Observations of heat and water vapor fluxes in unstable and convective boundary layers suggest that the fluxes vary more or less linearly from their maximum values at the surface to zero at the top of the PBL. For the total momentum flux, a linear profile is also expected under barotropic conditions. However, the presence of baroclinicity can cause large deviations from the linearity (Kaimal et al. [59]). Observations in the SBL also suggest simple profiles of fluxes decreasing with height, as represented by eqns. (66)-(68). The use of these similarity-scaled profiles of fluxes, variances and TKE should probably constitute the simplest parameterizations of these at interior levels in the PBL (see e.g., Arya [94]). The scaling parameters essentially depend on the surface fluxes, whose parameterization is discussed in the previous section, and the PBL height. These similarity scaling-based parameterizations may not be applicable, however, in the vicinity of steep topographical features or sharp changes in surface characteristics.

A more widely used scheme for parameterizing the interior turbulent fluxes is based on the gradient-transport relations in which eddy diffusivities or mixing lengths are specified as functions of height and stability (Businger and Arya [12]; Brost and Wyngaard [13]; Blackadar [15]; Pielke [92]). An implicit specification of eddy diffusivities in terms of mixing lengths might be preferable to an explicit eddy diffusivity profile, such às O'Brien's [95] cubic representation, especially in mesoscale models with high vertical grid resolution in which vertical gradients are well resolved. The dependence of subgrid scale mixing length on the vertical grid spacing can also be considered in such parameterizations.

In the above mentioned gradient-transport scheme, fluxes are always downgradient. Simple modifications to gradient-transport relations can also allow for some countergradient fluxes, as often observed in the convective mixed layer (Deardorff [96]). More satisfactory parameterizations of subgrid scale fluxes in the CBL would be those based on nonlocal mixing schemes, such as the asymmetric convective mixing (ACM) model (Pleim and Chang [17]). These have already been discussed in Section 2.2.2.

4.2.3 Height of the PBL or inversion

In most of the parametric relations discussed here, the height of the PBL (h) or the capping inversion (z_i) is an important length scale which must be calculated or parameterized. A variety of theoretical models have been proposed in the literature for predicting h or z_i during the daytime unstable and convective conditions (Deardorff [72]; Driedonks [20], Arya and Byun [21]; Stull [9]). Differences in the predictions of different types of models generally fall within the uncertainties of estimating the mean large scale vertical velocity $(\overline{w}_L)$ at the top of the PBL.

The prognostic equation for the inversion height is

$$\frac{d z_i}{d t} = w_e + \overline{w}_L, \tag{73}$$

where w_e is the rate of entrainment (velocity) of free atmospheric air into the mixed layer. The various height models differ primarily in the way w_e is represented or parameterized. For example, a simple thermodynamic model of the mixed-layer growth implies that

$$w_e = \frac{1.2}{\gamma z_i} \frac{Ho}{\rho c_p}, \tag{74}$$

where $\gamma = \partial\overline{\theta}/\partial z$ just above the top of the mixed layer.

Deardorff [72] has pointed out the limitations of the above model and proposed the following interpolation expression based on his large-eddy simulations of the PBL:

$$w_e = \frac{1.8\left(W_*^3 + 1.1 u_*^3 - 3.3 u_*^2 f z_i\right)}{\gamma z_i^2 \left(g/\overline{\theta}_o\right) + 9 W_*^2 + 7.2 u_*^2} \tag{75}$$

Note that in both eqns. (74) and (75), the rate of growth of z_i is directly proportional to the surface heat flux and inversely proportional to the stability of the overlying inversion layer. Unlike eqn. (74), (75) remains well behaved and approximately valid during each of the limiting cases of vanishing γ, u_*, or W_*.

The evolution of the nocturnal stable boundary layer (SBL) is rather slow, except immediately following the evening transition period, and it is not clear whether a simple diagnostic relation for h, such as eqn. (63), might be adequate or whether one should carry an appropriate rate equation for the same (see e.g., Nieuwstadt and Tennekes [75]). Since the SBL constitutes only a shallow surface inversion layer of weak turbulent transports and small TKE, its detailed parameterization in a mesoscale model may not be that important, unless the predictions of ground level concentrations and dry deposition are the primary objectives of the model.

Regional air quality models which are closely tied to mesoscale meteorological models are discussed elsewhere (Zwerver and van Ham [97]; Arya [3]). Pielke [92] describes the use of a boundary layer parameterization to estimate transport and diffusion in such a model with domain sizes of 50-500 km and horizontal grid resolutions of 1-10 km. Different methodologies and parameterizations are used in different models, which are very difficult to evaluate and compare with sparse and limited experimental data.

Acknowledgments

Partial support of our Cooperative Agreement No. NCC-1-188 with NASA, Langley Research Center, in preparing this manuscript is gratefully acknowledged.

References

[1] Arya, S. P., Unstable and convective boundary layers. Chapter 1, *Dynamics of Atmospheric Flows: Atmospheric Transport and Diffusion Processes*, eds. M.P. Singh and S. Raman, Computational Mechanics Publications, Southampton, pp. 1-38, 1998.

[2] Pasquill, F. and Smith, F. B., *Atmospheric Diffusion*, 3rd ed., Ellis Horwood, Chichester, 1983.

[3] Arya, S. P., *Air Pollution Meteorology and Dispersion*, Oxford University Press, New York, 1999.

[4] Holzworth, G. C., Estimates of mean maximum mixing depths in the contiguous United States. *Mon. Weath. Rev.*, **92**, pp. 235-242, 1964.

[5] Bradley, E. F., A micrometeorological study of velocity profiles and surface drag in the region modified by a change in surface roughness. *Quart. J. Roy. Meteorol. Soc.*, **94**, pp. 361-379, 1968.

[6] Stunder, M. and SethuRaman, S., A comparative evaluation of the coastal internal boundary-layer height equations. *Boundary-Layer Meteorol.,* **32**, pp. 177-204, 1985.

[7] Oke, T. R., *Boundary Layer Climates*, 2nd ed., Halsted Press, New York, 1987.

[8] Arya, S. P., *Introduction to Micrometeorology*, Academic Press, San Diego, 1988.

[9] Stull, R. B., *An Introduction to Boundary Layer Meteorology*, Kluwer Academic Pub., Boston, 1988.

[10] Holt, T. and Raman, S., A review and comparative evaluation of multilevel boundary layer parameterizations for first-order and turbulent kinetic energy closure schemes. *Rev. Geophys.*, **26**, pp. 761-780, 1988.

[11] Sorbjan, Z., *Structure of Atmospheric Boundary Layer*, Prentice Hall, Englewood Cliffs, NJ, 1989.

[12] Businger, J. A. and Arya, S. P., Height of the mixed layer in the stably stratified planetary boundary layer. *Advan. Geophys.*, **18A**, pp. 73-92, 1974.

[13] Brost, R. A. and Wyngaard, J. C., A model study of the stably stratified planetary boundary layer. *J. Atmos. Sci.*, **35**, pp. 1427-1440, 1978.

[14] Kao, S. K., Theories of atmospheric transport and diffusion. Chapter 6, *Atmospheric Sciences and Power Production*, ed. D. Randerson, U.S. Department of Energy, Technical Information Center, Oak Ridge, TN, pp. 189-239, 1984.

[15] Blackadar, A. K., Modeling pollutant transfer during daytime convection. *Preprints, Fourth Symp. on Atmospheric Turbulence Diffusion, and Air Quality*, American Meteorological Society, Boston, pp. 443-447, 1979.

[16] Stull, R. B., Review of transilient turbulence theory and non-local mixing. *Boundary-Layer Meteorol.*, **62**, pp. 21-96, 1993.

[17] Pleim, J. E. and Chang, J. S., A non-local closure model for vertical mixing in the convective boundary layer. *Atmos. Environ.*, **26A**, pp. 965-981, 1992.

[18] Fiedler, B. H. and Moeng, C.-H., A practical integral closure model for mean vertical transport of a scalar in a convective boundary layer. *J. Atmos. Sci.*, **42**, pp. 359-363, 1985.

[19] Alapaty, K., Pleim, J. E., Raman, S., Niyogi, D. S. and Byun, D. W., Simulation of atmospheric boundary layer processes using local and nonlocal-closure schemes. *J. Appl. Meteorol.*, **36**, pp. 214-233, 1997.

[20] Driedonks, A. G. M., Models and observations of the growth of the atmospheric boundary layer. *Boundary-Layer Meteorol.*, **23**, pp. 283-306, 1982.

[21] Arya, S. P. S. and Byun, D. W., Rate equations for the planetary boundary layer depth (urban vs. rural). *Modeling the Urban Boundary Layer*, American Meteorological Society, Boston, pp. 215-251, 1987.

[22] Byun, D. W. and Arya, S. P. S., A two-dimensional mesoscale numerical model of an urban mixed layer. I. Model formulation, surface energy budget, and mixed-layer dynamics. *Atmos. Environ.*, **24A**, pp. 829-844, 1990.

[23] Keyser, D. and Anthes, R. A., The applicability of a mixed-layer model of the planetary boundary layer to real data forecasting. *Mon. Weath. Rev.*, **105**, pp. 1351-1371, 1977.

[24] Deardorff, J. W., Ueyoshi, K., and Han, Y.-J., Numerical study of terrain-induced mesoscale motions and hydrostatic form drag in a heated, growing mixed layer. *J. Atmos. Sci.*, **41**, pp. 1420-1441, 1984.

[25] Byun, D. W., A two-dimensional mesoscale numerical model of St. Louis urban mixed layer, Ph.D. dissertation, North Carolina State University, Raleigh, 1987.

[26] Therry, G. and Lacarrere, P., Improving the eddy kinetic energy model for planetary boundary layer description. *Boundary-Layer Meteorol.*, **25**, pp. 63-68, 1983.

[27] Detering, H. W. and Etling, D., Application of the E-ε turbulence model to the atmospheric boundary layer. *Boundary-Layer Meteorol.*, **33**, pp. 113-133, 1985.

[28] Lacser, A. and Arya, S. P. S., A comparative assessment of mixing-

length parameterizations in the stably stratified nocturnal boundary layer (NBL). *Boundary-Layer Meteorol.*, **36**, pp. 53-70, 1986.

[29] Lumley, J. L., Second-order modelling of turbulent flows, Chapter 1, *Prediction Methods for Turbulent Flows*, ed. W. Kollmann, Hemisphere Pub., London, pp. 1-31, 1980.

[30] Mellor, G. L. and Yamada, T., A hierarchy of turbulence closure models for planetary boundary layers. *J. Atmos. Sci.*, **31**, pp. 1791-1806, 1974.

[31] Mellor, G. L. and Yamada, T., Development of a turbulence closure model for geophysical fluid problems. *Rev. Geophys.*, **20**, pp. 851-875, 1982.

[32] Donaldson, C. duP., Construction of a dynamical model of the production of atmospheric turbulence and the dispersal of atmospheric pollutants. Chapter 8, *Workshop on Micrometeorology*, ed. D.A. Haugen, American Meteorological Society, Boston, pp. 313-392, 1973.

[33] Wyngaard, J. C., Cote, O. R. and Rao, K. S., Modeling the atmospheric boundary layer. *Adv. Geophys.*, **18A**, pp. 193-211, 1974.

[34] Andre, J. C., deMoor, G., Lacarrere, P., Therry, G. and du Vachat, R., Modeling the 24-hour evolution of the mean and turbulent structures of the planetary boundary layer. *J. Atmos. Sci.*, **35**, pp. 1861-1883, 1978.

[35] Sun, W.-Y., Numerical study of dispersion in the convective boundary layer. *Atmos. Environ.*, **23**, pp. 1205-1217, 1989.

[36] Lumley, J. L., Computational modelling of turbulent flows. *Adv. Appl. Mech.*, **18**, pp. 123-176, 1979.

[37] Launder, B. E., Reece, G. J. and Rodi, W., Progress in the development of a Reynolds-stress turbulence closure. *J. Fluid Mech.*, **68**, pp. 537-566, 1975.

[38] Lewellen, W. S. and Teske, M. E., Second-order closure modelling of diffusion in atmospheric boundary layer. *Boundary-Layer Meteorol.*, **10**, pp. 69-90, 1976.

[39] Yamada, T., A numerical experiment on pollutant dispersion in a

horizontally-homogeneous atmospheric boundary layer. *Atmos. Environ.*, **11**, pp. 1015-1024, 1977.

[40] Monin, A. S. and Obukhov, A. M., Basic turbulent mixing laws in the atmospheric surface layer. *Tr. Geofiz. Inst. Acad. Nauk. SSSR*, **24**(151), pp. 163-187, 1954.

[41] Businger, J. A., Wyngaard, J. C., Izumi, Y. and Bradley, E. F., Flux-profile relationships in the atmospheric surface layer. *J. Atmos. Sci.*, **28**, pp. 181-189, 1971.

[42] Dyer, A. J., A review of flux-profile relationships. *Boundary-Layer Meteorol.*, **7**, pp. 363-372, 1974.

[43] Hicks, B. B., Wind profile relationships from the 'Wangara' experiment. *Quart. J. Roy. Meteorol. Soc.*, **102**, pp. 535-551, 1976.

[44] Hogstrom, U., Non-dimensional wind and temperature profiles in the atmospheric surface layer: A re-evaluation. *Boundary-Layer Meteorol.*, **42**, pp. 55-78, 1988.

[45] Hogstrom, U., Review of some basic characteristics of the atmospheric surface layer. *Boundary-Layer Meteorol.*, **78**, pp. 215-246, 1996.

[46] Garratt, J. R. and Hicks, B. B., Micrometeorological and PBL experiments in Australia. *Boundary-Layer Meteorol.*, **50**, pp. 11-29, 1990.

[47] Panofsky, H. A. and Dutton, J. A., *Atmospheric Turbulence*, Wiley-Interscience, New York, 1984.

[48] Web, E. K., Profile relationships: the log-linear range, and extension to strong stability. *Quart. J. Roy. Meteorol. Soc.*, **96**, pp. 67-90, 1970.

[49] Santoso, E. and Stull, R., Wind and temperature profiles in the radix layer: The bottom-fifth of the convective boundary layer. *J. Appl. Meteorol.*, **37**, pp. 545-558, 1998.

[50] Arya, S. P., Comments on wind and temperature profiles in the radix layer: The bottom-fifth of the convective boundary layer. *J. Appl. Meteorol.*, **38**, pp. 493-494, 1999.

[51] Kaimal, J. C. and Finnigan, J. J., *Atmospheric Boundary Layer Flows,* Oxford University Press, Oxford, 1994.

[52] Wyngaard, J. C. and Cote, O. R., The budgets of turbulent kinetic energy and temperature variance in the atmospheric surface layer. *J. Atmos. Sci.*, **28**, pp. 190-201, 1971.

[53] Kaimal, J. C., Wyngaard, J. C., Izumi, Y. and Cote, O. R., Spectral characteristics of surface-layer turbulence. *Quart. J. Roy. Meteorol. Soc.*, **98**, pp. 563-589, 1972.

[54] Hanna, S. R., Applications in air pollution modelling, Chapter 7, *Atmospheric Turbulence and Air Pollution Modelling*, eds. F.T.M. Nieuwstadt and H. van Dop, Reidel, Dordrecht, pp. 275-310, 1982.

[55] Garratt, J. R., *The Atmospheric Boundary Layer*, Cambridge University Press, Cambridge, 1992.

[56] Deardorff, J. W., Convective velocity and temperature scales for the unstable planetary boundary layer and for Rayleigh convection. *J. Atmos. Sci.*, **27**, pp. 1211-1213, 1970.

[57] Panofsky, H. A., Tennekes, H., Lenschow, D. H. and Wyngaard, J. C., The characteristics of turbulent velocity components in the surface layer under convective conditions. *Boundary-Layer Meteorol.*, **11**, pp. 355-361, 1977.

[58] Caughey, S. J. and Palmer, S. G., Some aspects of turbulence through the depth of the convective boundary layer. *Quart. J. Roy. Meteorol. Soc.*, **105**, pp. 811-827, 1979.

[59] Kaimal, J. C., Wyngaard, J. C., Haugen, D. A., Cote, O. R., Izumi, Y., Caughey, S. J. and Readings, C. J., Turbulence structure in the convective boundary layer. *J. Atmos. Sci.*, **33**, pp. 2152-2169, 1976.

[60] Caughey, S. J., Kitchen, M. and Leighton, J. R., Turbulence structure in convective boundary layers and implications for diffusion. *Boundary-Layer Meteorol.*, **25**, pp. 345-352, 1983.

[61] Quintarelli, F., A study of vertical velocity distributions in the planetary boundary layer. *Boundary-Layer Meteorol.*, **52**, pp. 209-219, 1990.

[62] Boers, R., A parameterization of the depth of the entrainment zone. *J. Appl. Meteorol.*, **28**, pp. 107-111, 1989.

[63] Nelson, E. R., Stull, R. and Eloranta, E., A prognostic relationship for entrainment zone thickness. *J. Appl. Meteorol.*, **28**, pp. 885-903,

1989.

[64] Tennekes, H., Similarity relations and scaling laws, Chapter 2, *Atmospheric Turbulence and Air Pollution Modelling*, eds. F.T.M. Nieuwstadt and H. van Dop, Reidel, Dordrecht, pp. 37-68, 1982.

[65] Kazanski, A. B. and Monin, A. S., On the dynamical interaction between the atmosphere and earth's surface. *Izv. Akad. Nauk. SSSR, Geophys. Ser.*, **5**, pp. 514-515, 1961.

[66] Zilitinkevich, S. S., Laikhtman, D. L. and Monin, A. S., Dynamics of the atmospheric boundary layer. *Izv. Atmos. Ocean. Phys.*, **3**, pp. 170-191, 1967.

[67] Arya, S. P. S., Geostrophic drag and heat transfer relations for the atmospheric boundary layer. *Quart. J. Roy. Meteorol. Soc.*, **101**, pp. 147-161, 1975.

[68] Nicholls, S., Aircraft observations of the Ekman layer during the Joint Air-Sea Interaction Experiment. *Quart. J. Roy. Meteorol. Soc.*, **111**, pp. 391-426, 1985.

[69] Zilitinkevich, S. S. and Chalikov, D. V., The laws of resistance and of heat and moisture exchange in the interaction between the atmosphere and an underlying surface. *Izv. Atmos. Ocean. Phys.*, **4**, pp. 438-441, 1968.

[70] Clarke, R. H. and Hess, G. D., Geostrophic departure and the functions A and B of Rossby-number similarity theory. *Boundary-Layer Meteorol.*, **7**, pp. 267-287, 1974.

[71] Deardorff, J. W., Numerical investigation of neutral and unstable planetary boundary layer. *J. Atmos. Sci.*, **29**, pp. 91-115, 1972.

[72] Deardorff, J. W., Three-dimensional numerical study of the height and mean structure of a heated planetary boundary layer. *Boundary-Layer Meteorol.*, **7**, pp. 81-106, 1974.

[73] Zilitinkevich, S. S., On the determination of the height of the Ekman boundary layer. *Boundary-Layer Meteorol.*, **3**, pp. 141-145, 1972.

[74] Garratt, J. R., Observations in the nocturnal boundary layer. *Boundary-Layer Meteorol.*, **22**, pp. 21-48, 1982.

[75] Nieuwstadt, F. T. M. and Tennekes, H., A rate equation for the nocturnal boundary-layer height. *J. Atmos. Sci.*, **38**, pp. 1418-1428,

1981.

[76] Arya, S. P. S., Parameterizing the height of the stable atmospheric boundary layer. *J. Appl. Meteorol.*, **20**, pp. 1192-1202, 1981.

[77] Koracin, D. and Berkowicz, R., Nocturnal boundary-layer height: Observations by acoustic sounders and predictions in terms of surface-layer parameters. *Boundary-Layer Meteorol.*, **43**, pp. 65-83, 1988.

[78] Arya, S. P. S., Suggested revisions to certain boundary layer parameterization schemes used in atmospheric circulation models. *Mon. Weath. Rev.*, **105**, pp. 215-227, 1977.

[79] Zilitinkevich, S. S., Resistance laws and prediction equations for the depth of the planetary boundary layer. *J. Atmos. Sci.*, **32**, pp. 741-752, 1975.

[80] Rao, K. S. and Snodgrass, H. F., Some parameterizations of the nocturnal boundary layer. *Boundary-Layer Meteorol.*, **17**, pp. 15-28, 1979.

[81] Zilitinkevich, S. S. and Deardorff, J. W., Similarity theory for the planetary boundary layer of time-dependent height. *J. Atmos. Sci.*, **31**, pp. 1449-1452, 1974.

[82] Arya, S. P. S. and Wyngaard, J. C., Effects of baroclinicity on wind profiles and the geostrophic drag law for the convective planetary boundary layer. *J. Atmos. Sci.*, **32**, pp. 767-778, 1975.

[83] Yamada, T., On the similarity functions A, B and C of the planetary boundary layer. *J. Atmos. Sci.*, **33**, pp. 781-793, 1976.

[84] Arya, S. P. S., Comparative effects of stability, baroclinicity and the scale-height ratio on drag laws for the atmospheric boundary layer. *J. Atmos. Sci.*, **35**, pp. 40-46, 1978.

[85] Garratt, J. R., Wyngaard, J. C. and Francey, R. J., Winds in the atmospheric boundary layer – prediction and observation. *J. Atmos. Sci.*, **39**, pp. 1307-1316, 1982.

[86] Lenschow, D. L., Li, X. S., Zhu, C. J. and Stankov, B. B., The stably stratified boundary layer over the Great Plains. I: Mean and turbulence structure. *Boundary-Layer Meteorol.*, **42**, pp. 95-121, 1988.

[87] Smedman, A., Some turbulence characteristics in stable atmospheric boundary layer flow. *J. Atmos. Sci.*, **48**, pp. 856-868, 1991.

[88] Rao, K. S. and Nappo, C. J., Turbulence and dispersion in the stable atmospheric boundary layer. Chapter 2, *Dynamics of Atmospheric Flows: Atmospheric Transport and Diffusion Processes*, eds. M.P. Singh and S. Raman, Computational Mechanics Pub., Southampton, pp. 39-91, 1998.

[89] Nieuwstadt, F. T. M., Turbulence structure of the stable, nocturnal boundary layer. *J. Atmos. Sci.*, **41**, pp. 2202-2216, 1984.

[90] Monin, A. S. and Yaglom, A. M., *Statistical Fluid Mechanics: Mechanics of Turbulence*, Vol. 1, MIT Press, Cambridge, MA, 1971.

[91] Wyngaard, J. C., On surface-layer turbulence. Chapter 3, *Workshop on Micrometeorology*, ed. D.A. Haugen, American Meteorological Society, Boston, pp. 101-149, 1973.

[92] Pielke, R. A., *Mesoscale Meteorological Modelling*, Academic Press, Orlando, 1984.

[93] Anthes, R. A., Seaman, N. L., and Warner, T. T., Comparisons of numerical simulations of the planetary boundary layer by a mixed-layer and multi-level model. *Mon. Weath. Rev.*, **108**, pp. 365-376, 1980.

[94] Arya, S. P. S., Parametric relations for the atmospheric boundary layer. *Boundary-Layer Meteorol.*, **30**, pp. 57-73, 1984.

[95] O'Brien, J. J., A note on the vertical structure of the eddy exchange coefficient in the planetary boundary layer. *J. Atmos. Sci.*, **27**, pp. 1213-1215, 1970.

[96] Deardorff, J. W., The countergradient heat flux in the lower atmosphere and in the laboratory. *J. Atmos. Sci.*, **23**, pp. 503-506, 1966.

[97] Zwerver, S. and van Ham, J., *Interregional Air Pollution Modelling*, Plenum Press, New York, 1985.

Chapter 4

Real time modeling and emergency response forecast

David P. Bacon
Center for Atmospheric Physics
Science Applications International Corporation
1710 SAIC Drive
McLean, VA 22102, USA

Abstract

Given an actual release of hazardous material into the atmosphere, time is of the essence in the response to the resultant emergency situation. Equally important, however, is accuracy, and the balance between these conflicting requirements is the challenge to emergency response teams. Instead of an idealized source, emergency response personnel must consider the detailed source configuration (spatial, temporal, and particle size distribution), the detailed dispersion in a potentially extremely complex urban terrain, and the human and health effects. The time required to obtain this information and to make it available to the analysis of the situation, however, reduces the time available to perform the actual analysis. Mesoscale dispersion modeling for emergency response thus consists of balancing the required level fidelity with speed of response.

1 Introduction

Emergency response to an atmospheric release of a hazardous material is an incredibly challenging problem. Some types of chemical releases may only be toxic in high concentrations and hence result in relatively short-range hazards under all atmospheric conditions; on the other hand, certain radiological releases may have nearly continental scale ramifications for human health and activity. This makes the problem of emergency response more encompassing than the already difficult issue of atmospheric dispersion, and a higher level of discussion is necessary.

An emergency response scenario begins with a **release**. This release may be natural (e.g., a volcanic eruption such as Monserrat), accidental (e.g., the Bhopal chemical disaster or the Chernobyl nuclear accident), or intentional (e.g., the releases of sarin in the Tokyo subway by the Aum Shinrikyo sect). The next phase in an emergency response scenario is **notification**. This may take the form of a detailed description of the release, or may be a sketchy news bulletin. At this point, a bifurcation takes place in the emergency response process. **First Responders** represent the first governmental agents on-scene, and may be police, fire department, or military personnel, who have access to a certain amount of local information (e.g., which way the wind is blowing now), but who are operating initially without detailed fundamental knowledge; the **analysis** team provides support to the first responders and has the mission of acquiring as much information as possible about the source details, the forcing meteorology, and the terrain, and determining the most likely dispersion of the released material *along with a measure of confidence in the final analysis*. The goal of the analysis team is to **disseminate** the results of their analysis forward to the first responders in as timely a fashion as possible.

In this chapter, we will explore each of the phases of an emergency response scenario in detail. We will discuss both the detailed scientific definition of each phase as well as the real-world requirements, which are often quite different.

2 Release

The scientific description of an atmospheric release involves quite a lot of detailed information about the spatial, temporal, and particle size distributions, the material physical and chemical properties, and any associated thermodynamic information that is source-specific. The *real-*

world description may be limited to "something happened over there". In the real world, source information is typically described in terms of *where, when, what* and *how much*.

2.1 Spatial, temporal, and particle size distribution

In the idealized world of research, atmospheric releases can be characterized as point sources, moving point sources, or areal releases. It is true that this does provide a rough representation of stack plumes, automobiles, boats and aircraft, and urban pollution, but it is not nearly enough to describe the spectrum of scenarios required by the emergency responder. Idealized releases are also either instantaneous or continuous. The real world, however, includes both industrial explosions that provide a nearly instantaneous release of a hazardous material and transportation accidents involving a tank car where the exact amount released will depend on the pressure in the tank; as the pressure changes due to material release, so does the rate of release. Finally, ideal releases involve massless passive tracers, or (at worst) releases of inert particles of a uniform size. The real world includes volcanic eruptions with particle sizes ranging from microns to meters.

Any function can be de-composed into a series of weighted basis functions. In the idealized description of Lagrangian dispersion models, the basis functions are Gaussian "puffs", hence it makes sense to use this as our frame of reference. In this framework, each puff, i, represents a specified mass of released material, $m_i(t)$, distributed over a spatial extent represented by a Gaussian distribution centered at the point ($x_i(t)$, $y_i(t)$, $z_i(t)$) with spatial variance ($\sigma_{xi}(t)$, $\sigma_{yi}(t)$, $\sigma_{zi}(t)$) and all of the material having a single particle size, a. This can be represented by:

$$c_i(x,y,z,t,a) = \frac{m_i(t)}{\sigma_{xi}(t)\sigma_{yi}(t)\sigma_{zi}(t)\sqrt{(2\pi)^3}}\delta(r-a) \times \exp\left\{-\left[\frac{(x-x_i(t))^2}{2\sigma_{xi}{}^2(t)}+\frac{(y-y_i(t))^2}{2\sigma_{yi}{}^2(t)}+\frac{(z-z_i(t))^2}{2\sigma_{zi}{}^2(t)}\right]\right\} \quad (1)$$

where c_i is the concentration associated with puff i.

Puffs are released at discrete points in time such that the released puff mass equals the amount of material that would have been released by the continuously varying source. In this fashion, a continuous process can be approximated by a series of discrete puff elements.

Using this paradigm, the release of sarin in a subway may involve, initially, a single spheroidal puff, while a volcanic eruption such as Monserrat may involve the specification of hundreds or thousands of puffs from the surface up to tens of kilometers and over an area of many square kilometers. The goal of the emergency response system, then, is to take a limited amount of information ("something happened over there") and to turn it into a set of puff specifications. This is in general a nearly impossible task due to the limited information available to the emergency response center and the time criticality of the analysis, and so it is necessary to have previously developed relocatable source models for certain baseline scenarios that take a limited amount of input and provide the detailed source puff specification.

2.2 Material and chemical properties

A tremendous amount of effort has been expended over the past five decades in studying the dispersion of passive, neutrally buoyant tracer gases. Real problems often involve volatile particulates; hence the density of the material as well as its temperature-dependent vapor pressure are important physical parameters. Another important parameter is its solubility in water; this impacts on the ability of the material to be scavenged by precipitation or absorbed on wet surfaces and bodies of water. Unfortunately, it is often true that the details of the released material are some of the last pieces of information available to the analyst; hence most initial analyses must be performed assuming passive gas or particulate releases.

2.3 Thermodynamic source considerations

One of the hardest issues to insert into the real-time response to an emergency is the thermodynamic forcing of the source. Chernobyl, for example, presented a massive thermal energy source such that the radiation was released in a strongly buoyant plume. In addition, the thermal output of the crippled reactor went through several surges creating an extremely complex source. At the other end of the spectrum, many hazardous materials, such as liquefied natural gas (LNG), may be strongly cooled in a tank rupture event due to the latent heat required for vaporization, resulting in an overdense plume.

Another, more unusual, source term is a momentum jet. Pyroclastic eruptions (c.f. Figure 1) such as occurred on May 18, 1980 at Mt. St. Helens can eject material at almost sonic velocities. These types of source terms, when coupled with broad particle size distributions, create a nearly impossible problem to approximate under the time constraints of a real emergency. The solution methodology is to estimate the initial stabilized plume height and then to develop an elevated source release at that altitude.

2.4 First approximation source terms

Given the range of complexity of real source terms, how does the emergency planner develop a system for building a suitable *representative* source term that incorporates the *known* information, *approximates* the unknown information, and guesses at the *unknowable* information? An example is provided by the National Atmospheric Release Advisory Center (NARAC) of the US Department of Energy (DOE) [2].

On the morning of July 26, 1993, NARAC responded quickly to requests from the State of California and San Francisco Bay Area agencies to provide needed guidance during a sulfuric acid gas leak from a rail tank car at the General Chemical plant in Richmond, California. In this case, the material released was *known* (sulfuric acid gas), the total amount released was *unknown*, but could be estimated from the total capacity of the tank car, but the time history of the release was *unknowable*. Since 1979, NARAC has responded to over seventy alerts, accidents, and disasters; based on these experiences, NARAC has developed a hierarchical questionnaire-based system for providing information on the source of a release (c.f. Figure 2). Other emergency response systems have also developed menu-based source information systems, often with intelligent defaults. One of the most extensive is the Hazard Prediction and Assessment Capability (HPAC) developed by the US Defense Threat Reduction Agency (DTRA) [3]. HPAC has

Figure 1: On May 18, 1980, Mt. St. Helens underwent a pyroclastic explosion that sent ash and rock over 10 miles into the sky [1].

developed rapid source determination models with intelligent defaults for nuclear, biological and chemical releases.

3 Notification

The first notice is obviously the most important, but often conveys the least quantitative information. This is the reason why the analysis team is often faced with conflicting courses of action - to begin the analysis in advance of

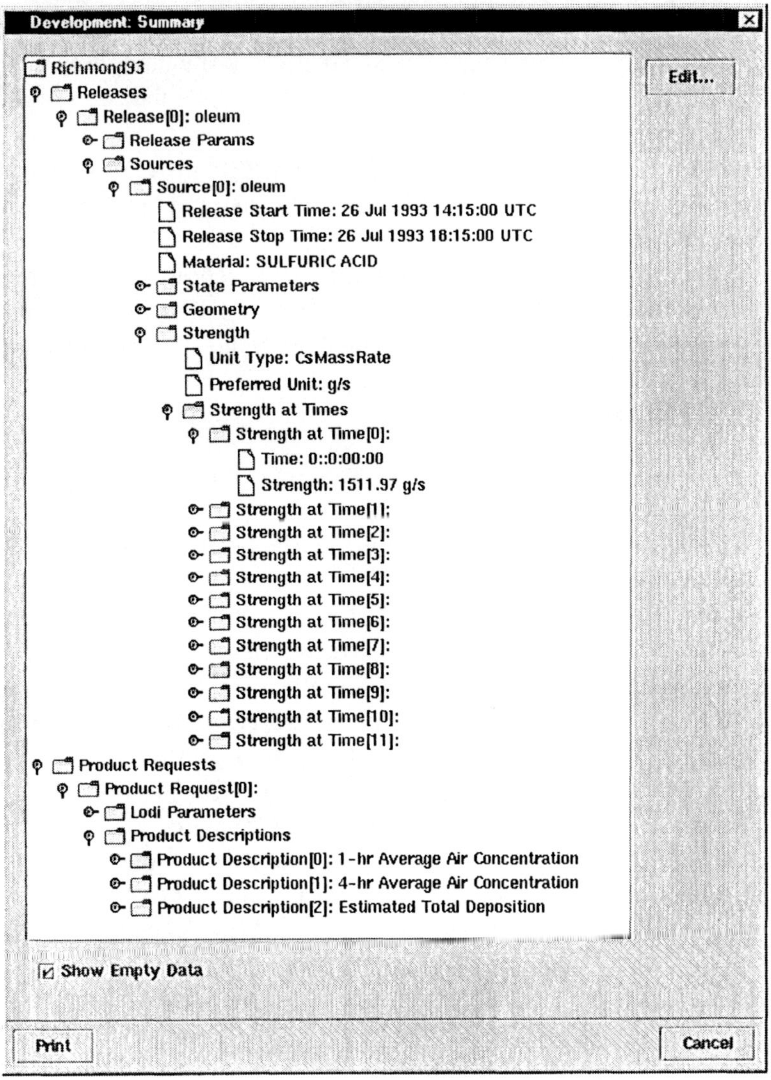

Figure 2: Hierarchical source question developed by NARAC.

complete information, or to wait until additional information is forthcoming. As we mentioned in the previous section, the real-world notification is rarely in scientific terms; it is more often couched in terms of *where, when, what* and *how much*. In an emergency situation, however, time is of the essence and most action begins therefore, once the *where* has been determined - even roughly.

Given a rough location, the first response team can start moving towards the area affected, and the analysis team can start acquiring whatever meteorological observations, analyses and forecasts (in that order) are available for the release point and its environs. The missing *when, what* and *how much* information can initially be overcome by using a continuous release and assuming a neutrally buoyant trace gas.

4 First responders

As we just mentioned, once the release location is known, even roughly, the first response team can begin to move towards the release point. The next critical pieces of information for them is the exact nature of the release - the *what* and *how much*.

Almost any substance is toxic if the concentration is high enough. Even oxygen, without which human life is impossible, is toxic at high concentration for extended periods of time leading to pulmonary oxygen toxicity and oxygen convulsions (Council for National Cooperation in Aquatics, [4]). Dealing with the plethora of potential material spills would be nearly impossible without computer-based tools. One of the most commonly used is the Computer Aided Management of Emergency Operations (CAMEO) system developed jointly by the US National Oceanographic and Atmospheric Administration (NOAA) and the US Environmental Protection Agency (EPA). CAMEO's chemical database provides ready access to response recommendation for over 4,000 chemicals, including fire fighting, physical properties, health hazards, and first aid guidance. The system contains over 60,000 chemical synonyms and identification numbers in a searchable format so that unknown substances can be quickly identified during an incident.

The CAMEO system also includes the Areal Location Of Hazardous Atmospheres (ALOHA) straight-line dispersion model [5]. ALOHA uses current local wind information, assumes that it is horizontally uniform and constant, and predicts a simple downwind dispersion hazard.

CAMEO and ALOHA were designed to be fast-running on PC-based systems and are valid for ranges from a few tens of meters to a few kilometers. Fortunately this covers the bulk of small-scale industrial accidents such as an accidental spill or small tank rupture. So great is the need of the first responders, however, that it is often used for problems with much greater ranges and hence a more detailed analysis team is needed to back up the quick simple answer with something more substantial.

5 Analysis

In the world of emergency response, the great bulk of events are the accidental spill of a limited quantity of a hazardous chemical, which results in a hazardous area of a kilometer or less. A large number of models have been designed to handle this type of accident. A recent compendium of models by the Office of the Federal Coordinator for Meteorological Services and Supporting Research (OFCM) [6] listed over sixty atmospheric transport and diffusion models, the great majority of which were simple puff dispersion models such as ALOHA.

Rapidly changing meteorological conditions or regions of complex terrain cannot be handled by such systems. In addition, the potential consequences of a major release of nuclear, biological or chemical material with hazardous footprints of tens to hundreds of kilometers are so severe that they must also be considered. Finally, a release in an urban environment with the dispersion dominated by the urban canyon flow and thermally driven circulations poses its own set of problems.

High resolution weather forecasting is the key to accurate dispersion modeling; neither single point surface data nor large scale analysis information is sufficient. To improve the fidelity of hazardous dispersion models, it is essential that the meteorological forecast itself be improved. This is because the modeling of atmospheric dispersion involves virtually all scales of atmospheric motion from microscale turbulence to planetary scale waves. The Operational Multiscale Environment model with Grid Adaptivity (OMEGA) was designed to improve the state-of-the-art in atmospheric dispersion. OMEGA is a multiscale simulation tool which simulates fine scale flows without the need for multiple nested grids, and incorporates an embedded dispersion model [7]. An example of an OMEGA simulation in a data sparse region is shown in Figure 3. This figure shows the predicted plume from the Kuwait oil fires for a specific date in March 1991 and the actual observation.

In this section we explore the determination of the weather conditions for an emergency response situation, including the level of confidence therein, and discuss high resolution dispersion modeling, including the estimate of uncertainty.

5.1 The need for analysis – Bhopal

A classic example of the need for rapid analysis is seen in the events that occurred in Bhopal, India, in the early morning of December 2, 1984. At 12:20 local time, a pressure relief valve on a storage tank opened, venting 20 tons of methylisocyanate (MIC) into the atmosphere.

The fate of this heavier-than-air gas has been the subject of many studies including one by Boybeyi et al. [8]. Bhopal was a city of several hundred thousand and at the time of the release, the mesoscale circulation was dominated by the stable air mass caused by a large lake in the area and the urban heat island that ducted the light winds over and around the city. Within two hours, those closest to the release were affected with eye and lung irritation, but no general alert had gone out. During this time the dense MIC pooled, increasing its concentration above the lethal threshold. As the urban heat island dissipated, and the winds picked up, the now lethal plume moved over the city killing 5,000 people.

5.2 Weather analysis

The amount of global *in situ* weather data is abysmally small, roughly 25 MB of data in total per day for both the worldwide surface and upper air data. The surface observations are taken hourly at roughly 8000 points on the globe; the upper air observations are sampled from less than 1000 locations twice a day. Given this limited amount of data, the probability that an observation will be close to a random release point is extremely small – *unless the site has been instrumented prior to the release.* This latter approach was the basis of the NARAC system previously mentioned.

(a)

(b)

Figure 3: A comparison of (a) OMEGA simulated oil-field fire plumes and (b) AVHRR image data. The OMEGA simulation is valid at 1030Z on March 11, 1991; the AVHRR image was taken at 1020Z.

NARAC was created to serve as an emergency management tool for releases at nuclear facilities. Since the facilities were known, it was possible to instrument each site with a local mesonet. While this approach has many advantages (e.g., high quality local data, known calibration of the instrumentation), it is obviously not appropriate for the general case.

In the general case, the data is just a starting point. If raw data is obtained, it must undergo quality control prior to use due to the large number of observations that possess instrument, transcription or transmission errors. The raw data is processed by the national weather services of the various countries to create gridded *analyses* and *forecasts*. Unfortunately for many dispersion problems, the spatial and temporal resolution of these analyses are not sufficient.

Most global weather forecast models are run at a spatial resolution of 50-100 km. In addition, most are spectral models, which assume a spherical Earth and thus possess extremely limited information about the local terrain. For a dispersion problem near a coastal region or in complex terrain, this lack of spatial resolution is intolerable. Many operational forecast centers are now running regional or mesoscale models down to resolutions of 30-50 km. Although these models have more terrain information, even this resolution is not sufficient for the detailed dispersion analysis necessary in an emergency response situation. This is part of the reason why even finer scale models with grid resolution down to 1 km (e.g., OMEGA) have been developed.

Resolution alone is not the answer. Increased resolution permits the resolution of fine scale atmospheric features that are important for dispersion. But the physics that drives these fine scale features must be modeled. In addition, these fine scale features must also be allowed to affect the larger scale circulation. Thus a model such as OMEGA can resolve microscale events while being fully consistent and providing the large scale forcing which is required to ensure the proper simulation of the atmosphere. In order to accomplish this, however, it is necessary to include all of the physical parameters and processes, which affect the local flows. These include not only the topography, but also the land use, the land/water composition, the vegetation, the soil moisture, the snow cover (if appropriate), and the surface moisture and energy budgets.

Much of this additional physics is only appropriate because of the increased spatial resolution.

5.3 Weather uncertainty

Given a weather observation, analysis, or forecast, how sure are we of the validity of that data to the problem at hand? Rarely, if ever, will the data be recorded at the precise time and location of the actual event, hence we have to make assumptions as to the representativeness of the observation some distance removed from the event. For example, if a *perfect* observation is taken in the midwestern US with relatively flat terrain and a meteorological condition that is relatively constant, then the observation may be valid for some considerable distance and time. On the other hand, if it is taken in mountainous terrain with a front moving through, then the observation may be immaterial to the problem at hand. The same can be said for analyses and forecasts. Their representativeness for dispersion studies depends on their own spatial resolution and that of the databases that support them. A grid point analysis or forecast at 100 km spacing is a useful qualitative tool, but it may result in an incorrect dispersion analysis.

The errors in these systems can be grouped into observation, analysis, or forecast errors, any of which can lead to errors in calculating the dispersion of a hazardous aerosol or gas. Quality control can eliminate bad observations prior to use leaving analysis and forecast errors. Of these, given equivalent surface information and grid resolution, the latter is the more significant.

Forecast errors can be *estimated* by a variety of validation runs. Unfortunately, numerical weather prediction (NWP) models cannot be validated in the mathematical sense. In mathematical terms, NWP is plagued by a lack of data (data sparse), has poorly defined lateral and top boundary conditions, has incomplete surface boundary conditions, and contains motions spanning many temporal and spatial scales. For this reason, while parts of NWP models may be rigorously validated (in the mathematical sense), the only true test is by comparison with observations.

Forecast errors arise from a combination of limited grid resolution and the approximations that resolution imposes, actual errors in model physics, and in the initial and boundary conditions of the model. The errors in the initial conditions are often termed the "butterfly effect", but this is more true of synoptic motion than local terrain forced flow. Many operational forecast centers have developed *ensemble* forecast systems for their global forecast models. Typically this involves running a degraded resolution model with slightly differing initial conditions or with particular spectral components (the "breeding" modes) enhanced. By running a set of variations, they can determine whether the baseline forecast of the large scale flow of the atmosphere is robust – and for how long it will continue to be robust. An example is given in Figure 4 which shows the results of just 5 of 12 runs of the Medium Range Forecast (MRF) model used at the National Centers for Environmental Prediction [9]. This figure shows the intersection of the 500 mb pressure level with the 5640 m geopotential height. The figure indicates that the 3-day forecast is relatively robust for North America (top) while the 5-day forecast has considerably more scatter.

A very real question is the validity of this analysis as a basis for quantifying the forecast error. The twelve standard MRF runs used for the NCEP ensemble are the baseline MRF forecast result (T126), a control run at degraded resolution (T62), and five (5) pairs (plus / minus) of perturbations with different breeding modes run at the degraded resolution. The result is an exercise in *model* variability, not true forecastability. For those cases where agreement exists, however, some level of robustness can be assumed.

Another valid question is the impact of these synoptic forecast errors on dispersion calculations. Ensemble forecasts are produced for only a very few products due to the high computational cost of production. Wind speed and direction are not typically driving needs for most operational weather forecast centers, hence this is not a high priority for ensemble forecasting. (This is another example of why it is necessary to improve the state of weather forecasting to improve the state of dispersion forecasting.) But even if it was, it is valid to ask whether this is a reasonable methodology for dispersion purposes.

Ensemble forecasts rely on multiple forecasts at degraded spatial resolution. The resolution **must** be degraded due to the large amount of computational cycles that must be spent on these multiple forecasts. But dispersion is a problem which is critically dependent on the *details* of the

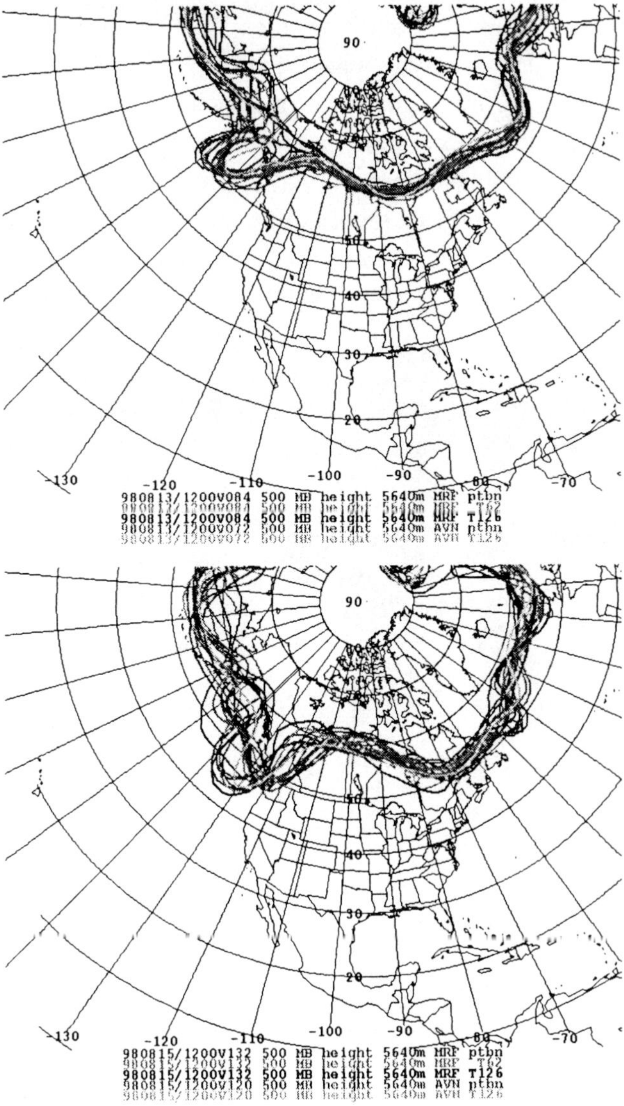

Figure 4: Ensemble forecast products produced by NCEP showing the increase in forecast variability from 3 to 5 days. The top panel shows a collection of simulation results for a 3-day forecast; the bottom panel for a 5-day forecast. (Courtesy of NOAA / NCEP.)

atmospheric circulation. In addition, most dispersion problems occur in the planetary boundary layer where they are forced by fixed surface features that may be quite small (terrain, coastlines, vegetation or land use variations). Thus an argument can be made that it is preferable to utilize all the available computational cycles to create a *single, high fidelity* forecast than multiple low fidelity ones.

Figure 5 shows a situation that demonstrates the problems of ensembles of low resolution simulations. This figure shows a part of the

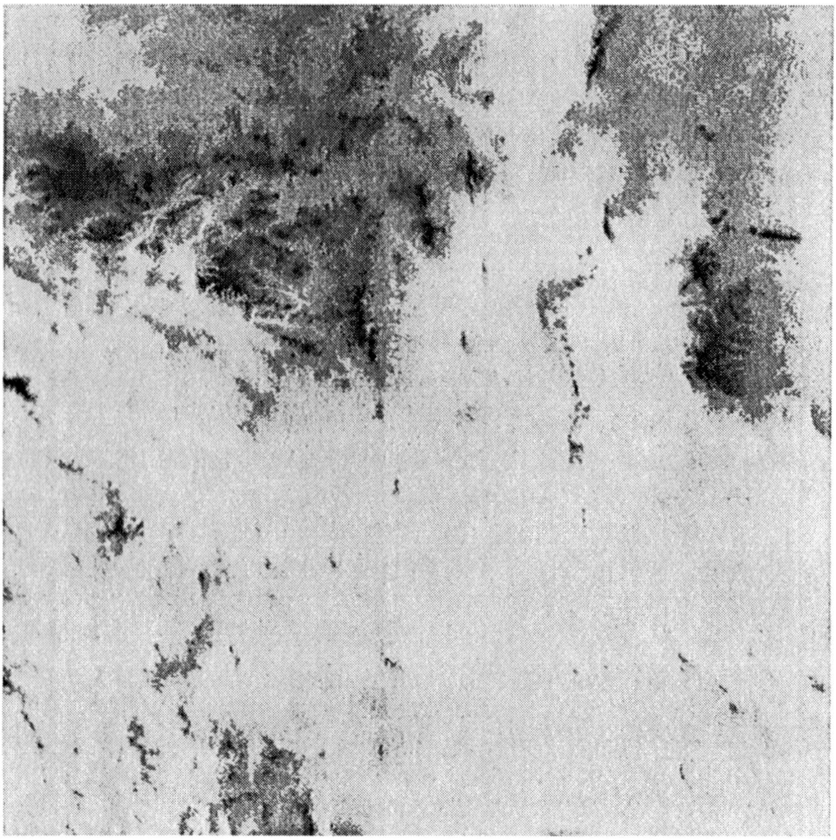

Figure 5: The terrain in southeastern New Mexico. Near the top of the figure the terrain is quite complex with many features on the order of a few kilometers.

terrain of south-eastern New Mexico. As can be seen at the top of the figure, some of this terrain is extremely complex with many features at the few kilometers scale. If this terrain is resolved only at 10-30 km resolution, then all of the critically important terrain features will be missed by the NWP simulations. Hence any ensemble of these results will be worthless as they would **all** have failed to resolve the critical feature – the terrain.

Situations such as this demonstrate the natural separation of operational weather forecasting and high resolution forecasting for dispersion. The low resolution ensembles currently in vogue at national and regional forecast centers provide a means of measuring the synoptic forecastability but provide no guidance for dispersion. Since dispersion simulations *require* the highest possible resolution of surface feature, all available computational cycles should be devoted to a single, high resolution simulation.

The use of a single high resolution simulation, however, does not eliminate the need to understand the limitations of the model and to try to quantify the forecast error. Without a true mathematical means of establishing the accuracy of an NWP model, the best that can be done is to develop model output statistics (MOS) for the model by comparing the forecasts against the observations or the analyses. Comparing against the observations is preferable, but only allows MOS development for the observation locations; analysis-derived MOS have the advantage of total coverage (over the forecasted domain), but are removed somewhat from a direct comparison with the data. Whichever method is used, the MOS can then be utilized to provide the dispersion calculation with a measure of the uncertainty in the forecasted products directly applicable to dispersion: temperature, dewpoint, wind speed, and wind direction.

5.4 High fidelity dispersion modeling

Improving the forecast of the weather is only the first step in improving the forecast of dispersion. A critical step is providing this weather information to the dispersion model. Most dispersion models are designed to ingest weather information produced at some spatial and temporal resolution and then to calculate the dispersion. (e.g., the EPA Models-3 program has designed their air quality modeling system around this paradigm.) Unfortunately, as the

NWP model resolution increases, the input/output (I/O) requirements of the dispersion model increases as the *cube* of the NWP model resolution. Increasing the resolution from 100 km (the norm at the beginning of this decade), to 1 km increases the data handling requirement by 10^6. The alternative to this is to embed the dispersion model into the NWP model; this is the paradigm used by OMEGA.

OMEGA contains both puff and particle dispersion models as part of its embedded Atmospheric Dispersion Model (ADM) routines [7]. These models are run as part of the overall simulation and have access to the full NWP model results at each timestep. Since the ADM routines are embedded inside OMEGA, there is no additional I/O constraint. An example of the ADM particle output is shown in Figure 3.

While the use of embedded transport is ideal for high fidelity simulations, it has a significant drawback for emergency response support: the lack of ability to change source location and characteristics and/or aerosol or gas characteristics. For this reason, it is still necessary to provide a means to export the NWP output for use in post-processed dispersion models. The embedded transport results, however, can significantly improve the data export by providing guidance on the exact domain for which the NWP results are required.

5.5 Mean and variance

Turbulence in the atmosphere produces fluctuations in measured concentration *even if the source is absolutely constant.* Most dispersion models were developed to simulate just such a situation: stack emissions from power plants or chemical facilities. The data taken from situations such as this is typically reduced to provide the mean value of the effluent concentration as a function of range and atmospheric conditions, and this is the standard against which most dispersion models are judged. Emergency response situations, however, often involve chemicals for which the *peak* rather than the mean concentration is the critical value. This requires understanding of the variance associated with the mean concentration.

Several variance models have been developed over the years. The first is due to Wilson et al. [10, 11] and uses conventional first-order turbulence closure; the second has been developed by Sykes et al. [12]

and is based on a second-order closure turbulence model. Both have their strengths and weaknesses. Upon initial inspection, the Second-Order Closure Integrated PUFF (SCIPUFF) model appears to have an edge due to its higher order closure of the turbulence terms in the Navier-Stokes equations; however, it relies on knowledge of the initial and boundary conditions for the Reynolds stresses, which are not standard observations. SCIPUFF gets around this limitation using a climatological database for Reynolds stresses. Hence what should be a clear advantage becomes a philosophical issue as to the benefit of using current observations with a lower order closure scheme (as in Wilson), or a higher order closure scheme with climatological initial and boundary conditions (as in SCIPUFF).

Regardless of the approach, for emergency response it is the availability of a quantitative measure of the variance that is critical to the response teams.

5.6 Operational considerations

The ideal situation aside, the real-time operational constraints of an emergency response center lead to a fine balance between what *should* be done and what *can* be done. (Remember, the analysis team is often forming as the first responders are en route; time is critical.) Retrieving the closest observations and the most recent large scale analysis is often the best that can be done before the first responders have arrived on site. This information can then be compared with their on-site observation. At this time, it may also be possible to make a first estimate of the hazard given the meteorological observations since the release and assuming that the current or on-site meteorological observations hold constant (c.f., Figure 6). This is the most current methodology, and while it works under many situations, it is extremely dangerous in coastal situations or changing weather conditions. In these situations, it is necessary to access the operational forecast products and to generate a high resolution forecast to support further analysis.

Once the analysis has reached this point, a great deal of judgement comes into play. If a high resolution forecast is planned, then the model domain, resolution, and physics must be balanced against the available computational resources in order to achieve a sufficient time compression for the operational need.

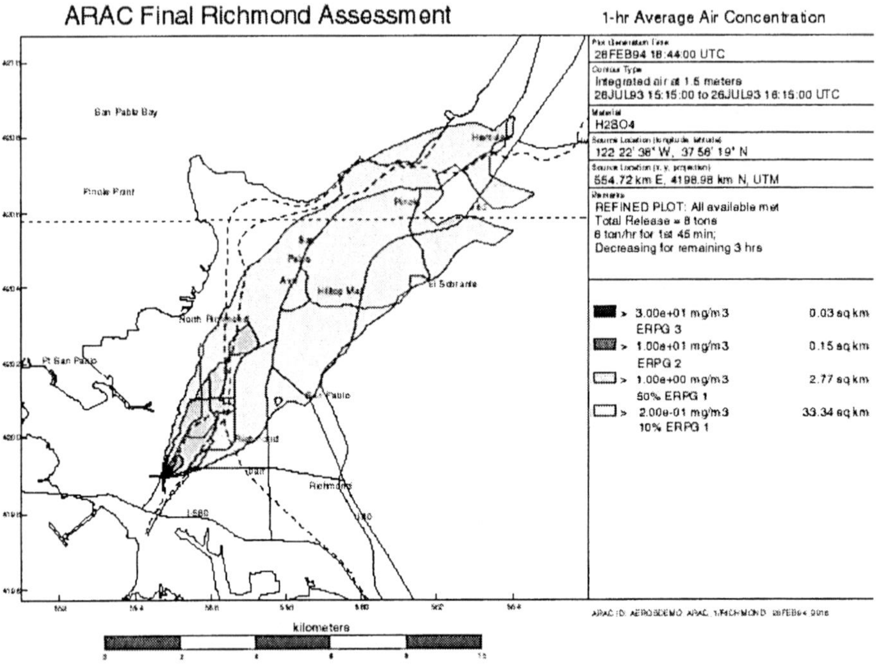

Figure 6: The 1-hour average air concentration for the second hour after release for the event described in Section 2.4 (Courtesy of NARAC).

6 Results dissemination

The best possible analysis is worthless unless it can be transmitted to those making decisions. In the past, this data dissemination was achieved via dedicated circuits to central command facilities; today's mobile communications are rapidly placing the ability to provide graphical products to first responders on site. Cellular phones would appear to be ideal for this purpose, but in fact tend to saturate in emergency situations and may not be available in remote locations. Satellite based communication, on the other hand, is ideal for this purpose.

Satellite communication bandwidth depends primarily on the power of the transmission unit. This leads to a natural asymmetry in the communications path with higher bandwidth available *from* the analysis cell to the field and lower bandwidth available *from* the field to the analysis cell. Fortunately, the amount of information that needs to be passed back to the analysis cell from the field is primarily text based information while the analysis cell is sending primarily graphical products to the field, and so this natural asymmetry does not have negative implications.

As the power of mobile communications and computing increases, we will see an increasing move towards on-demand graphical distribution of dispersion analysis products to the first responders.

7 Summary

Emergency response is a problem that requires a delicate balance between accuracy and speed. The most accurate dispersion forecast possible is useless if it cannot be delivered in time to save lives or prevent injury. A timely forecast that is incorrect, however, because it failed to consider important details in the source, the terrain, or in the ambient meteorology can be worse than no answer, however, as it may guide emergency personnel and the local populace into harm's way.

In any application of numerical methods to solve a physical problem, it is of paramount importance that sufficient spatial and temporal resolution is provided to resolve the dominant physical processes. Dispersion, especially mesoscale dispersion, is a problem that depends critically on *details*. Modeling mesoscale dispersion requires the resolution of all the significant local terrain features (coastlines, elevation, land use, vegetation, etc.) as well as the details of the ambient meteorology (horizontal and vertical). All of this attention to detail comes at the cost of *time*, and the key to dispersion analysis for emergency response is to understand those features that **must** be included and those that can be left to later analysis.

References

[1] United States Geological Survey / Cascades Volcano Observatory.

[2] Sullivan, T. J., Ellis, J. S., Foster, C. S., Foster, K. T., Baskett, R. L.,

Nasstrom, J. S. and Schalk, W. W. III, Atmospheric Release Advisory Capability: Real-Time Modeling of Airborne Hazardous Materials. *Bull. Am. Met. Soc.*, **74**, No. 12, 1993.

[3] Defense Special Weapons Agency, HPAC Version 1.3 User's Guide, *Reference Manual Version 7.5*, 1996.

[4] Council for National Cooperation in Aquatics, *The New Science of Skin and SCUBA Diving*, pp 94-97, Association Press, NY, 1974.

[5] EPA/CEPPO, *ALOHA User's Manual*. National Safety Council, Washington, DC, 1992.

[6] Federal Coordinator for Meteorological Services and Supporting Research, Directory of Atmospheric Transport and Diffusion Consequence Assessment Models. FCM-I3-1999, OFCM, Suite 1500, 8455 Colesville Rd., Silver Spring, MD 20910, 1999.

[7] Bacon, D. P., Ahmad, N. N., Boybeyi, Z., Dunn, T. J., Hall, M. S., Lee, P. C. S., Sarma, R. A., Turner, M. D., Waight, K. T., Young, S. H. and Zack, J. W., A Dynamically Adapting Weather and Dispersion Model: The Operational Multi-scale Environment model with Grid Adaptivity (OMEGA). *Mon. Wea. Rev.*, in publication, 2000.

[8] Boybeyi, Z., Raman, S. and Zannetti, P., Numerical investigation of possible role of local meteorology in Bhopal gas accident. *Atmospheric Environment,* **29**, 479-496, 1995.

[9] Toth, Z. and Kalnay, E., Ensemble forecasting at NCEP and the breeding method. *Mon. Wea. Rev.*, **125**, 3297-3319, 1997.

[10] Wilson, D. J., Robins, A. G. and Fackrell, J. E., Predicting the spatial distribution of concentration fluctuations from a ground level source. *Atmospheric Environment*, **16**, 497-504, 1982.

[11] Wilson, D. J., Fackrell, J. E. and Robins, A. G., Concentration fluctuations in an elevated plume: a diffusion-dissipation approximation. *Atmospheric Environment*, **16**, 2581-2589, 1982.

[12] Sykes, R. I., Lewellen, W. S. and Parker, S. F., A Gaussian Plume Model of Atmospheric Dispersion Based on Second-Order Closure. *Journal of Climate and Applied Meteorology*, **25**, 322-331, 1986.

Chapter 5

Urban parameterizations for mesoscale meteorological models

Michael J. Brown
Los Alamos National Laboratory
Energy and Environmental Analysis Group
Group TSA-4, MS F604
Los Alamos, NM 87545, USA

Abstract

Urban parameterizations developed for use in mesoscale meteorological models are described. These parameterizations attempt to account for the area-average effect of drag, turbulence production, heating, and surface energy budget modification induced by buildings and urban land use. Some insights garnered through field observations regarding the urban influence on local meteorology are given, including short descriptions of the urban heat island and urban roughness. A brief survey of prior work on mesoscale modeling of urban areas is presented. Finally, problems that arise when implementing or validating the urban canopy parameterizations are discussed.

1 Introduction

Cities affect the local weather by perturbing the wind, temperature, moisture, turbulence, and surface energy budget fields. Buildings alter the wind, produce turbulent eddies, create shade, and trap heat. The urban fabric – made up of such materials as concrete, asphalt and steel –

stores and releases heat differently than rural areas. Energy consumption related to home and office heating and cooling, manufacturing and transportation releases heat to the urban environment. Cities in arid environments may be wetter than their surroundings due to high water use, while cities in humid environments may be drier due to replacement of natural vegetation with urban materials. Urban-rural thermal differences can lead to the generation of winds.

Numerous investigations have shown that buildings and urban land use significantly modify the micro- and mesoscale flow fields (e.g., see reviews by Bornstein [1] and Hosker [2]). Accounting for the urban impact on atmospheric dynamics and thermodynamics is important for many applications, e.g., urban photochemical modeling, plume transport and dispersion, wind loading on buildings studies, urban design and energy usage studies, thermal comfort level evaluations, global warming assessments. For example, a plume trajectory may be modified by urban heat island circulations, the transport speed may be reduced due to building-induced drag, and vertical mixing might be enhanced as a result of heat island convection or building-created turbulence. For air quality applications and accidental release scenarios, mesoscale numerical models are often used to provide meteorological fields to air chemistry and puff dispersion models or boundary conditions to higher resolution models. Since mesoscale models do not have the spatial resolution directly to simulate the fluid dynamics and thermodynamics in and around urban structures, urban canopy parameterizations are sometimes used to approximate the drag, heating, radiation attenuation and enhanced turbulent mixing produced by the sub-grid scale urban elements.

In this chapter, we will focus on current methods for incorporating urban effects into mesoscale models. We will cover techniques for incorporating drag and turbulence production into the flow equations and modifications to the surface energy budget and heat equation to account for urban influences. We will distinguish between methods intended for use *above* the urban canopy from those intended to be used *within* the canopy. By necessity, we will present a short review of urban effects on mesoscale flows and give references to more thorough reviews. We will end with a section on implementation and practical difficulties associated with the urban parameterizations, namely parameter specification and model validation issues. We will not cover the specifics of atmospheric dispersion in urban environments. We point the interested reader to very good reviews of urban impacts on dispersion by Hanna and Chang [3], Hanna et al. [4], Hosker [5], Yamartino et al. [6], and Brown and Streit [7].

2 Urban canopy impact on mesoscale flow

In the 1960s and 70s, the atmospheric science community began looking seriously at how cities affect the natural climate system (e.g., Chandler [8], Daigo and Nagao [9], Landsberg [10], Oke [11]). In the 1970s a number of groundbreaking urban field experiments were conducted, partially in response to air quality concerns in large metropolitan cities (e.g., Bornstein [12], Clarke [13], Ludwig [14], Oke and East [15], Angel et al. [16], Ackerman [17]). Likewise, in the late 60s and early 70s, computer models were first being used to understand the dynamics of urban-induced circulations and heating patterns (e.g., Myrup [18], Atwater [19], McElroy [20], Bornstein [21]). Based on these field experiments and numerical model results, the urban climate system was found to be multi-dimensional and complex with numerous feedback mechanisms between components. Explanations were hypothesized as to why many cities were warmer at night than the surrounding rural areas (coined the "urban heat island"), and urban-scale flow patterns were found that were associated with these thermal differences. In addition, it became apparent that the drag and turbulence created by the "roughness" of buildings were large enough to reduce the strength of the mesoscale wind and enhance boundary-layer-scale mixing. In the next two sub-sections, we give short reviews of the so-called urban heat island and urban roughness effects. Our main purpose here is to identify the general flow features and mechanisms that are important to simulate and/or account for in mesoscale models. More complete accounts can be found in the excellent reviews by Oke [22] on the urban surface energy budget, by Bornstein [1] on urban circulation and thermodynamic evolution, and by Oke [23] on urban climate modification.

2.1 Urban heat island

The well-known urban heat island phenomenon is characterized by warmer temperatures in the city as compared to the surrounding rural area. Generally, the heat island occurs at night and results because the rural area cools faster than the urban area. Urban heat islands can induce thermodynamically driven urban-scale flows. In calm or low wind conditions, the warmer air in the city core rises, pulling air near the surface radially inwards (Figure 1). A radially outward return flow may develop aloft. A dome of heated air often forms above the city. For slightly stronger ambient winds, a plume of heated air may extend downstream of the city.

Figure 2 shows temperature measurements at the surface for Okayama City, Japan, and reveals the urban heat island signature. Temperatures are highest near the core of the city, where buildings are tall and urbanization is dense. Temperature differences of up to 10-12 K have been measured across large cities (Oke [23]). The vertical structure of the urban heat island often shows a several hundred meter well-mixed layer, as indicated in Figure 3. Here, a vertical profile of the temperature outside Sendai, Japan shows a deep stable layer, while the profile over the city reveals that the air temperature is uniform up to 50 meters and is warmer near the surface relative to the rural profile. Vertical profiles of the average of many urban-rural temperature difference measurements are given in Figure 4 for the cities of New York, Christchurch, and Montreal. The profiles show a heated region extending from the surface to between 300 and 800 meters above the cities.

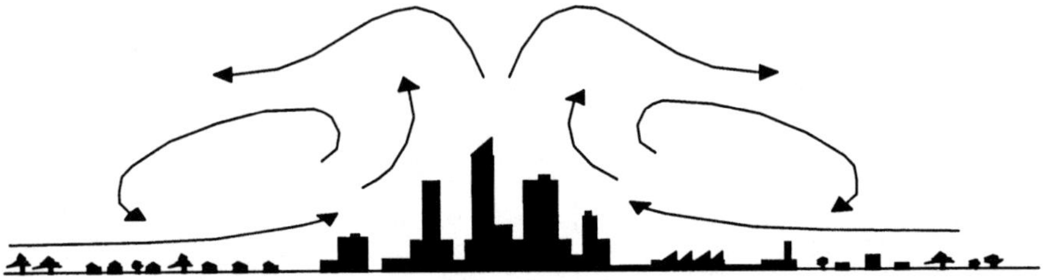

Figure 1: Illustration of urban heat island circulation during calm wind conditions. Pressure differences resulting from warmer temperatures in the city and cooler temperatures in the surrounding rural area lead to thermally-driven flows. (Adapted from Lowry [24] and Liu et al. [25].)

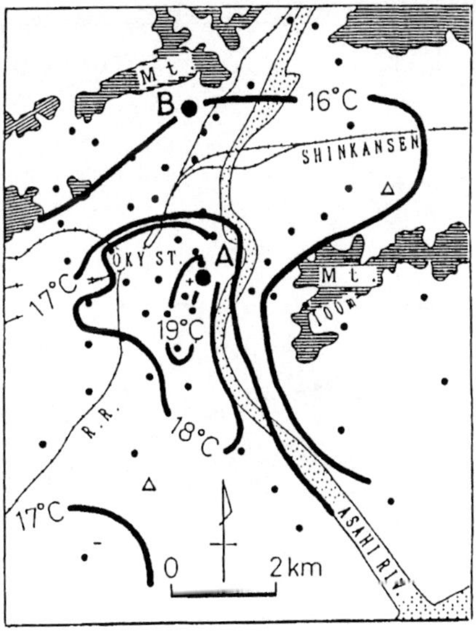

Figure 2: Surface temperatures observed at 2100 local time for Okayama City, Japan. The hotter core is centered over the downtown area. (Adapted from Sahashi et al. [26].)

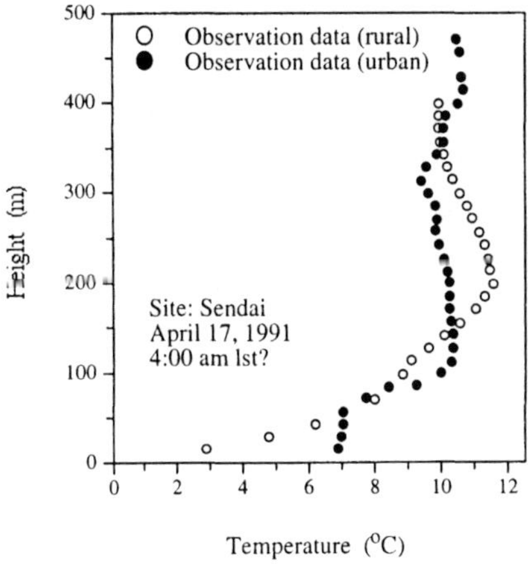

Figure 3: Night time temperature profiles at urban and rural sites reveal the heat island well-mixed layer. (Adapted from Saitoh et al. [27].)

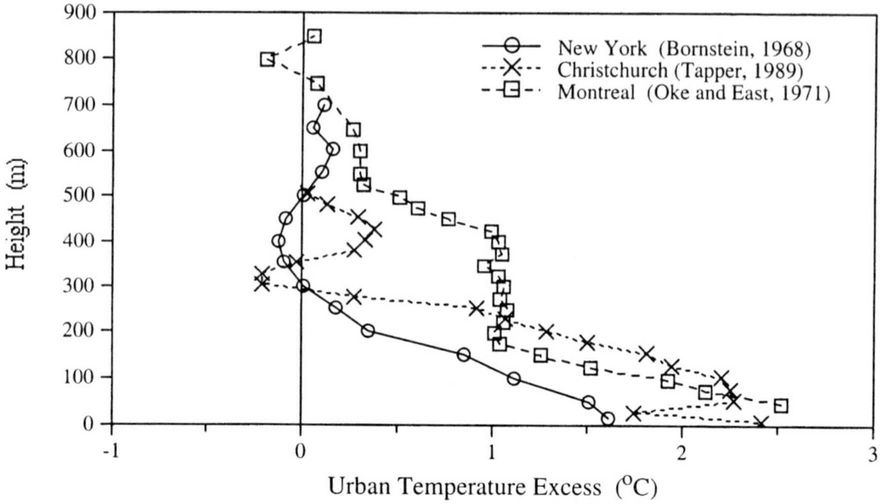

Figure 4: Average urban-rural temperature difference as function of height for 3 cities near sunrise. (Adapted from Tapper [28].)

There have been fewer field measurements of the thermo-dynamically-driven winds, in part because of the difficulty of separating the large and small scale components of the wind circulation. Climatological averages of surface wind sensor data in St. Louis, USA (Shreffler [29]) and Bochum, Germany (Kuttler and Romberg [30]) reveal radial inward motion (Figure 5). These climatological averages were obtained during low wind speed conditions and in the former case by subtracting the assumed prevailing wind components.

Although urban heat island intensity as measured by the maximum temperature difference between urban and rural sites correlates well with population, Oke's [31] review of existing measurements showed that European and North American cities collapsed onto two different curves (Figure 6). Further analysis suggested that the difference was caused by the relatively taller buildings in the core of American cities. Figure 7 shows the urban heat island intensity plotted as a function of building height-to-width ratio, one measure of urban density. It should be pointed out that these data are for calm wind and cloudless conditions.

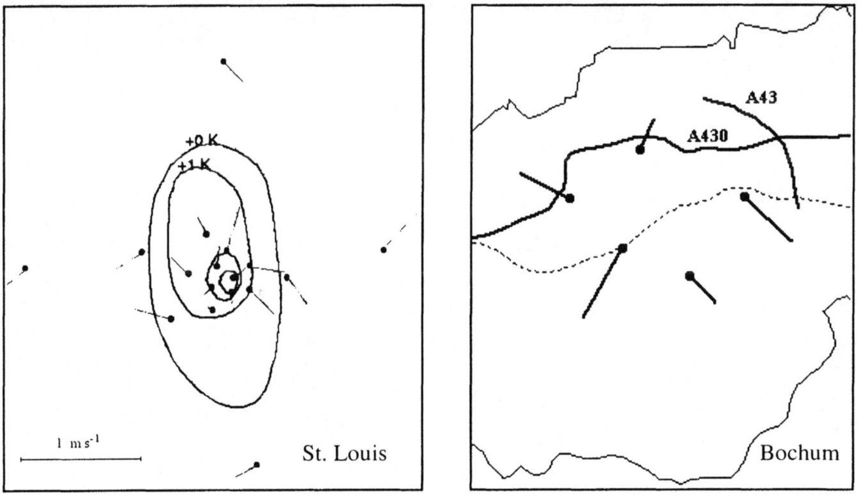

Figure 5: Urban heat island induced winds over St. Louis, USA and Bochum, Germany. (Adapted from Shreffler [29] and Kuttler and Romberg [30].)

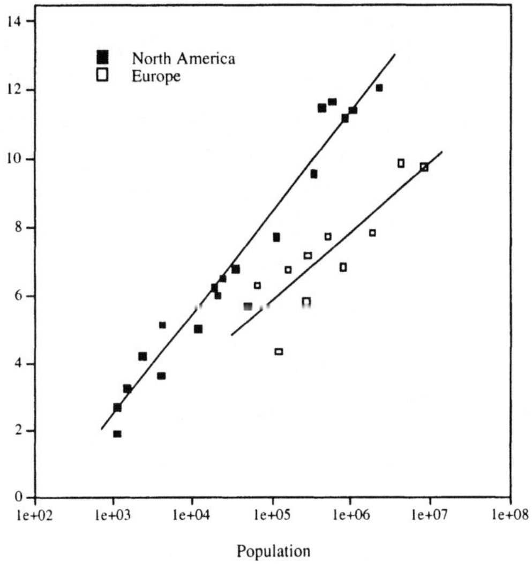

Figure 6: Measurements of urban heat island intensity as function of population showing differences in European and N. American cities. (Adapted from Oke [31].)

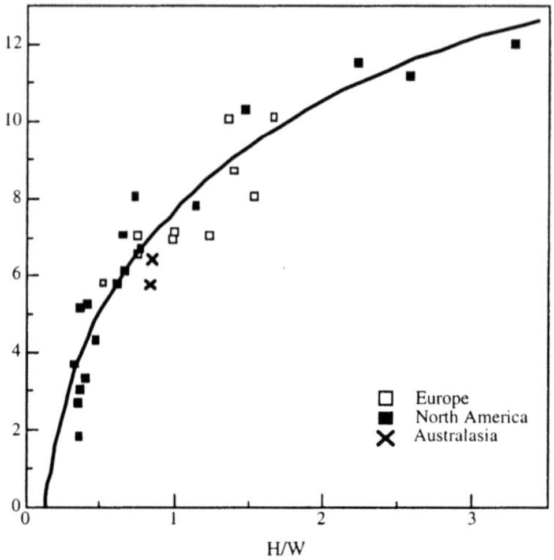

Figure 7: Measurements of urban heat island intensity as function of building height-to-width ratio. (Adapted from Oke [31].)

Although the measurements tend to collapse fairly well, results from numerous studies suggest that the formation and evolution of the urban heat island is complicated, dependent on a number of competing factors. For example, Figure 8 indicates that the urban–rural temperature difference is a strong function of wind speed. If the winds are too strong, the heat can be advected away faster than it can be replenished by the city. In fact, Oke and Hannell [32] proposed that there is a critical wind speed above which urban heat islands do not form. They found that the critical wind speed U_c was a function of population P:

$$U_c = 3.4 \log P - 11.6. \tag{1}$$

Relationships such as those shown in Figures 7 and 8 and eqn. (1) will be useful for direct testing and validation of mesoscale meteorological model results and indirect testing of the urban parameterizations.

Differences in the surface energy budget between urban and rural locales are suspected of being primary factors in the formation of the urban heat island. The energy balance in a control volume containing the urban canopy can be written as (Oke [22])

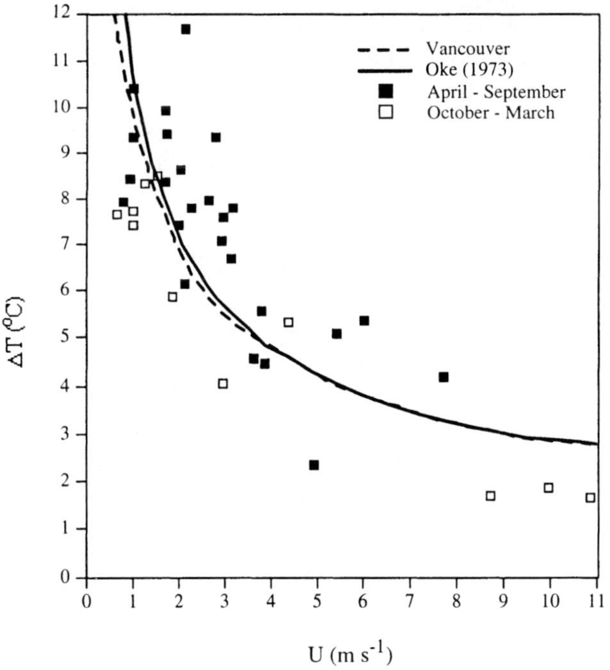

Figure 8: Measurements of heat island intensity plotted as a function of wind speed. (Adapted from Oke [33].)

$$
\begin{aligned}
Q^* + Q_F &= \left(R_{L\downarrow} - R_{L\uparrow}\right) + \left(R_{S\downarrow} - R_{S\uparrow}\right) + Q_F \\
&= \Delta R_L + (1-\alpha) R_{S\downarrow} + Q_F \\
&= Q_H + Q_E + \Delta Q_S + \Delta Q_A
\end{aligned}
\quad , \qquad (2)
$$

where Q^* is the net radiation, Q_F is the anthropogenic heat flux, $R_{L\downarrow}$ and $R_{L\uparrow}$ are the downward and upward long wave radiation, respectively, $R_{S\downarrow}$ and $R_{S\uparrow}$ are the downward and upward short wave radiation, respectively, ΔR_L is the net long wave radiation, $(1-\alpha)R_{S\downarrow}$ is the net short wave radiation, α is the surface albedo, Q_H is the sensible heat flux, Q_E is the latent heat flux, ΔQ_S is the energy storage in the canopy, and ΔQ_A is the advection of energy into and out of the control volume. The net radiation represents the amount of energy coming into ($Q^* > 0$) or out of ($Q^* < 0$)

the canopy from short and long wave sources. The energy associated with the net radiation and the anthropogenic heat flux is partitioned into the sensible heat flux (heats or cools the air), latent heat flux (evaporates water or condenses water vapor), energy storage (heats or cools urban surfaces), and advective heat flux (represents the energy transported by the wind into or out of the canopy volume). Later in Section 3.5 we will discuss eqn. (2) in more detail and present a slightly different form that we feel is more useful for mesoscale meteorological model implementation.

Figure 9 shows a diurnal cycle of surface energy fluxes for a suburban and a rural site in Vancouver. A number of major differences are apparent. First, the latent heat flux, Q_E, is much larger during the daytime at the rural site, indicating that the rural surface has more moisture than the suburban surface. Second, the canopy storage term, ΔQ_S, is much larger during the daytime at the suburban site, implying that the suburban canopy has more energy storage capacity. Third, the sensible heat flux, Q_H, remains positive for a few hours after sunset at the suburban site and the canopy storage term, ΔQ_S, reverses sign at night (i.e., the canopy mass gives off energy to the atmosphere), becoming relatively large in magnitude at the suburban site. This additional input of energy into the atmosphere helps to explain why the rate of cooling at night is smaller in the urban area and why temperatures would be warmer there. Additionally, the lack of a strong heat island during the daytime,

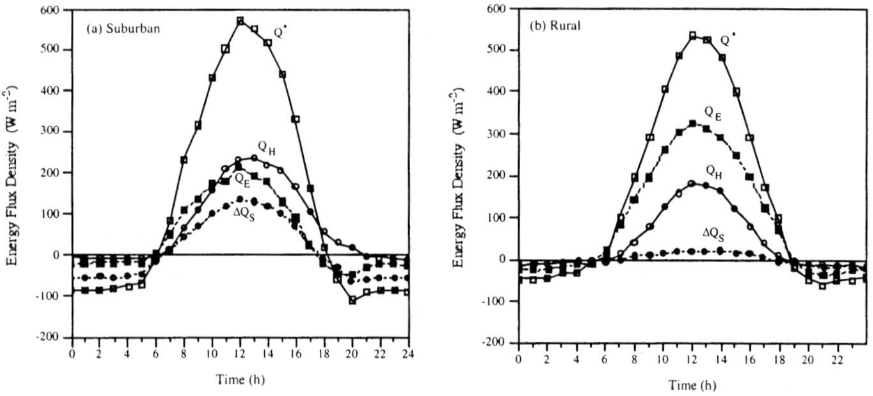

Figure 9: Surface energy balance measured over 30 days during the same time period at suburban and rural sites in Vancouver, BC. (Adapted from Oke [22].)

which one might expect due to the small fraction of the net radiation, Q^*, that goes into latent heating in the suburban area, is partially explained by the relatively large fraction of net radiation that goes into heating the canopy elements (ΔQ_S), thereby reducing the amount of energy that goes into heating the air (Q_H) in the suburban area.

Oke [22] lists seven (sometimes competing) causes for why cities may become warmer than the surrounding rural areas: 1) decreased long wave radiation loss due to reduced sky factor (i.e., the building walls trap, or intercept, infrared radiation trying to escape up into the sky); 2) increased downward long wave radiation from the warmer air above the city (this may be due to trapping and re-emission from polluted layers aloft and/or from heat-island-induced vertical advection of warm surface air above the city); 3) increased short wave absorption (the "bulk" albedo of urban areas is usually smaller than that of rural areas, possibly resulting from a combination of more surface area due to building walls and the trapping of reflected solar radiation onto other urban surfaces due to canyon geometry); 4) decreased evapotranspiration due to less vegetation and moisture availability (this results in more energy going into heating the air and the canopy elements and less into latent heating of moisture); 5) anthropogenic heat input; 6) increased heat storage by canopy elements and 7) reduced heat transport (within the urban canopy the wind speeds and turbulent mixing are generally smaller).

Although decreased urban evapotranspiration, the anthropogenic heat flux, and the urban heat storage are major factors in the development of the urban heat island, it is difficult to draw general conclusions and the exact nature is site and time specific. For example, Oke [22] indicates that urban heat island development is a function of season (e.g., the anthropogenic heat flux increases in the winter for many northern latitude cities), the weather and local mesoscale flows (for example, sea breezes will interact with heat island development in coastal cities), local construction materials, local watering practices (for example, in downtown Mexico City it is common for the sidewalks to be cleaned every morning by hosing them down with water), and surrounding rural land use (e.g., rural areas surrounding arid southwestern U.S. cities will cool at a different rate than moist forest-covered Midwestern and Atlantic seaboard cities).

Many of the factors cited above by Oke [22] are site and time dependent and have feedbacks with each other and with the flow field.

Hence, it would be difficult to obtain universal functions for urban heat island intensity except for idealized cases. For modeling purposes, the individual factors need to be addressed in a robust way so that they interact reasonably. Several of the factors cited above are a strong function of the building height-to-width ratio (e.g., long wave radiation loss, short wave absorption, wind speed, turbulent mixing) which helps to explain the strong dependence of urban heat island intensity on urban density (see Figure 7). For broad application in different urban environments, a mesoscale meteorological model should have urban canopy parameterizations capable of capturing the effects of urban density as a function of land use type, for example. In Sections 3.4 and 3.5, we present methods that have been used to approximate the effect of the urban canopy on heat transport and the surface energy budget in mesoscale models. Next we look at another manifestation of the city on the atmosphere: urban roughness.

2.2 Urban roughness

From the macroscopic viewpoint, cities can be thought of as rough surfaces. When flying high over a city and looking down, one can understand why buildings are often considered to be surface roughness elements. Mesoscale models "see" the world similarly. With grid sizes on the order of kilometers, buildings are not resolved and hence are often parameterized as surface roughness.

Increased surface roughness, generally associated with urban areas, leads to greater frictional momentum loss and increased turbulent fluxes of heat, momentum, and moisture. The review by Bornstein [1] reports that several field studies show wind speed deficits generally exist in urban areas, while turbulence levels are generally elevated. For example, the climatological annual mean wind speed for an expanding city in Russia decreased over time from 3.9 m/s in 1945 to 2.6 m/s in 1971. An elegant study by Hogstrom et al. [34] illustrates the impact of the urban area on the vertical profile of wind speed (Figure 10). By placing an instrumented tower at the urban-rural interface and doing conditional sampling based on wind direction, average wind speed profiles were obtained for both urban and rural fetch wind directions. It is expected that the reduction in wind speed by surface roughness will be offset somewhat

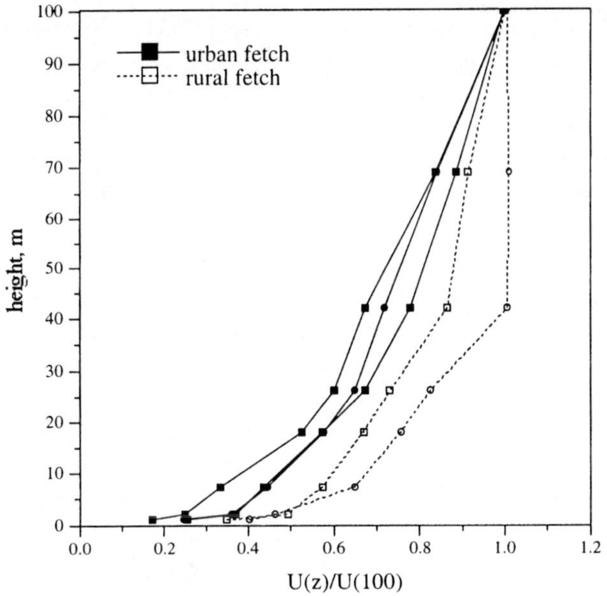

Figure 10: Average of many wind profiles measured on a tower during near-neutral conditions at an urban-rural interface. (Hogstrom et al. [34].)

by the speeding up of the wind associated with the urban heat island circulation.

Surface measurements analyzed by Bowne and Ball [35] revealed that urban turbulence levels in Fort Wayne, Indiana, were 30 to 50% higher than rural levels. Bornstein [1] found that surface measurements in and around New York City of the standard deviation of the horizontal wind direction σ_θ, a measure of turbulence intensity, indicated that values in the city were from 2 to 2.5 times greater than those in rural areas. Analysis of horizontal traverses over St. Louis by Godowitch [36] showed that the vertical velocity variance was about 50% larger as compared to outlying rural areas. It should be pointed out that these results include both the effects of urban heat island and roughness induced turbulence.

Wind tunnel studies can isolate the surface roughness effect on wind and turbulence profiles. For example, Pendergrass and Arya [37] found that the mean velocity and Reynolds shear stress profiles for smooth surface and block roughness boundary layers were significantly different (Figure 11). Theurer et al. [38] have performed a number of experiments over different configurations and shapes of building roughness elements. They found that wind speed and turbulence profiles above building rooftops are strongly impacted by the particular arrangement of buildings.

The studies cited above will be helpful in evaluating the wind and turbulence fields produced over cities. However, to understand the details of the parameterizations associated with urban roughness, it will be helpful to "zoom" down to the microscale and look explicitly at the flow fields around groups of buildings.

Numerous examples exist showing that wind flow patterns and turbulence mixing are dramatically altered around groups of buildings (e.g., Hosker [2]). Figure 12 depicts wind vectors and turbulence levels measured in a wind tunnel around a rather simplified array of wide buildings. The measurements reveal vortices that form between the buildings in the street canyons, a jet region and recirculating flow above the first building rooftop, elevated turbulence levels above the rooftops, and low turbulence levels in the street canyons. For rectangular buildings

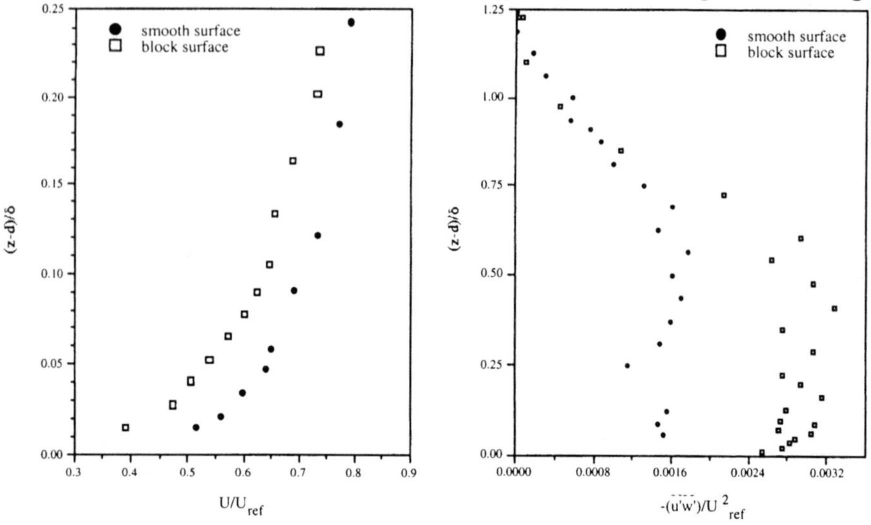

Figure 11: Comparison of non-dimensionalized mean wind profiles (left) and Reynolds shear stress profiles (right) for flow over a smooth surface and a block (urban) surface. (Pendergrass and Arya [37].)

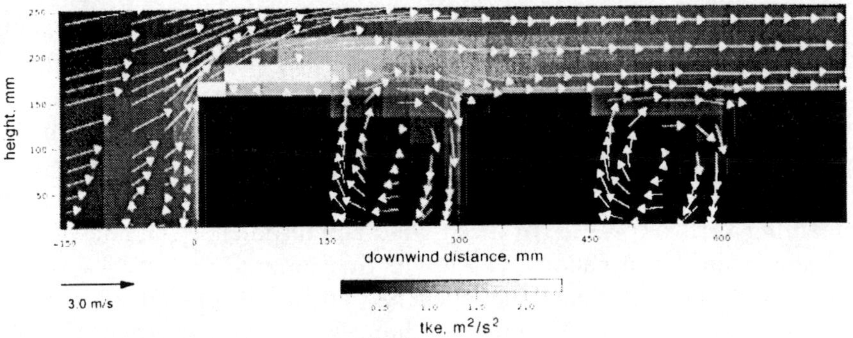

Figure 12: Wind vector and turbulent kinetic energy fields measured along centerline around a 2-d building array in the USEPA meteorological wind tunnel. (Brown et al. [39].)

of equal height, the nature of the flow around the buildings is a function of the building width-to-height ratio. As summarized by Oke [22], a single vortex develops between buildings for skimming flow (*w/h* < 1), two counter-rotating vortices may develop for wake interference flow (*w/h* ~ 1.5), and for isolated roughness flow (*w/h* > 3) the flow field looks similar to the single building case (Figure 13). Significantly more complicated flows can develop for groups of narrow buildings, for variable spacing between buildings, for buildings of different heights and shapes, and for different approach flow angles.

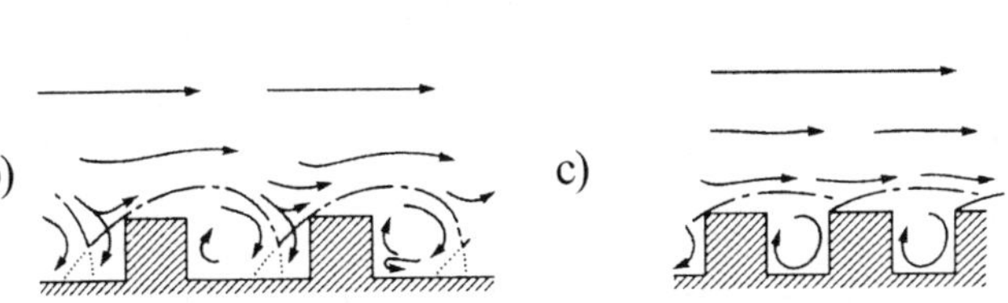

Figure 13: Flow regimes as function of width-to-height ratio: a) isolated roughness flow (*w/h* > 3); b) wake interference flow (*w/h* ~ 1.5); c) skimming flow (*w/h* < 1). (From Oke [22].)

Few detailed mean and turbulence flow measurement campaigns have been performed within the real urban canopy. Roth [40] summarized results above the canopy layer in the so-called roughness sub-layer from twelve field experiments. He found that relationships derived from the logarithmic wind profile (see Section 3.2.2) described the data fairly well above the canopy height. Rotach [41] obtained a unique dataset containing vertical profiles of mean wind and turbulence statistics within and above an urban canopy in Zurich, Switzerland. Figure 14 shows mean wind and turbulent kinetic energy profiles averaged over many days of measurements. An inflection point is apparent in the mean wind profile, a signature of many vegetative canopy velocity profiles. Oikawa and Meng [42] measured mean wind and turbulence profiles in a suburban area in Sapporo, Japan that were in qualitative agreement with the measurements of Rotach [41]. They found that the Reynolds shear stress, a measure of the vertical turbulent momentum flux, peaked at about 1.5 times the canopy height.

As we have shown in this section, buildings act as a sink for momentum and result in a net loss of wind speed (see Section 3.2.1). The

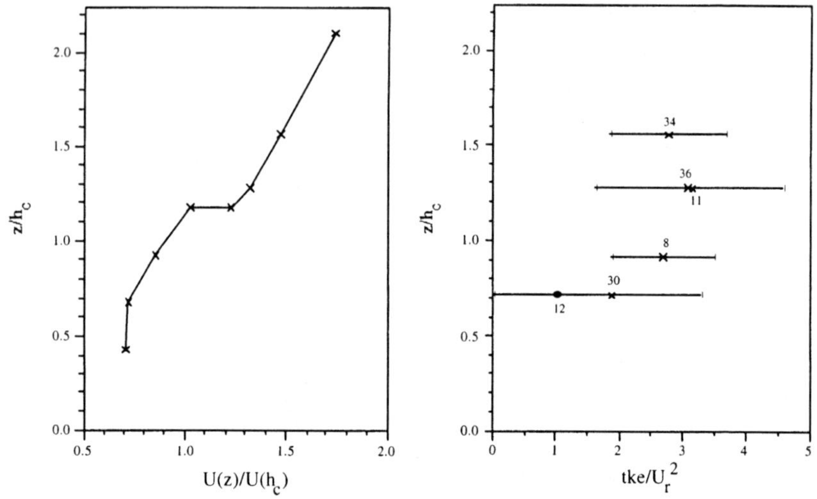

Figure 14: Average wind (left) and turbulent kinetic energy (right) profiles measured in a street canyon in Zurich (Rotach [41]). h_c is canopy height. Range and number of measurements are included on the TKE plot.

strong gradients in the wind produced by the buildings results in enhanced mechanical production of turbulence in the shear zones. The vertical variation of the area-average wind speed and TKE is of course highly dependent on the arrangement, relative heights, and shapes of the buildings. For some applications, urban roughness parameterizations developed for mesoscale models should account for area-averaged drag and turbulent mixing effects and possibly be a function of the building geometry and configuration. In the next section, we discuss methods that have been developed for incorporating urban effects into mesoscale models.

3 Accounting for urban canopy effects in mesoscale models

In this section, we will describe several different methods for including urban canopy effects in mesoscale models. Although we will cover the more traditional methods (e.g., increasing the aerodynamic roughness length, altering the land-use properties), we will bias our discussion towards slightly more detailed approaches. However, we will not be focusing on microscale models in which the effects of buildings are explicitly accounted for, e.g., Arnfeld's [43] energy budget and radiation balance model, Murakami's [44] computational fluid dynamics models for flow around buildings. Rather we are looking at methods that attempt to parameterize the bulk, or area-average, effects of the urban canopy layer for use in mesoscale models. The "area-average" concept is a key one. We are not interested in simulating the details of what happens on particular street corners, but rather we want to compute the bulk effects of buildings and urban land-use over many square blocks, i.e., over a mesoscale grid cell which is typically on the order of kilometers in the horizontal dimension. In addition, we are assuming that the mesoscale model will have adequate resolution to resolve the canopy layer (if a model's first layer is hundreds of meters above ground level, then detailed urban canopy parameterizations would probably not be as beneficial).

Of course, the utility of the urban parameterizations will depend on the model and the model physics being used, so that some of the parameterization methods will not be applicable. For example, a diagnostic wind model may not be able directly to utilize the drag coefficient approach to account for the reduction in wind speed due to buildings. Or, a prognostic mesoscale code using empirical equations for the eddy diffusivity would most likely not be able to incorporate the turbulent kinetic energy production methods described below. Components of these parameterizations should be useful, however, for many of the operational and research mode mesoscale models that are in use today, e.g., COAMPS, FITNAH, HOTMAC, MEMO, MERCURE, MM5, OMEGA, RAMS. Below, we will discuss methods for incorporating urban effects into the momentum, turbulent transport, heat, and surface energy budget equations. First we give a brief overview of past research work where urban parameterizations have been incorporated into mesoscale and boundary-layer models.

3.1 Summary of urban parameterization modeling efforts

Myrup [18] was one of the first to model numerically the details of the urban energy budget. Surface energy balance equations similar to eqn. (2) were solved numerically, while atmospheric conditions were prescribed (i.e., prognostic equations for the atmospheric flow field were not solved). Urban values of the roughness length z_0, the ground albedo α, the soil heat capacity C_p, the soil conductivity k_s, and relative humidity were required to specify the surface energy fluxes. Using this simple model, Myrup [18] was able to simulate reasonable urban-rural temperature differences and found that reduced evaporation and increased roughness in cities, the thermal properties of urban materials, and the wind speed were major factors in determining the strength of the heat island. Without prognostic equations for the atmospheric dynamics, however, he was not able to look at the feedback mechanisms between the flow field and the surface energy budget. More recent work (e.g., Oke et al. [45], Mills [46]) has incorporated urban canyon influences on the urban energy budget, including shadowing, multiple long and short wave reflections off walls and ground surfaces, and heat flux from building interiors. Oke et al. [45] found that the building spacing (width-to-height ratio) and the thermal properties of the urban canopy were controlling variables for urban heat island intensity.

Delage and Taylor [47], Bornstein [21] and Vukovich et al. [48] numerically simulated the urban heat island circulation using the atmospheric conservation equations for mass, momentum and heat. They did not solve a surface energy budget equation, but rather defined different surface temperatures or cooling rates for the urban and rural areas in the modeling domain. Urban effects were captured through the roughness length z_0 and the specified urban-rural temperature differences. Model simulations were found to reproduce many of the characteristics of the urban heat island circulation. For example, in the 2-d simulations performed by Bornstein [21], a critical upstream wind speed was found that resulted in either faster or slower winds over the urban area, a result of the dual competing effects of urban roughness-induced drag and thermally induced winds. In 3-d simulations, Vukovich et al. [48] found cyclonic inflow around the heat island for low wind speeds. They found that as the ambient wind speed increased, the urban heat island intensity decreased and the length of the downwind thermal plume increased. Although much was learned about the development of urban heat island circulation, they were not able to study the feedback mechanisms between different components of the energy balance and the flow field.

Some of the first simulations performed using linked atmospheric flow and surface energy budget equations were performed by Atwater [19] and McElroy [20]. Although the simulations were 2-d, steady-state in the case of McElroy [20], and included no advection in the case of Atwater [19], they were able to look at how the surface energy budget modified the wind fields, how the modified wind fields changed the surface energy budget terms, and so forth. In both cases, anthropogenic heat flux terms were added to the surface energy budget and urban-specific ground albedo, roughness length, thermal diffusivity, density, and heat capacity were defined. The effect of urban-induced drag and mixing were accounted for through the roughness length parameter. McElroy [20] found reasonable agreement between the model-computed thermal structure in the boundary layer and experimental measurements. Atwater [19] also included the effects of radiatively active pollutants in the surface energy budget and introduced a moisture availability term for computing the latent energy flux term. Although the pollution layer was shown to reduce incoming solar radiation, Atwater [19] found that the physical properties of the surface were more important in the creation of the urban heat island.

2-d and full 3-d mesoscale atmospheric simulations utilizing surface energy budget equations and the roughness length approach have been performed over urban areas by a number of modelers, e.g., Hjelmfelt [49], Schultz and Warner [50], Byun and Arya [51], Kimura and Takahashi [52], Saitoh et al. [27], Mochida et al. [53], Carissimo [54], Kitada et al. [55], Hafner and Kidder [56], Taha [57], Perez-Garcia and Nickerson [58]. They have performed simulations in St. Louis, Los Angeles, El Paso, Tokyo, Paris, Nagoya, Atlanta and Mexico City with reasonable success. Using a larger roughness length to account for the urban influence on drag and eddy diffusivity, and incorporating the anthropogenic heat flux into the surface energy budget, several of the investigators were able to simulate qualitatively reasonable looking urban heat islands. As will be discussed below in Section 3.2, these simulations did not capture the flow dynamics within the urban canopy. In addition, they did not account for the effects of radiation attenuation, trapping, and emission due to buildings.

Sorbjan and Uliasz [59] were among the first to incorporate urban-canopy induced drag and turbulence production directly into the fluid dynamics equations (as opposed to the roughness length approach, which is incorporated through boundary conditions). They introduced a drag term into the horizontal momentum equations and a turbulent generation term into the eddy diffusivity profile formula. Using this approach they were able to compute the flow fields *within* the urban canopy. Also unique to their study was the introduction of a turbulent generation term associated with vehicles and the adaptation of Atwater's [19] radiation scheme to account for pollution on the development of the urban heat island.

Based on forest canopy parameterizations of Yamada [60], Williams et al. [61] included urban-induced drag and turbulent production terms into a mesoscale model with a q^2-l turbulence closure. Like Sorbjan and Uliasz [59], they included a drag term in the horizontal momentum equations, but also included a turbulence production term in the turbulent kinetic energy ($q^2/2$) equation. In addition, as outlined in Brown and Williams [62], the radiation attenuation due to the urban canopy was crudely accounted for in the surface energy budget while the heat flux from rooftops and vertically-distributed anthropogenic heat sources were included in the heat conservation equation.

Recently, a number of researchers have developed similar urban canopy parameterizations with different levels of complexity for k-ε turbulence closure models. Maruyama [63], Urano et al. [64], and Ca et al. [65] have all included the urban canopy drag term in the momentum equations. Maruyama [63] included urban-induced drag terms in the turbulent kinetic energy (k) and the turbulent dissipation rate (ε) equations as well. The surface energy balance was ignored however. Urano et al. [64] included drag and turbulence production terms in their k-ε mesoscale model and solved an energy balance equation at the ground surface. Like Brown and Williams [62], they added a vertically-varying anthropogenic heat flux term to the heat equation as well. Ca et al. [65] did not add urban production terms to the turbulent kinetic energy or dissipation terms, but like Maruyama [63] did include an effective air volume term in the mass, momentum, heat, turbulent kinetic energy, and dissipation equations. The effective air volume term accounts for the reduced air volume in the grid cells within the urban canopy. Ca et al. [65] included a detailed radiation balance model accounting for shading and multiple reflections of long and short wave radiation off building walls and streets in their surface energy budget model. They also included a building air conditioning model explicitly to account for heat emissions and feedback between urban temperature and air conditioning usage.

3.2 Drag

The drag force induced by buildings and other obstacles in the urban environment results in a transfer of momentum out of the bulk flow and a reduction in the area-average wind speed (see Section 2.2). A number of different methods have been used in mesoscale and boundary-layer modeling in order to account for the drag effect. In this section, we will look at three methods that cover a range of complexity: the popular roughness length approach in which the wind profile is modified in the lowest grid cell through boundary conditions, a straightforward attenuation approach in which the wind profile is modified through the canopy layer using empirically-derived formulae, and a drag formulation which gets incorporated into the prognostic momentum equations. First we begin with a definition and derivation of the drag term.

3.2.1 Drag force definition

The drag force is the force that a fluid exerts on a solid (e.g., Whitaker [66]). The solid may be the earth's surface, buildings, trees, a car or an airplane, among other things. The drag force on an object is in the same direction as the mean flow (in addition, there may also be a lift force that develops off-angle to the mean flow depending upon the geometry of the obstacle). The drag is typically divided into form (or pressure) drag and frictional (or viscous) drag. Form drag is a result of the pressure differential that forms around objects when they are enveloped in a flow field (usually the front face of an obstacle has a high pressure associated with the flow impacting the front surface, and the back face has a low pressure that is a function of whether or not flow separation occurs on the back side). The frictional drag is a result of mixing of momentum in a direction normal to the surface of the object. The fluid elements closest to the object "stick" to the surface, shear layers develop, and either molecular or turbulent diffusion transports the slow moving fluid away from the obstacle resulting in momentum loss around and downstream of the solid object.

By definition, the drag force F_D on an obstacle is the integral of the pressure and viscous forces over the surface of the obstacle (e.g., Panton [67]):

$$F_{D_x} = \int n_x P - n_i \tau_{ix} dS\,, \tag{3}$$

where x is the direction of the flow, n_x is the unit normal vector parallel to the flow, P is the pressure (or perturbation pressure), n_i is the unit normal vector perpendicular to the surface, and τ_{ix} is the shear stress term. For laminar flows, τ_{ix} represents molecular viscous forces while for turbulent flows it also includes the turbulent Reynolds shear stresses.

As a body force, the drag force term can be added to the right hand side of the atmospheric fluid flow momentum conservation equation:

$$\rho\frac{DU_i}{Dt} = -\frac{\partial P}{\partial x_i} - \rho g \delta_{i3} - 2\rho\Omega\varepsilon_{ijk}\eta_j U_k + \frac{\partial \tau_{ji}}{\partial x_j} - \frac{\rho F_{D_i}}{\delta V}\,, \tag{4}$$

where ρ is the fluid density, U_i is the mean velocity in the ith direction, $D(\ )/Dt = \partial(\ )/\partial t + U_j\partial(\ )/\partial x_j$ is the total derivative, P is the pressure, g is the gravitational acceleration term, δ_{i3} the Kronecker delta function,

Ω is the angular velocity of the earth's rotation, ε_{ijk} is the permutation tensor, η_j is the unit vector parallel to the earth's axis of rotation, τ_{ij} the shear stress tensor, and δV a unit volume. F_{Di} is defined to be positive in the direction opposite to the mean flow U_i. For incompressible flows, the shear stress tensor τ_{ij} can be written out in turbulent and viscous components, i.e.,

$$\tau_{ij} = -\rho < u_i u_j > + \mu \frac{\partial U_i}{\partial x_j}, \tag{5}$$

where u_i is the pertubation from the mean velocity, the < >s represent an ensemble average, $<u_i u_j>$ is the Reynolds turbulent shear stress term, and μ is the molecular viscosity.

The non-dimensionalized form of the drag force, called the drag coefficient C_D, is obtained by dividing the drag force by the mean kinetic energy of the flow and the cross-sectional area of the obstacle:

$$C_D = \frac{F_D}{\frac{1}{2}\rho U_0^{\,2} A}. \tag{6}$$

For idealized flow conditions and obstacle shapes, analytical solutions for C_D can be obtained (e.g., Panton [67]). For more realistic flow conditions and obstacle shapes, C_D is determined experimentally (e.g., Hoerner [68]). Assuming one knows C_D for a given obstacle and flow condition, the drag force imparted by the obstacle on the flow can be computed with additional knowledge of only a reference velocity U_0 and the cross-sectional area of the obstacle A.

3.2.2 Aerodynamic roughness length approach

Drag due to the urban surface is commonly accounted for in mesoscale models by modifying the surface roughness parameter associated with the lower boundary conditions of the model. Many mesoscale codes use Monin-Obukhov similarity relationships for defining the turbulent fluxes and/or the mean flow variables at the lowest grid cell above the surface. One of the parameters in the Monin-Obukhov similarity relationships is the aerodynamic roughness parameter z_0 which influences the gradients of the flow variables and the magnitude of turbulent mixing at the surface.

Monin-Obukhov similarity states that the wind shear in the surface layer is a function of a velocity scale and two length scales (e.g., Arya [69]), i.e.,

$$\frac{\partial U}{\partial z} = \frac{u_*}{\kappa z}\phi_m(z/L), \tag{7}$$

where U is the mean wind velocity in the direction of the wind, z is the height above ground, u_* is the friction velocity ($u_*^2 = <uw>_0$), u and w are the horizontal and vertical velocity pertubations, respectively, k is von Karman's constant (0.4), ϕ_m is a stability-dependent, empirically-determined function that relates the turbulent shear stress to the mean velocity gradient, and L is the Obukhov length that pertains to the relative importance of shear and buoyancy effects in the surface layer.

Integrating eqn. (7) from $U=0$ to $U=U(z)$ and assuming the turbulent momentum flux is constant within the surface layer, one obtains the logarithmic velocity profile:

$$U(z) = \frac{u_*}{\kappa}\left[\ln\frac{z}{z_0} - \psi_m(z/L)\right], \tag{8a}$$

where z_0 is the roughness length and ψ_m are stability-dependent corrections to the mean velocity profile. The roughness length z_0 can be thought of as the height above the surface at which U goes to zero. Figure 15 shows how the mean velocity profile changes with z_0 given a constant u_* and neutral atmospheric stability conditions.

For very rough surfaces, such as urban and forested areas, a displacement height d is needed for the logarithmic profile method to work. That is, the surface reference plane should be moved up off the ground to some fraction of the roughness element height h:

$$U(z) = \frac{u_*}{\kappa}\left[\ln\frac{z-d}{z_0} - \psi_m(\frac{z-d}{L})\right]. \tag{8b}$$

Arya [69] showed how z_0 relates to the drag coefficient by equating the surface stress τ_0 to the mean wind speed at a reference height z_r through a drag relation:

$$\tau_0 = \rho C_D U(z_r)^2 \tag{9}$$

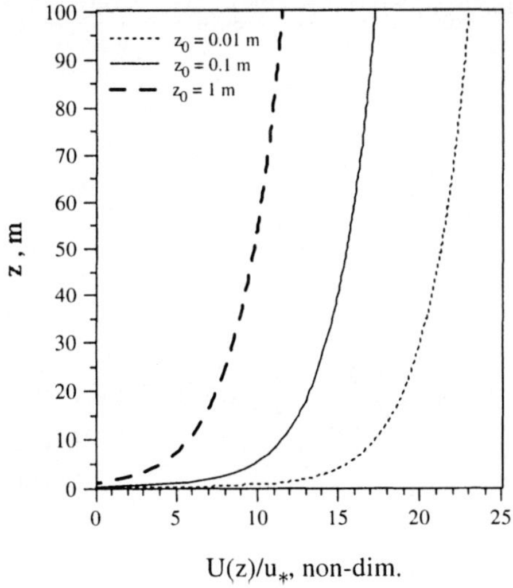

Figure 15: Logarithmic velocity profile for neutral conditions as function of z_0.

Assuming neutral stability, substituting u_*^2 for τ_0/ρ and eqn. (8) for $U(z_r)$, and rearranging, Arya [69] obtained:

$$C_D = \frac{\kappa^2}{\{\ln[(z_r - d)/z_0]\}^2} \,. \tag{10}$$

Hence, as the roughness length increases, the frictional drag increases as well (Figure 16).

In practice, the roughness length is incorporated into mesoscale simulations in one of two ways. In many mesoscale codes the velocity at the first grid level above the surface is specified as a boundary condition using the logarithmic velocity profile eqn. (8). In this approach, the effect of surface roughness is transmitted directly through the boundary condition to the velocity field. Alternatively, some codes prognostically compute the velocity at the first grid level using the momentum conservation equations, and use eqn. (7) as a boundary condition to specify the wind shear in the Reynolds shear stress computation at the lowest grid cell. The roughness length, z_0, is used to compute u_* following eqn. (8), which is then used to determine the wind shear

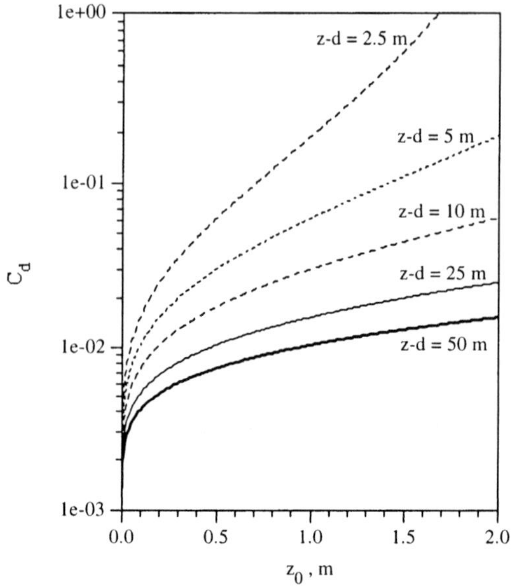

Figure 16: Variation of drag coefficient C_D with roughness length z_0 at different heights according to eqn. (10).

boundary condition. In a similar way, the roughness length also influences the boundary condition computations of the heat and moisture fluxes, which are derived from Monin-Obukhov similarity as well (see Section 3.5).

The advantage of the aerodynamic roughness approach is that it is easy to implement and considerable information is available for specifying z_0 and d as functions of land-use type. Simple formulae have been developed that relate z_0 and d to the canopy element height h_c (e.g., Plate [70]):

$$z_0 = 0.15\, h_c \tag{10a}$$

and

$$d = h_c\,. \tag{10b}$$

In reality, z_0 and d are functions of canopy element (e.g., building, tree, bush) spacing, shape, stiffness, relative heights, etc. For example, wind-tunnel experiments by Hall et al. [71] show that closely spaced buildings have a larger displacement height than widely spaced buildings, and the

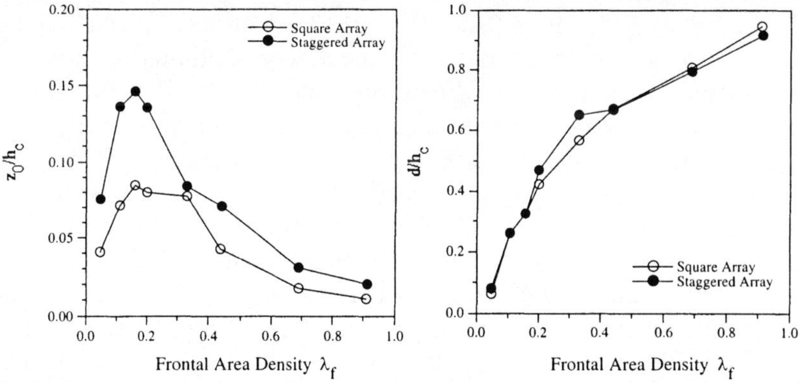

Figure 17: Wind-tunnel measurements of roughness length and displacement height as function of frontal area density for cubes in square and staggered arrays. (Adapted from Hall et al. [71].)

roughness length initially increases as the building density increases and then decreases beyond a critical density (Figure 17). Letau [72] provided a relationship for z_0 that considered the frontal area density λ_f:

$$z_0 = 0.5\, h_c\, A_f / A_d = 0.5\, h_c\, \lambda_f , \tag{11}$$

where A_f is the total frontal area of the obstacles and A_d is the total area covered by the obstacles. This expression was found to be valid for small λ_f (< 20 to 30%), i.e., for sparsely populated building arrays. Raupach [73], Bottema [74], McDonald et al. [75] and Duijm [76] derived relationships for z_0 and d incorporating the sheltering effect of building obstacles, which are valid over a wider range of λ_f. For example, McDonald et al. [75] proposed the following:

$$z_0 = \frac{h_c - d}{h_c} \exp\left[-\frac{0.5 C_D \lambda_f}{k^2} (1 - \frac{d}{h_c}) \right] \quad \text{and} \tag{12a}$$

$$d = h_c \cdot \left\{ 1 + A^{-\lambda_p} (\lambda_p - 1) \right\}, \tag{12b}$$

where λ_p is the plan area density (plan view area of obstacles divided by total plan area) and A is a constant that depends on obstacle arrangement.

McDonald et al. [75] found that A=4.4 and 3.6 worked well for staggered and aligned arrays of square blocks, respectively. Although it is not clear which formulae are "best", we direct the reader to careful evaluations that have been carried out by Grimmond and Oke [77] and Duijm [76].

There are a number of disadvantages to the aerodynamic roughness approach which might be problematic depending on the application. First, the lowest height at which winds can be computed with the logarithmic formula is at the displacement height, d (typically a large fraction of the urban canopy height). This provides problems when one is interested in what happens at street level (e.g., street-level dispersion, surface heat fluxes). So, in fact, when a modeler presents results of winds at the lowest level using a roughness length approach as described above, these winds are actually at or above the building rooftops. A second disadvantage is that the horizontally-homogeneous condition used in the Monin-Obukhov similarity theory is not satisfied within and for some distance above the urban canopy. Plate [70] reported that field experiments above vegetative canopies showed that the logarithmic profile was valid for $z > 2h_c$. For an urban canopy, Roth [40] found that above $2.5h_c$ Monin-Obukhov similarity appeared to work well. Lastly, traditional Monin-Obukhov similarity cannot account for the inflection point in the velocity profile (or the elevated peak in tke) as was shown in Figure 14. The methods described in the next two sections attempt to overcome these limitations.

3.2.3 Attenuation approach

Much work has been done accounting for drag within the vegetative canopy layer. Numerous mean wind profiles have been measured within real and artificial plant canopies such as corn fields, wheat fields, orchards, forests, wooden pegs, and plastic strips (e.g., see Cionco [78], Plate [70], Arya [69]). Early work by Cionco [78] included the following formula for flow within the canopy:

$$U(z) = U(h_c)\exp\left[b(z)\cdot\left(\frac{z}{h_c}-1\right)\right], \tag{13a}$$

where $U(z)$ is an area-average wind velocity, $U(h_c)$ is the wind velocity at canopy height, and $b(z)$ is an attenuation coefficient dependent on the

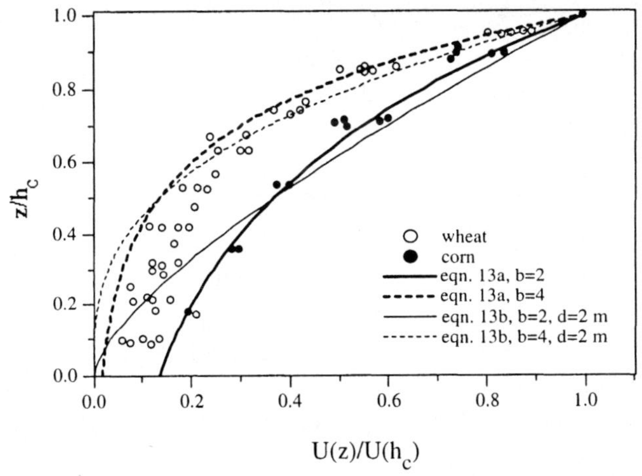

Figure 18: Comparison of eqns. (13a) and (13b) to measurements made in vegetative canopies. (after Arya [69].)

plant canopy. In Figure 18, eqn. (13a) is plotted for several values of b and yields the expected sharp decay with height in the canopy, but does not approach zero at the surface. As presented in Pielke [79], a slightly different formulation has been proposed that does go to zero at the surface:

$$U(z) = U(D)\exp\left[b \cdot \left(\frac{\ln z}{\ln D} - 1\right)\right]. \tag{13b}$$

Eqn. (13b) is also plotted in Figure 18 assuming that $D=h_c$. Note that this formulation does not work for D less than or equal to one.

As described in Cionco and Ellefsen [80], these attenuation methods can be extended to urban canopies. eqn. (13) could be used as a boundary condition in place of the logarithmic velocity profile (eqn. (8)) or could be used to extrapolate below the lowest grid level in coarse vertical resolution simulations. In addition, this approach could be easily adapted for use with diagnostic wind models (whereas the drag approach we describe below in the next section would be more difficult to implement). The method is straightforward, but does require specification of the attenuation coefficient which is a function of the canopy type and also may be a function of height.

The functional form of *b(z)* can be derived for several idealized cases. Following the derivation presented in Arya [69] for steady-state, horizontally-averaged flow within a canopy, the conservation equation for momentum (eqn. (4)) reduces to:

$$\frac{\partial <uw>}{\partial z} = -\frac{1}{\rho}\frac{\partial P}{\partial x} - \frac{1}{2}C_D a(z)U^2, \tag{14}$$

where *a(z)* is the effective aerodynamic canopy surface area per unit volume of canopy ($A/\delta V$). If one assumes that the horizontal pressure gradient is negligible (perhaps valid near the top or above the canopy), then eqn. (14) reduces to a balance between the gradient of the turbulent shear stress and the drag force. Using the mixing length hypothesis to describe the Reynolds shear stress, i.e.,

$$<uw> = -l_m^{\,2}(\partial U / \partial z)^2, \tag{15}$$

and assuming that $C_d a(z)$ and the mixing length l_m are constant, eqn. (14) can be integrated to yield eqn. (13a) with

$$b = h_c \cdot \left(\frac{C_d a}{4 l_m^{\,2}}\right)^{1/3}. \tag{16}$$

Although attenuation coefficients have been experimentally determined for a number of plant canopies (e.g., Cionco [78]), we are unaware of *b(z)* values for urban canopies. In addition, the assumptions that were used to simplify eqn. (14), namely that the pressure gradient is negligible, are not valid within the urban canopy where form drag is dominant. However, the form of eqns. (13a-b) may still describe the horizontally averaged urban canopy velocity profile adequately for many model applications.

3.2.4 Drag approach

As shown in Section 3.2.1, a drag force term can be added to the atmospheric momentum conservation equations to account for obstacle drag. This approach has been successfully used to parameterize the frictional drag due to forest canopies (e.g., Wilson and Shaw [81], Yamada [60], Liu et al. [82]) and has been used in several cases for urban canopies (e.g., Sorbjan and Uliasz [59], Brown and Williams [62], Urano et al. [64]). Using the drag coefficient approach, modifications can be made to the horizontal components of the momentum equations to account for the area-average effect of the sub-grid urban canopy elements:

$$\frac{DU}{Dt} = -\frac{1}{\rho}\frac{\partial P}{\partial x} + fV - \frac{\partial < uu_j >}{\partial x_j} - f_{urb} C_d a(z) U|U|, \tag{17a}$$

$$\frac{DV}{Dt} = -\frac{1}{\rho}\frac{\partial P}{\partial y} - fU - \frac{\partial < vu_j >}{\partial x_j} - f_{urb} C_d a(z) V|V|, \tag{17b}$$

where f is the Coriolis parameter, the $<u_i u_j>$ are the Reynolds shear stresses, f_{urb} is the fraction of the grid cell covered by urban canopy, $a(z)$ is the canopy area density (surface area of buildings perpendicular to the wind per unit volume of urban canopy), and the other variables are as defined in eqn. (4). The drag force will be non-zero within the canopy and "turn-off" at some height above the canopy in response to the canopy area density $a(z)$ term. The drag formulation by Urano et al. [64] and Sorbjan and Uliasz [59] is nearly identical to eqn. (17), except that in the latter case they lumped C_d and $a(z)$ together and did not explicitly state that it could vary with height.

In these approaches, it is assumed that the sub-grid buildings affect the flow field, but do not take up any volume within the grid cell. Maruyama [63] and Ca et al. [65] included a fractional volume term G in the conservation equations to account for the space occupied by buildings. If one assumes that the building canopy actually occupies volume within the computational domain, then this approach ensures that mass fluxes are conserved across grid cells. In principle, even without the drag terms added, the flow will slow down or speed up as it travels through grid cells with more or less air volume. It is not clear what the relative impact of the G and drag terms are on simulations. Some authors (e.g., Hirt [83]) have used the fractional volume approach to account for the drag effect of sub-grid forest canopies and assumed that the G term accounted for the drag, i.e., no drag force term was needed. Porous flow approaches are somewhat similar to the drag approach as well, adding a frictional term to the equations that is similar in behavior to the C_d term (e.g, Ingram et al. [84]). Bottema and Sini [85] found that traditional porous flow methods did not work well for urban simulations, basically because the frictional force was too small, i.e., porous flow dynamics did not well represent the flow within an urban canopy.

This drag method has the advantage over the roughness length approach that the effects of the canopy can be applied to the flow interior, not just as a boundary condition. As shown in Fig. 19, the model can account for the height variation of the wind profile within the canopy. Figure 20 shows the impact on the mesoscale flow field. The simulation, performed over Dallas, illustrates the wind speed reduction over the urban area due to the drag force. The major disadvantage with the drag approach includes specifying the new terms that need to be incorporated into the model, including the drag coefficient C_d, the canopy area density $a(z)$, and the urban grid cell fraction f_{urb}. We discuss their specification in Section 4.1.

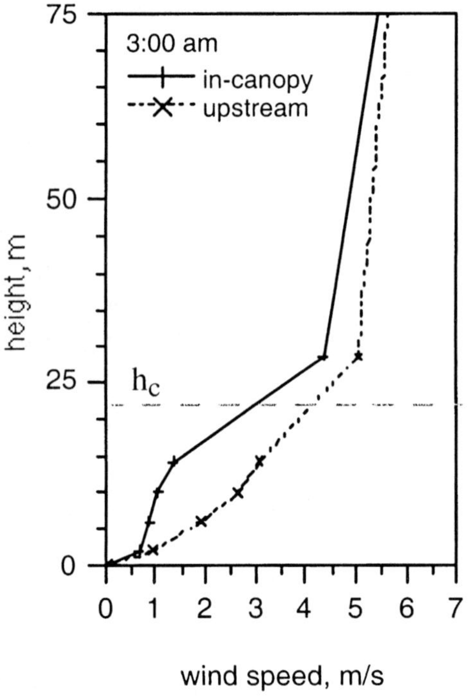

Figure 19: Model-computed wind speed profile at non-urban (upstream) and urban (in-canopy) sites. (Brown and Williams [62].) Canopy height = 22 m.

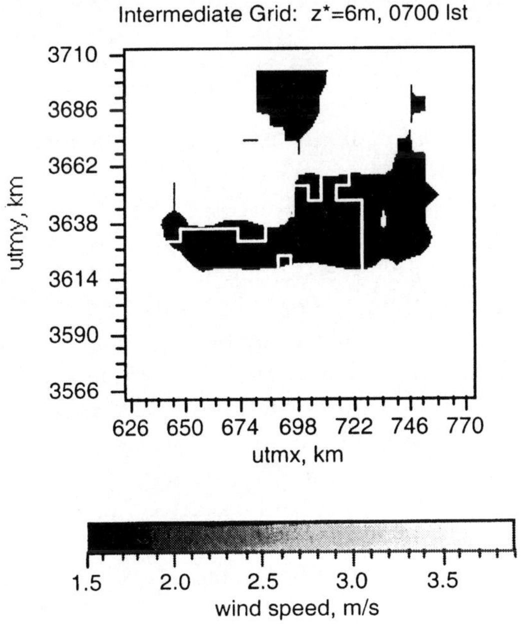

Figure 20: Mesoscale simulation of Dallas/Ft. Worth area (white outline) showing wind speed reduction over the urban area due to drag parameterizations. (Brown and Muller [86].)

3.3 Turbulence production/destruction

Buildings result in enhanced turbulence production in the urban environment. The flow blockage, isolated regions of jetting, and adverse pressure gradients associated with the building obstacle result in flow separation, recirculation zones, high shear, and production of turbulent eddies. Within the street canyon, turbulent mixing may be suppressed due to a disconnection with the ambient airflow above. In this section, we will look at three methods that have been used in mesoscale modeling for incorporating enhanced turbulent mixing due to building obstacles: the roughness length approach, a local scaling method, and a drag formulation. All three methods either directly or indirectly modify the Reynolds shear stress term, often modeled as a first-order gradient diffusion term:

$$< u_i u_j >= -K_m \left(\frac{\partial U_i}{\partial x_j} + \frac{\partial U_j}{\partial x_i} \right), \tag{18}$$

where K_m is the turbulent eddy diffusivity for momentum.

3.3.1 Roughness length approach

As discussed in Section 3.2, the roughness length influences the friction velocity (i.e., the surface layer turbulent momentum flux) through the surface boundary conditions. For urban surfaces, the larger roughness length results in a larger friction velocity u_* through eqn. (8). Depending on the turbulent closure scheme and the particulars of the boundary condition formulation, the eddy diffusivity, the Reynolds shear stress, and/or the velocity gradient at the lowest vertical grid level are enhanced due to their dependence on the friction velocity. In this way, the surface roughness associated with urban land use affects turbulent mixing.

Although this approach is easy to implement, it is valid only at some height above the canopy where the flow is quasi-horizontally homogeneous. As mentioned in Section 3.2.2, Roth [40] found that Monin-Obukhov similarity appeared to work well over an urban area above $2.5h_c$. Using the roughness length approach, one cannot account for the expected peak in turbulent mixing at or above the canopy height and the reduced mixing within the street canyons. The approach has worked well, however, for defining the general characteristics of the urban heat island.

3.3.2 Empirical local scaling approach

Similar to the approach proposed by Cionco [78] for accounting for canopy drag on the wind profile (Section 3.2.3), an equation describing the vertical profile for turbulent kinetic energy, eddy diffusivity, or Reynolds shear stress could be formulated within and/or above the urban canopy. Based on an extensive review of field data, Roth [40] proposed the following formulae for a locally-scaled friction velocity u_* and the turbulence intensity in the roughness sublayer above the urban canopy for neutral conditions (see Fig. 21):

$$\begin{aligned} (-< uw(z)>)^{1/2} &= u_*(z) \\ &= U \cdot \{0.095 + 0.389 \exp[-0.971 \cdot (z-d)/h_c]\}, \end{aligned} \tag{19a}$$

and

$$I_i = \sigma_i / U = \kappa A_i \left[\ln\left(\frac{z-d}{z_0} \right) \right]^{-1}, \quad i = u,\ v,\ w, \tag{19b}$$

where σ_u, σ_v, and σ_w are the standard deviations of the horizontal and vertical components of the wind velocity components, and A_i is an empirical constant. Rotach [87] proposed an alternative expression for the friction velocity within the roughness sublayer:

$$u_*(z) = u_* \cdot C_1 \cdot \{1 - \exp[-C_2 \cdot (z-d)]\}^{1/3}, \tag{20}$$

where u_* is the friction velocity above the roughness sublayer within the inertial sublayer and C_1 and C_2 are experimentally-determined constants. Rotach [87] proposed equations for the velocity variances that were non-dimensionalized by the local friction velocity eqn. (20). In both cases, the equations can be used to extrapolate down to slightly below the canopy element height $(z/h_c = 0.7)$ using wind velocities or friction velocities computed by the mesoscale model at the top of the roughness sublayer. These methods could be used to provide finer details within the first grid cell for vertically coarse resolution mesoscale simulations or to

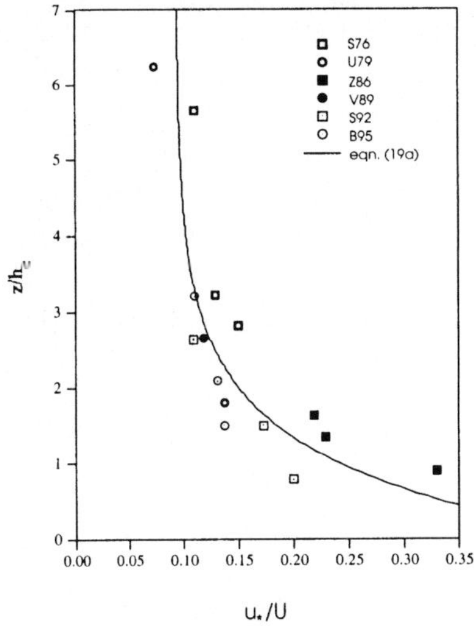

Figure 21: Eqn. (19a) compared to measurements of the friction velocity over urban areas. (Roth [40].)

complement diagnostic wind models by providing turbulence parameters for dispersion modeling applications. We are not aware of any area-average formulations for the turbulent kinetic energy, eddy diffusivity, or Reynolds shear stress *below* canopy height.

3.3.3 Drag approach

In mesoscale models using the k-ε or q^2-*l* turbulence closure approach to specify the eddy diffusivity, a turbulent kinetic energy production term can be added to the tke equation to account for urban-induced turbulence, (e.g., Brown and Williams [62], Maryuma [63], Urano et al. [64])

$$\begin{aligned}\frac{Dtke}{Dt} &= \delta_{i3}\frac{g}{\Theta} < u_i\theta > - < u_i u_j > \frac{\partial U_i}{\partial x_j} - \frac{\partial < u_j \cdot tke >}{\partial x_j} \\ &- \frac{1}{\rho}\frac{\partial < u_i p >}{\partial x_i} - \varepsilon + f_{urb} C_d a(z)\left(|U|^3 + |V|^3\right)\end{aligned}, \quad (21)$$

where $tke = k = q^2/2 = 1/2(<u^2>+<v^2>+<w^2>)$ and ε is the dissipation term. This is the exact form of the tke equation without the closure assumptions used in k-ε or q^2-*l* modeling applied yet. The last term is the tke production term due to sub-grid urban canopy elements, where f_{urb} is the urban fraction of the grid cell, C_d is a drag coefficient, $a(z)$ is the canopy area density, and U and V are the horizontal components of the wind velocity. The absolute value signs on the velocities ensure that the drag term is always a source of turbulent kinetic energy. Little justification has been given for the form of the tke production term, except that it is dimensionally correct. Figures 22 and 23 show the impact of the drag term on tke levels in a mesoscale simulation over an urban area.

Similarly, drag terms have been added to the modeled dissipation equation (e.g., Maruyama [63])

$$\begin{aligned}\frac{D\varepsilon}{Dt} &= \frac{c_{\varepsilon 1} P_k}{\rho}\frac{\varepsilon}{tke} - c_{\varepsilon 2}\frac{\varepsilon^2}{tke} + \frac{\partial}{\partial x_j}\left(\frac{K_m}{\sigma_\varepsilon}\frac{\partial \varepsilon}{\partial x_j}\right) \\ &+ f_{urb} C_d a(z)\frac{\varepsilon}{tke}\left(|U|^3 + |V|^3\right)\end{aligned}, \quad (22)$$

and the modeled q^2-*l* equation (e.g., Yamada [60])

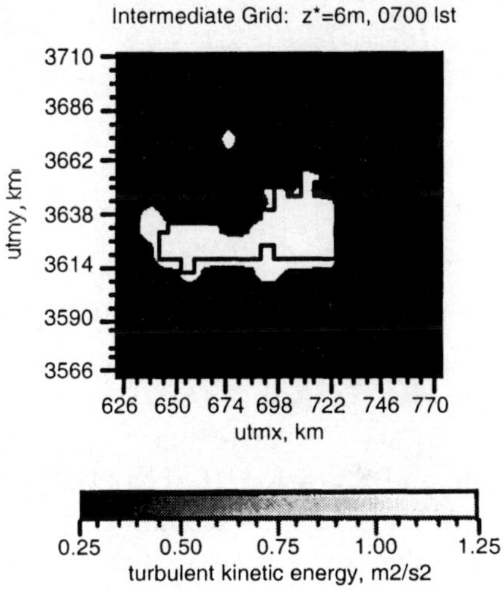

Figure 22: Model-computed tke over the Dallas/Ft. Worth area (black outline) showing high tke values over the urban area (Brown and Muller [86]).

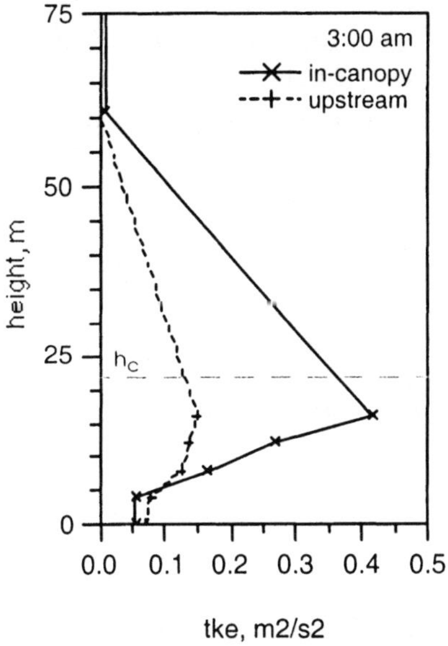

Figure 23: Comparison of tke profiles computed within and upstream of urban canopy (Brown and Williams [62]).

$$\frac{Dtke \cdot l}{Dt} = \frac{\partial}{\partial z}\left[\frac{tke^{1/2} \cdot l \cdot \widetilde{S}_q}{2}\frac{\partial tke \cdot l}{\partial z}\right] + \frac{l \cdot E_1}{2}\left[<u_i u_j>\frac{\partial U_i}{\partial x_j} + \beta g <w\theta>\right]$$
$$-\frac{tke^{3/2}}{2B_1}\left[1 + E_2\left(\frac{l}{kz}\right)^2\right] + f_{urb} \cdot C_d \cdot a(z) \cdot l \cdot \left(|U|^3 + |V|^3\right), \quad (23)$$

where $c_{\varepsilon 1}$, $c_{\varepsilon 2}$, σ_e, E_1, E_2, and B_1 are empirically-determined coefficients and S_q is a stability-dependent factor. Note that the dissipation eqn. (22) does not have a buoyancy term included.

Eqns. (21) and (22) are similar to those proposed by forest canopy modelers (e.g., Wilson and Shaw [81])). Liu et al. [82] proposed adding a sink term to eqns. (21) and (22) to account for the accelerated cascade of tke to small scales due to small-size leaf foliage. It is not clear whether a sink term is needed for urban canopies. Ayotte et al. [88] incorporated drag terms into a 2nd order turbulence closure model for vegetative canopy flow. From this type of modeling, one might be able to learn how to better parameterize the first-order closure models described above.

For mesoscale models using empirical formulae for determination of the eddy diffusivity, K_m can be modified as a function of height using the drag approach to account for building obstacle effects. Based on an approximated solution to a simplified form of the turbulent kinetic energy equation, Sorbjan and Uliasz [59] obtained the following formula for the eddy diffusivity within the urban canopy for neutral conditions:

$$K_m = l^2 \left\{ \left| \frac{dU}{dz} + \frac{dV}{dz} \right|^3 + \frac{E_c}{l^2} \right\}^{1/3}, \quad (24)$$

where E_c is the rate of urban-generated turbulent kinetic energy defined by

$$E_c = 1/2 \cdot C_d \cdot a(z) \cdot \left(U^2 + V^2\right)^{3/2}. \quad (25)$$

This formulation results in a larger eddy diffusivity above urban surfaces with precise values determined by the specification of the drag coefficient

C_d and the canopy area density *a(z)*. Sorbjan and Uliasz [59] did not modify *l* within the urban canopy.

3.4 Heat equation

In this section, two possible modifications to account for urban effects in the atmospheric conservation equation for heat are discussed. The first is radiative heating or cooling aloft due to pollution associated with urban activities. The second is related to ambiguities in how to define the urban surface and whether it is best to account for urban canopy heating terms in the surface energy budget or in the atmospheric heat conservation equation.

3.4.1 Heating aloft associated with urban pollution

Clearly, urban haze and smoke can impact the transference of short and long wave energy through the atmosphere and thus impact the surface radiation budget. In addition, pollutant layers above the city can absorb energy resulting in warming of the air aloft and can emit energy resulting in local cooling. The modification of the thermal structure could lead to changes in the heat island circulation and intensity.

In a 2-d numerical modeling study, Atwater [19] found that carbonaceous aerosols resulted in a small amount of warming (0 to 0.5 K) aloft over urban areas during the daytime. Interestingly, however, he found that over rural areas a 0.5-1.0 K heating aloft occurred due to absorption of solar radiation. The increased stability reduced the growth of the mixed layer over the rural area by one-half. The small amount of heating over the urban area did not affect the mixed-layer growth and Atwater [19] hypothesized that this was because the heat flux associated with the pollutant layers was negligible compared to the anthropogenic and other urban heat flux terms. Using a more sophisticated Mie-scattering radiation scheme, Yoshida and Kunitomo [89] found 0.5 – 1.5 K daytime heating aloft in a 1-d simulation of the atmosphere. The amount of warming was determined to be a function of aerosol diameter.

3.4.2 Urban canopy heating terms

The Reynolds-averaged form of the conservation equation for heat is given by

$$\frac{D\Theta_V}{Dt} = -\frac{\partial}{\partial x_j} < u_j \theta_V > + \frac{1}{\rho C_p} \frac{\partial R_N}{\partial z} , \tag{26}$$

where Θ_v is mean virtual potential temperature, $<u_j\theta_v>$ is the turbulent heat flux, ρ is the density of air, C_p the specific heat capacity of air, and R_N is the net radiation ($R_{S\downarrow} - R_{S\uparrow} + R_{L\downarrow} - R_{L\uparrow}$). The last term represents the flux divergence of net radiation and can be either a source or sink of heat.

For a mesoscale model with adequate vertical resolution to resolve the urban canopy layer, but with horizontal grid spacing on the order of kilometers, it becomes somewhat ambiguous as to where to define the surface and how to spatially attribute heat sources due to buildings and urban activities. Rather than add the anthropogenic heating term to the surface energy budget equation, Ca et al. [65], Urano et al. [64], Brown and Williams [62], and Taha [57] assumed that the heat is released directly to the air. In addition, Brown and Williams [62] assumed that the rooftop heat flux was emitted directly to the air. As shown in Figure 24, the model grid cell can be divided into an urban fraction (f_{urb}) and non-urban fraction ($1-f_{urb}$), and then the urban canopy fraction can be further subdivided into roof fraction (f_{roof}) and "between-building" fraction (f_{cnyn}). A modified heat equation is then obtained:

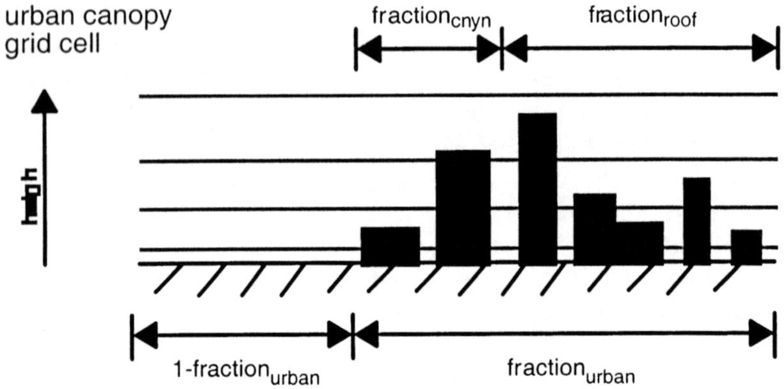

Figure 24: Illustration of how each urban grid cell is divided up into urban and non-urban fractions. The urban fraction is further sub-divided into roof and canyon fractions. (Brown and Williams [62].)

$$\frac{D\Theta_V}{Dt} = -\frac{\partial}{\partial x_j} < u_j \theta_V > + \frac{(1-f_{urb})}{\rho C_p}\frac{\partial R_N}{\partial z} + \left(1+\frac{1}{B}\right)^{-1}\left(\frac{f_{cnyn}}{\rho C_p}\frac{\partial R_{N_c}}{\partial z} + \frac{f_{roof}}{\rho C_p}\frac{q_{roof}}{\Delta z}\right) + \frac{f_{urb} q_{urb}}{\rho C_p \Delta z} \tag{27}$$

where R_{Nc} is the net radiation within the canopy, B is the Bowen ratio (sensible heat to latent heat flux ratio), $(1+1/B)^{-1}$ is the fraction of energy that goes into heating the air, q_{roof} is the amount of energy emanating from the rooftop, and q_{urb} is the anthropogenic heat flux in a particular height interval. In most prior atmospheric modeling studies, the roof and anthropogenic heating terms have been accounted for in the surface energy budget equation, not in the heat equation.

Urano et al. [64] included a height-varying anthropogenic heat flux in the heat equation; however, they sub-divided the anthropogenic heat flux into the heat and surface energy budget equations. They split the anthropogenic heat flux from vehicles equally between the heat and surface energy budget equations, and allocated all the anthropogenic heat flux from the first floor of buildings to the surface energy budget equation and above the first floor to the heat equation. They found that the vertical distribution of heat significantly impacted surface temperatures in their model simulations of the Tokyo urban heat island. In principal, the anthropogenic heat flux could be multiplied by the $(1+1/B)^{-1}$ term, however, in most studies it has been assumed that all the anthropogenic energy goes into heating the air.

Brown and Williams [62] calculated the rooftop heat flux from

$$q_{roof} = R_{N_{hc}} - R_{L_{roof}} = (1-\alpha_{roof})R_{S\downarrow} + \Delta R_{L_{hc}} - \varepsilon_{roof}\sigma T_{roof}^{4}, \tag{28}$$

where α_{roof} is the rooftop albedo, ε_{roof} is the rooftop emissivity, and T_{roof} is the rooftop temperature. Implied in eqn. (28) is the assumption that the rooftop is infinitely thin and all solar radiation absorbed by the roof is immediately reemitted as longwave and sensible heat, i.e., the roof has no heat storage capacity. Without further information, they assumed that the rooftop longwave radiated at air temperature, clearly an underestimation during the daytime. For a better approximation, one could solve a "surface" energy budget equation at rooftop and solve for the rooftop temperature T_{roof}. In addition, inclusion of the heat flux from building walls into eqn. (23) or at a minimum accounting for their extra surface area in the surface energy budget equation might be important as well (e.g., Voogt and Oke [90]).

The net radiation within the canopy is assumed to exponentially decay as the ground is approached:

$$R_{N_C}(z)=R_{N_{hc}}\exp[-k\cdot bai(z)] \,, \tag{29}$$

where R_{Nhc} is the net radiation at canopy top, k is an extinction coefficient, and *bai(z)* is the cumulative "building" area index (analogous to the leaf area index) defined as

$$bai(z)=\int_z^{h_c} a(z')dz' \,, \tag{30}$$

where $a(z)$ is the canopy area density (note that the canopy area density for drag, e.g., eqn. (17), is the surface area of the buildings perpendicular to the mean wind per unit volume, while here the surface area is the plan area of the buildings per unit volume). The extinction coefficient k should be a function of time of day as the sun angle changes. The net radiation approach described here is a modification of the forest canopy parameterizations proposed by Yamada [60]. For that reason, it is considered to be only a crude approximation for the net radiation attenuation within the urban canopy. Methods based on sky-view factor or derived from explicit radiation balance modeling around clusters of buildings should be considered in order to improve the radiation attenuation formulation (e.g., Oke et al. [45], Mills [46], Ca et al. [65]).

Figure 25 shows near-surface temperatures over the El Paso/Ciudad Juarez region from a simulation that utilizes the urban heating parameterizations described above. A several degree urban heat island intensity is apparent. Figure 26 shows the vertical potential temperature profile of an idealized urban heat island simulation. The profile shows a well-mixed region within the canopy.

3.5 Surface energy balance

The surface energy balance is solved in order to define the surface temperature, a boundary condition for the mesoscale model. Typically,

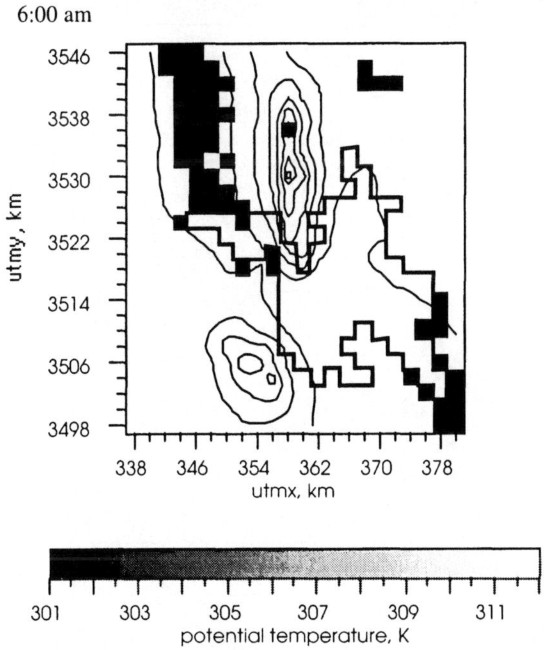

Figure 25: Near-surface temperatures computed over the El Paso/Ciudad Juarez region showing urban heat island. (Brown and Williams [91].) Urban area denoted by black outline.

the surface energy budget is defined at the ground plane by making a balance equation for the amount of incoming and outgoing energy. For urban areas, it is difficult to specify exactly where the ground is located: is it at street level running under the buildings, a surface that follows the contours of the building surfaces, a horizontal plane that is defined at rooftop? In principal, separate energy balance equations could be solved at rooftop level, on building sidewalls, and at street level as in the microscale models of Arnfeld [43], Mills [46] and Ca et al [65]. This is not practical for typical mesoscale modeling applications, as the level of detail needed to define the building layout for an entire city is typically not readily available (if one could define "typical" building configurations for different types of urban land use (e.g., downtown, high density residential, low density residential, commercial, residential)

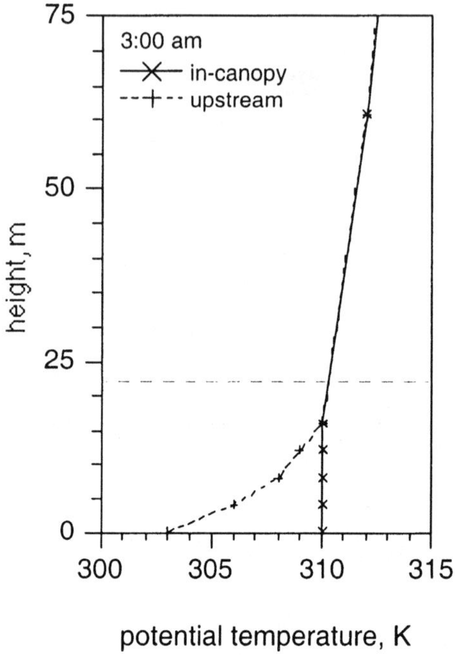

Figure 26: Comparison of model-computed profiles of potential temperature within and upstream of the urban canopy. (Brown and Williams [62].)

and/or as a function of height-to-width ratios or canopy area density, then this approach might be more practical for mesoscale applications).

Equation (2) has been used by many mesoscale modelers for defining the surface energy budget over urban areas, with and without the canopy storage and anthropogenic heat flux terms. This method places a control volume around the canopy elements, computes the heat fluxes at the top and sides of the box, and assumes that there is a canopy heat storage term within the volume. A simple, straightforward approach for accounting for the urban impact on the surface energy budget eqn. (2) has been outlined by Taha [57] and Perez-Garcia and Nickerson [58]. They incorporated the objective hysteresis model of Grimmond et al. [92]. This model defines the urban canopy storage term ΔQ_S in eqn. (2) as a function of the net radiation Q* based on experimental measurements:

$$\Delta Q_S = \sum_{i=1}^{n} \left(a_i Q^* + b_i \frac{\partial Q^*}{\partial t} + c_i \right) A_i \,, \qquad (31)$$

where *a*, *b*, and *c* are empirical coefficients associated with land use, *n* is total number of landuse types, and A_i is the area fraction of the particular land use within the area of interest (e.g., model grid cell). Taha [57] incorporated the anthropogenic heat flux term into the lowest grid cell of the heat equation, however, and added a soil heat flux term Q_G in the surface energy budget and scaled it by the non-urban grid cell fraction $(1-f_{urb})$.

However, although this method of enclosing the urban canopy elements within a control volume is conceptually appealing and useful for understanding the partitioning of the energy budget, it is not particularly amenable to mesoscale model applications where one wants to solve for the meteorological fields *within* the urban canopy. In Brown and Williams [62], the surface energy budget equation is defined at ground-level and is slightly different than that given by eqn. (2):

$$\begin{aligned} Q^* &= (R_{L\downarrow} - R_{L\uparrow}) + (R_{S\downarrow} - R_{S\uparrow}) \\ &= (1\text{-}f_{urb})\{(1\text{-}\alpha_G)R_{S\downarrow} + \Delta R_L\} + f_{cnyn} R_{N_{hc}} \exp[-k \cdot bai(0)] \,, \qquad (32) \\ &= Q_H + Q_E + Q_G \end{aligned}$$

where α_G is the ground albedo in the non-urban area, and ΔR_L is the net long wave radiation, R_{Nhc} is the net radiation at canopy height, *k* is an extinction coefficient, and $bai(z{=}0)$ is the cumulative building area index as defined in eqn. (30). As discussed in Section 3.4, the model grid cell was divided into urban and non-urban fractions, and then the urban fraction was subdivided into canyon and rooftop fractions. The first term on the right hand side on the second line represents the net long wave and net short wave radiation in the non-urban fraction of the grid cell, while the last term represents the fraction of the solar and long wave radiation that reaches ground-level in the urban canyon fraction of the grid cell. Following the forest canopy work of Yamada [60], it is assumed that the grid-cell averaged net radiation falls off exponentially in the urban canopy. The solar radiation that is intercepted by the rooftops is assumed to be converted to heat and then directly emitted into the air through the heat eqn. (27); therefore the rooftop fraction of the solar radiation is not put into the surface energy budget eqn. (32).

As in many of the modeling studies cited in Section 2, the advective term ΔQ_A was omitted from surface energy budget eqn. (32) because the mesoscale model explicitly accounts for heat advection. In addition, the anthropogenic heat flux Q_f was put into the heat eqn. (27) rather than into the surface energy budget eqn. (32). The reasoning for this was that the anthropogenic heat flux is emitted into the air (e.g., heat loss from buildings, vehicle-related heat emissions) and its distribution with height could be better accounted for in the heat conservation equation. The anthropogenic flux term will still impact the surface energy budget, but indirectly through heating of the air aloft and changing the magnitude of the flux terms, most notably the sensible heat flux Q_H.

The soil or ground heat flux can be approximated by

$$Q_G = K_s \left. \frac{\partial T_s}{\partial z_s} \right|_G , \tag{33}$$

where K_s is the soil diffusivity, and the soil temperature gradient at the ground can be obtained solving a one dimensional heat conduction equation in the ground. K_s is typically modified to account for the diffusivity properties of urban materials. One might also consider variation with depth in the urban materials; for example, a street would have an asphalt layer overlying a road bed of packed soil, or a wall might have an outer plaster layer and then wood, insulation, and/or dead air space underneath.

Here we wish to clarify that Q_G is the energy flux into and out of ground and urban surfaces (asphalt, concrete, masonry, etc.), while Q_f is the energy flux from human activities, primarily the heat released from energy use. So Q_f includes heated air escaping from buildings to the atmosphere, while Q_G also includes energy fluxes from the building surfaces, but just that component associated with the heat received from the atmosphere (in the form of short and long wave radiation). In practice, it may be difficult to distinguish between the two fluxes when measured experimentally. In addition, it might be prudent to subtract the roof fraction from the soil flux term in eqn. (32), so that only the ground heat flux occurs for the non-urban and street canyon ground surfaces.

However, as pointed out by Voogt and Oke [90], the extra surface area associated with the non-anthropogenic heat flux into and out of building walls should be considered in the heat balance eqn. (32). Hence, the overestimation of the surface area in the first case and the underestimation in the second case may offset one another.

As an alternative to solving the heat conduction equation, the simpler force-restore method can be used to compute the surface temperature (e.g., Kimura and Takahashi [52], Carissimo [54]):

$$\frac{\partial \Theta_G}{\partial t}=\frac{2\omega}{k \cdot C_p}(\Delta \mathrm{R_S}+\Delta \mathrm{R_L}-\mathrm{Q_H}-\mathrm{Q_E})-\omega(\Theta_G-\Theta_D), \tag{34}$$

where Θ_G is the ground surface temperature, $\omega = 2\pi/24$, k is soil heat conductivity, C_p is soil heat capacity, and Θ_D is the deep soil temperature.

The final terms in eqn. (32) that need to be evaluated are the sensible and latent heat fluxcs

$$Q_H=-\rho C_p u_* \Theta_* \quad \text{and} \tag{35a}$$

$$Q_E=-\rho L_v u_* q_* . \tag{35b}$$

These are typically evaluated using Monin-Obukhov similarity relationships to define the friction velocity u_*, the temperature scale Θ_*, and the water vapor scale q_* , i.e., eqn. (8) and

$$\Theta(z)-\Theta_G=\frac{\Theta_*}{\kappa}\left\{\ln\left(z/z_{o_t}\right)-\Psi_h(z/L)\right\}, \tag{36a}$$

$$Q_W(z)-Q_{W_G}=\frac{q_*}{\kappa}\left\{\ln\left(z/z_{o_v}\right)-\Psi_v(z/L)\right\}, \tag{36b}$$

where L_v is the latent heat of vaporization, Θ_G and Q_{WG} are ground level temperature and mixing ratio of water vapor, respectively, z_{0t} and z_{0v} are the roughness lengths for temperature and water vapor, respectively, and Ψ_h and Ψ_v are stability correction functions. The heat balance eqn. (32) can be used iteratively with eqn. (36a) to solve for the ground-level temperature Θ_G.

Determination of the latent heat flux is problematic since its magnitude depends on Q_{WG}, which depends on the underlying moisture content of the surface. Oke [23] gives equations for the surface water balance, but since the moisture content of urban areas is a function of recent weather, watering practices, and urban cover, it is very difficult to simulate the water balance. Often rather simplifying assumptions are used to specify the latent heat flux. For example, Brown and Williams [62] and Ca et al. [65] specify a Bowen ratio for the urban canopy, so that the latent heat flux is calculated directly from the sensible heat flux:

$$Q_L = Q_H / B \,. \tag{37}$$

Mochida et al. [53], Kimura and Takahashi [52], and Atwater [19] utilized similar moisture availability equations to compute Q_{Wg}. Mochida et al.'s [53] version is

$$Q_{wg} = q_g = q(z) + \beta(q_{sat} - q(z)) \,, \tag{38}$$

where β, the moisture availability constant, is specified for different land use types and varies from 0 for dry surfaces to 1 for water bodies and q_{sat} is the saturation water vapor mixing ratio at the surface temperature. Basically eqn. (38) says that the ground can act as a source of moisture (q_{sat}) and that the flux of moisture is proportional to the difference between the amount of moisture in the ground and the amount in the air (q_{sat}-$q(z)$). In general, β is small for urban land use and larger for vegetated and moisture laden land use.

An important item to note is that Monin-Obukhov similarity is being used to specify boundary conditions at the surface *within* the urban canopy. As discussed earlier, Monin-Obukhov similarity falls apart in the canopy layer. However, without alternative "theories" there is no other recourse as the surface boundary conditions must be specified in order to run the numerical model. Research on alternative approximations for how to specify area-averaged surface fluxes within the canopy layer would be extremely valuable.

4 Practical difficulties

4.1 Parameter specification

The drawback of potentially more universal, in-depth urban parameterizations is that there is often a corresponding need to specify more or new parameters. A major challenge lies in determining the different coefficients that are needed for the urban canopy parameterizations. Data is often sparse and may contain large uncertainties. As outlined in Sections 3.2 and 3.3, the advantage of using the urban roughness approach to account for some of the effects of the urban canopy is that there are numerous references on the roughness length z_0 and displacement height d for urban areas. However, there is still quite a bit of uncertainty in the specification of even these parameters (e.g., Grimmond and Oke [77]) and the method cannot account for many of the within-canopy effects of buildings. In this section, we give typical ranges for some of the parameters needed to implement the urban canopy drag, heating, and surface energy budget parameterizations as given in eqns. (17), (21), (27) and (32).

Land use Urban land use needs to be specified as a function of space within the mesoscale modeling domain. In the approach of Brown and Williams [62], the fraction of the horizontal grid cell area occupied by urban land use f_{urb} needs to be defined. In the approach of Ca et al. [65] and Maruyama [63] the fractional volume occupied by buildings needs to be determined for each grid cell. The area fractions can be evaluated from land use datasets which are often freely available online (e.g., the USGS Land use/Land Cover dataset). These datasets, however, can be outdated and there is often an ambiguity in determining the urban fraction based on the given land use categories. Determination of the building volume fractions would require specialized datasets or some mapping of typical land use categories into volume fractions.

Many of the urban parameters are a function of the urban land use type or building density and therefore it may be useful to break down urban land use into finer sub-divisions. Brown and Williams [62] have divided urban landuse into four classes: 1) downtown / city center, 2) industrial/commercial, 3) residential with mature trees, and 4) residential without mature trees. The principal characteristics distinguishing these categories are built vs. green space fractions, canopy height, and building density. The canopy height is utilized in defining the depth through which the urban canopy impacts the flow field through such parameters as the canopy area density $a(z)$. In addition, the surface energy budget and heat conservation equation parameterizations given by eqns. (27) and (32) require specification of the building rooftop and street canyon fraction of the grid cell area. Although these quantities are site-dependent, Table 1 provides typical ranges found in the literature.

Surface properties A number of land-use-dependent surface properties are needed for the urban surface energy balance eqn. (32). For example, an extinction coefficient or sky view factor should be utilized in determining the net radiation attenuation within the urban canopy. Specification of the Bowen ratio or moisture availability coefficient is especially problematic as their value depends on weather and city watering policy, but generally a larger Bowen ratio and a smaller moisture availability coefficient is expected as the urban fraction increases. Residential areas, in general, should have more moisture availability than downtown, industrial or commercial areas. A range of measured values is given in Table 1. One could try to model the urban water balance, but this is often beyond the realm of atmospheric mesoscale modeling.

Urban and rooftop albedo and emissivity measurements cover a wide range of values as well and generally have not been broken down by urban land use type. However, albedos specific to different types of

Table 1: Urban land use properties.

	downtown/city center	industrial / commercial	residential
f_{urban}	0.8 – 1.0	0.9 – 1.0	0.5 – 0.7
f_{roof}	0.3 – 0.4	0.3 – 0.4	0.15 – 0.25
f_{cnyn}	0.5 – 0.6 (est.)	0.6 – 0.7 (est.)	0.15 – 0.4
h_c [m]	15 – 100's	5 – 25	5 – 15
Bowen Ratio	1.5 – ∞	1.5 – ∞	0.5 – 1.0
urban albedo		0.10 – 0.27 (avg. 0.15)	
urban emissivity		0.85 – 0.96 (avg. 0.95)	
roof albedo		0.08 – 0.35	
roof emissivity		0.90 – 0.92	

building rooftops have been published in the literature (e.g., Oke [23]). There can be some ambiguity in what measured albedos represent, i.e., do the measurements include the effect of building wall reflections (a sort of area-average urban albedo) or are they representative of a particular surface (such as an asphalt road or a tile roof)? It is often stated that the albedo of urban areas is lower due to multiple reflections off building walls which effectively "capture" more of the incoming solar radiation. The specified albedo needs to be compatible with the particular urban parameterization. For example, an area-average urban albedo will not be appropriate for a method that tries to incorporate wall reflection implicitly or that computes rooftop albedo separately.

Molecular properties Urban values for density, heat capacity, and thermal conductivity are needed to compute the ground flux in the surface energy balance. Values for specific building materials (e.g., steel, brick, concrete, asphalt) can be found in the literature (e.g., Oke [23]). However, in general, area-average values are needed that consider the fraction of urban and vegetated land use within a grid cell, and the representative fractions of building materials. Brown and Williams [62] have used ρ =2300 kg/m^3, C_p = 880 J/(kg K), and $k = 1.2\times10^{-06}$ m^2/s for all urban land use types in the surface energy budget balance. When converted to thermal admittance ($\rho C_p k^{1/2}$), these values fall within the 800 – 3000 range given by Oke [23]. One might consider having separate molecular property values for walls, roofs and streets, for example, if the fractions of each within each grid cell can be determined.

Drag coefficient The drag force in the momentum eqn. (17) and the turbulent production term in the tke eqn. (21) are linearly dependent on the magnitude of the drag coefficient C_D. Hoerner [68] lists C_D values ranging from 0.7 to 1.5 for buildings of different sizes and shapes. Although there should be dependence on number, spacing, height, and shape of buildings, as well as approach flow angle, Brown and Williams [62] use an average value of 1.0 for all urban land use categories. One might also expect that the drag coefficient will decrease in regions where buildings are closely spaced and buildings are all about the same height, and increase when building heights change relative to one another. For comparison, Yamada [60] used a C_D = 0.2 for trees. Irvine et al. [93] and Shaw and Schumann [94] deduced values of 0.20 and 0.15, respectively, for forested areas.

Canopy area density The height dependence of the drag force and turbulent kinetic energy production is controlled through the canopy area density term $a(z)$. There are numerous measurements of $a(z)$ for different vegetative canopies (e.g., Arya [69], Watanabe and Kondo [95]), but we have not found $a(z)$ profiles in the literature for the urban canopy. However, in earlier studies Brown and Williams [62] hypothesized a pyramid-shaped canopy area density profile for a mixed distribution of buildings (see Figure 27). They use a mix of the pyramid and fir tree shapes for residential areas with mature trees which several authors have indicated may in fact be better described as a forest canopy (e.g., Oke [96]). The canopy area density profiles are described by analytical functions so that the cumulative canopy (leaf or building) area index eqn. (30) can be easily computed.

There is clearly a lot of uncertainty in the values of the above parameters. More analysis of urban datasets needs to be performed in order better to prescribe these parameters. In addition, high resolution modeling could be performed to determine some of the coefficients (e.g., the drag and extinction coefficients) as functions of building

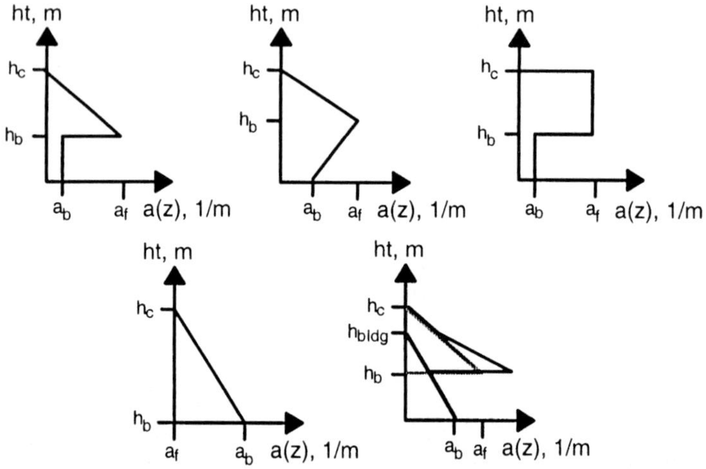

Figure 27: "Shape profiles" used for describing the canopy surface area density a(z) $[m^2/m^3]$. Four parameters are needed for describing the profiles: canopy height, base height, base surface area density, and foliage surface area density.

configuration, for example. Sensitivity tests, such as those being done by Leach and Chen [97], need to be performed as well in order to determine what coefficients need to be specified more precisely.

4.2 Validation

One of the major difficulties in testing the urban canopy parameterizations is in the comparison of area-averaged model results to measurements that are typically representative of a point in space. From point-to-point within the urban environment, the local impact of buildings and land use can vary dramatically. For example, the wind speed can be very strong on one street due to channeling, but on a side street be very weak and in the opposite direction due to street canyon vortex formation. Or, on the sunny side of a street overlying asphalt, the near-surface air temperature could be very warm, whereas on the opposite side of the street in the shade overlying a strip of vegetation the near-surface air temperature could be dramatically lower. Hence, individual measurements can be misleading when assessing the impact of the urban canopy and can prove difficult to use for model validation.

For urban canopy parameterization validation, field experiments need to be designed such that enough measurements are taken properly to represent an area average, or sensors should be ideally placed to represent an upstream area rather than a local microscale feature. However, in practice this is difficult to do due to limited resources.

Another option is to numerically simulate at high resolution the fluid flow, thermodynamics, and/or surface energy budget around groups of buildings. For example, a detailed radiation balance could be performed using models of the type developed, for example, by Arnfeld [43] and Mills [46] and then area-averaged results could be obtained for different building configurations. Figure 28 shows the area-averaged flow field produced by a computational fluid dynamics model around a 3-d array of obstacles. Velocity profiles computed with a mesoscale model incorporating urban parameterization could then be compared to the CFD result in order to test, for example, the drag coefficient value.

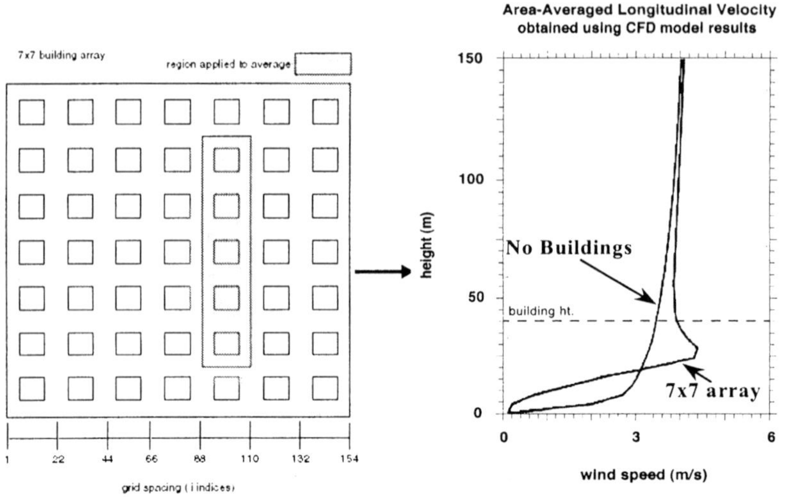

Figure 28: Top view of obstacle layout (left) and area-averaged velocity profiles computed by computational fluid dynamics model with and without obstacles (right). (Smith [98].)

5 Conclusions

Urban parameterizations developed for use in mesoscale meteorological models to account for the area-average effect of drag, turbulence production, heating, and surface energy budget modification induced by buildings and urban land use have been described. It was shown that the city affects the surface energy budget and the dynamic and thermodynamic flow fields. Urban areas are typically warmer than the surrounding countryside at night and the heat island may result in thermodynamically-driven winds. The buildings in urban areas also create drag and produce turbulence, effectively changing the flow dynamics within and above the urban canopy. Being able to simulate these urban effects in mesoscale models is important for many applications.

Modifications to the surface energy budget and heat equation to account for urban influences were described, including approaches that accounted for radiation attenuation, anthropogenic heating, and urban surface properties. Drag and turbulence parameterizations with different levels of sophistication were covered as well, including the surface roughness, attenuation, and the drag coefficient approaches. It was argued that the traditional surface roughness approach is not appropriate for some applications, specifically when one is interested in computing the area-averaged flow fields within the urban canopy. The attenuation approaches were found suitable as a method for diagnostic wind models or vertically coarse resolution prognostic simulations. Additional work, however, needs to be done in determining the coefficients for urban areas in order to make this approach practical. The drag coefficient approach was described as being useful for mesoscale modeling when the urban canopy layer is resolved by the model. In theory, it is capable of correctly simulating the area-average wind and tke profile variation with height within the canopy layer. The drawback to this approach is the number of parameters that need to be specified, some of which are not known with high precision. Uncertainties in parameter specification and difficulties with testing the parameterizations were covered. Future work using high resolution CFD and surface energy balance models was proposed in order to test the parameterizations and specify some of the parameters.

Acknowledgements

The author wishes to thank Mike Williams for insightful discussions on urban canopy parameterizations, Cathrin Muller for help with digitizing and plotting many of the figures, and Gerald Streit, Steven Burian, Matthias Roth, and Mathias Rotach for their editorial comments.

References

[1] Bornstein, R., Mean diurnal circulation and thermodynamic evolution of urban boundary layers. *Modeling the Urban Boundary Layer*, Am. Met. Soc., Boston, 1987.

[2] Hosker, R., Flow and diffusion near obstacles. *Atmospheric Science and Power Production*, USDOE, 1984.

[3] Hanna, S. and Chang, J., Boundary-layer parameterizations for

applied dispersion modeling over urban area. *Bound. Layer Meteor.*, **58**, pp. 229-259, 1992.

[4] Hanna, S., Ramsdell, J. and Cramer, H., Urban Gaussian diffusion parameters. *Modeling the Urban Boundary Layer*, Am. Met. Soc., Boston, 1987.

[5] Hosker, R., The effects of buildings on local dispersion. *Modeling the Urban Boundary Layer*, Am. Met. Soc., Boston, 1987.

[6] Yamartino, R., Strimaitis, D. and Messier, T., Modification of highway air pollution models for complex site geometries, vol. 1, Data analyses and development of the CPB-3 model. *Rep. No. FHWA-RD-89-112*, 1989.

[7] Brown, M. and Streit, G., Emergency Responders' 'Rules-of-Thumb' for Air Toxics Releases in Urban Environments. *Los Alamos Report LA-UR-98-4539*, 1999.

[8] Chandler, T., *The Climate of London*, Hutchison, London, 1968.

[9] Daigo, M. & Nagao, T., *Urban Climatology*, Asakura-Shoten, Tokyo, 214 pp., 1972.

[10] Landsberg, H., Inadvertent atmospheric modification through urbanification. *Weather and Climate Modification*, W. Hess, ed., John Wiley and Sons, New York, pp. 726-763, 1974.

[11] Oke, T., Review of Urban Climatology, 1969-1973, *WMO Publ. Tech. Note 134*, 132 pp., 1974.

[12] Bornstein, R., Observations of the urban heat island effect in New York City. *J. Appl. Meteor.*, **7**, pp. 575-582, 1968.

[13] Clarke, J., Nocturnal urban boundary layer over Cincinnati, Ohio. *Mon. Wea. Rev.*, **97**, pp. 582-589, 1969.

[14] Ludwig, F., Urban temperature fields. *WMO Publ. Tech. Note 108*, pp. 80-107, 1970.

[15] Oke, T. and East, C., The urban boundary layer in Montreal. *Bound.-Layer Meteor.*, **1**, pp. 411-437, 1971.

[16] Angel, J., Pack, D., Dickson, C. and Hoecker, W., Urban influence

on nighttime airflow estimated from tetroon flights. *J. Appl. Meteor.*, **10**, pp. 194-204, 1971.

[17] Ackerman, B., METROMEX: Wind fields over the St. Louis metropolitan area. *Bull. Amer. Meteor. Soc.*, **55**, pp. 93-95, 1974.

[18] Myrup, L., A numerical model of the urban heat island. *J. Appl. Meteor.*, **8**, pp. 896-907, 1969.

[19] Atwater, M., Thermal effects of urbanization and industrialization in the boundary layer: a numerical study. *Bound.-Layer Meteor.*, **3**, pp. 229-245, 1972.

[20] McElroy, J., A numerical study of the nocturnal heat island over a medium-sized mid-latitude city (Columbus, Ohio). *Bound.-Layer Meteor.*, **3**, pp. 442-453, 1973.

[21] Bornstein, R., The two-dimensional URBMET urban boundary layer model. *J. Appl. Meteor.*, **14**, pp. 1459-1477, 1975.

[22] Oke, T., The surface energy budgets of urban areas. *Modeling the Urban Boundary Layer*, Am. Met. Soc., Boston, 1987.

[23] Oke, T., *Boundary-Layer Climates.* Routledge, London, 1987.

[24] Lowry, W., The climate of cities. *Scientific American*, 1967.

[25] Liu, J., Arya, S., Snyder, W. and Lawson, R., A laboratory simulation of turbulent convection over an urban heat island. *J. Appl. Meteor,* **36**, pp1392-1402, 1997.

[26] Sahashi, K., Hieda, T. and Yamashita, E., Nitrogen-oxide layer over the urban heat island in Okayama City. *Atmos. Env.*, 30, pp. 531-535, 1996.

[27] Saitoh, S., Shimada, T. and Hoshi, H., Modeling and simulation of the Tokyo urban heat island. *Atmos. Env.*, **30**, pp. 3431-3442, 1996.

[28] Tapper, N., Urban influences on boundary layer temperature and humidity: results from Christchurch, New Zealand. *Atmos. Env.*, **24B**, pp. 19-27, 1990.

[29] Shreffler, J., Detection of centripetal heat-island circulations from tower data in St. Louis. *Bound.-Layer Meteor.*, **15**, pp. 229-242,

1978.

[30] Kuttler, W. and Romberg, E., The occurrence and effectiveness of country breezes by means of wind tunnel and in-situ measurements. *Proc. 9th World Clean Air Congr.*, Montreal, Quebec, 1992.

[31] Oke, T., Canyon geometry and the nocturnal urban heat island: comparison of scale model and field observations. *J. Climatol.*, **1**, pp. 237-254, 1981.

[32] Oke, T. and Hannell, F., The form of the urban heat island in Hamilton, Canada. *Urban Climates, WMO Tech. Note 108*, pp. 113-126, 1970.

[33] Oke, T., The distinction between canopy and boundary layer urban heat islands. *Atmos.,* **14**, pp. 268-277, 1976.

[34] Hogstrom, U., Taesler, R., Karlsson, S., Enger, L. and Hogstrom, A., The Uppsala Urban Meteorology Project. *Bound.-Layer Meteor.*, **15**, pp. 69-80, 1978.

[35] Bowne, N. and Ball, J., Observation comparison of rural and urban boundary layer turbulence. *J. Appl. Meteor.*, **9**, pp. 862-873, 1970.

[36] Godowitch, J., Characteristics of vertical turbulent velocities in the urban convective boundary layer. *Bound.-Layer Meteor.*, **35**, pp. 387-407, 1986.

[37] Pendergrass, W. and Arya, S., Dispersion in neutral boundary layer over a step change in surface roughness – I. Mean flow and turbulence structure. *Atmos. Env.*, **18**, pp. 1267-1279, 1984.

[38] Theurer, W., Baechlin, W. and Plate, E., Model study of the development of boundary layers above urban areas. *J. Wind Eng. & Ind. Aerodyn.*, **41-44**, pp 437-448, 1992.

[39] Brown, M., Lawson, R., DeCroix, D. and Lee R., Mean flow and turbulence measurements around a 2-d array of buildings in a wind tunnel. *11th AMS Appl. of Air Poll. Meteor. Conf.*, Long Beach, Ca., 2000.

[40] Roth, M., Vertical structure of turbulence over cities. *2nd AMS Urban Env. Symp.*, Albuquerque, NM, 1998.

[41] Rotach, M., Profiles of turbulence statistics in and above and urban street canyon. *Atmos. Env.*, **29**, pp. 1473-1486, 1995.

[42] Oikawa, S. and Meng, Y., A field study of diffusion around a model cube in a suburban area. *Bound.-Layer Meteor.*, **84**, pp. 399-410, 1997.

[43] Arnfeld, A., An approach to the estimation of the surface radiative properties and radiation budgets of cities. *Phys. Geog.* **3**, pp. 97-122, 1982.

[44] Murakami, S., Current status and future trends in computational wind engineering. *J. Wind Eng. & Ind. Aerodyn.*, **67** & **68**, pp. 3-34, 1997.

[45] Oke, T., Johnson, G., Steyn, D. and Watson, I., Simulation of surface urban heat islands under "ideal" conditions at night, Part 2: Diagnosis of causation. *Bound.-Layer Meteor.*, **56**, pp. 339-358, 1991.

[46] Mills, G., An urban canopy-layer climate model. *Theor. Appl. Climatol.*, **57**, pp. 229-244, 1997.

[47] Delage, Y. and Taylor, P., Numerical studies of heat island circulations. *Bound.-Layer Meteor.*, **1**, pp. 201-226, 1970.

[48] Vukovich, F., Dunn, J. and Crissman, B., A theoretical study of the St. Louis heat island: the wind and temperature distribution. *J. Appl. Meteor.*, **15**, pp. 417-440, 1976.

[49] Hjelmfelt, M., Numerical simulation of the effects of St. Louis in mesoscale boundary layer airflow and vertical air motion: simulations of urban vs. non-urban effects. *J. Appl. Meteor.*, **21**, pp. 1239-1257, 1982.

[50] Schultz, P. and Warner, T., Characteristics of summertime circulations and pollutant ventilation in the Los Angeles basin. *J. Appl. Meteor.*, **21**, pp. 672-682, 1982.

[51] Byun, D. and Arya, S., A two-dimensional mesoscale numerical model of an urban mixed layer - I. Model formulation, surface energy budget, and mixed layer dynamics. *Atmos. Env.*, **24A**, pp. 829-844, 1990.

[52] Kimura, F. and Takahashi, S., The effects of land-use and anthropogenic heating on the surface temperature in the Tokyo metropolitan area: a numerical experiment. *Atmos. Env.*, **25B**, pp. 155-164, 1991.

[53] Mochida, A., Murakami, S., Ojima, T., Kim, S., Ooka, R. and Sugiyama, H., CFD analysis of mesoscale climate in the greater Tokyo area. *J. Wind Eng. Ind. Aerodyn.*, **67** & **68**, pp. 459-477, 1997.

[54] Carissimo, B., Numerical simulation of meteorological conditions for peak pollution in Paris. *22nd NATO/CCMS Conf.*, Clermont-Fernand, France, 1997.

[55] Kitada, T., Okamura, K. and Tanaka, S., Effects of topography and urbanization on local winds and thermal environment in the Nohbi Plain, coastal region of central Japan: a numerical analysis by mesoscale meteorological model with a k-ε turbulence model. *J. Appl. Meteor.*, **37**, pp. 1026-1046.

[56] Hafner, J. and Kidder, S., Urban heat island modeling in conjunction with satellite-derived surface/soil parameters. *J. Appl. Meteor.*, **38**, pp. 448-465.

[57] Taha, H., Modifying a mesoscale meteorological model to better incorporate urban heat island storage: a bulk parameterization approach. *J. Appl. Meteor.*, **38**, pp. 466-473.

[58] Perez-Garcia, I. and Nickerson, E., The applicability of a mesoscale model in the Valley of Mexico during extreme air pollution episodes, personal communication, 1996.

[59] Sorbjan, Z. and Uliasz, M., Some numerical urban boundary-layer studies. *Bound.-Layer Meteor.*, **22**, pp. 481-502, 1982.

[60] Yamada, T., A numerical study of turbulent airflow in and above a forest canopy. *J. Met . Soc. Japan*, **60**, 439-454, 1982.

[61] Williams, M., Brown, M., Cruz, X., Sosa, G. and Streit, G., Development and testing of meteorology and air dispersion models for Mexico City. *Atmos. Env.*, **29**, pp. 2929-2960, 1995.

[62] Brown M. and Williams, M., An urban canopy parameterization for

mesoscale meteorological models. *AMS 2nd Urban Env. Symp.*, Albuquerque, NM, 1998.

[63] Maruyama, T., Surface and inlet boundary conditions for the simulation of turbulent boundary layer over complex rough surfaces. *J. Wind. Eng. Ind. Aerodyn.*, **81**, pp. 311-322, 1999.

[64] Urano, A., Ichinose, T. and Hanaki, K., Thermal environment simulation for three dimensional replacement of urban activity. *J. Wind. Eng. Ind. Aerodyn*., **81**, pp. 197-210, 1999.

[65] Ca, V., Asaeda, T. and Ashie, Y., Development of a numerical model for the evaluation of the urban thermal environment. *J. Wind Eng. Ind. Aerodyn*., **81**, pp. 181-196, 1999.

[66] Whitaker, S., *Introduction to Fluid Mechanics*, Robert Krieger Pub. Co., Malabar, FL, 1984.

[67] Panton, R., *Incompressible Flow*, John Wiley and Sons, New York, NY, 1984.

[68] Hoerner, S., *Fluid-Dynamic Drag*, self-published, Lib. Congress Cat Card #64-196666, 1965.

[69] Arya, S., *Introduction to Micrometeorology*, Academic Press, San Diego, CA, 1988.

[70] Plate, E., *Aerodynamic Characteristics of Atmospheric Boundary Layers*, AEC Critical Reviews, USDOE, TID-25465, 1971.

[71] Hall, D., McDonald, R., Walker, S. and Spanton, A., Measurements of dispersion within simulated urban arrays – a small scale wind tunnel study, *BRE Client Report CR 178/96*, Building Research Establishment, Garston, Watford, 1996.

[72] Letau, H., Note on aerodynamic roughness parameter estimation on the basis of roughness element description. *J. Appl. Meteor*., **8**, pp. 828-832, 1969.

[73] Raupach, M., Drag and drag partition on rough surfaces. *Bound.-Layer Meteor*., **60**, pp. 375-395, 1992.

[74] Bottema, M., Urban roughness modeling in relation to pollutant dispersion. *Atmos. Env*., **31**, pp. 3059-3075, 1997.

[75] McDonald, R., Griffiths, R. and Hall, D., An improved method for the estimation of surface roughness of obstacle arrays. *Atmos. Env.*, **32**, pp. 1857-1864, 1998.

[76] Duijm, N., Estimates of roughness parameters for arrays of obstacles. *Bound.-Layer Meteor.*, **91**,

[77] Grimmond, S. and Oke, T., Aerodynamic properties of urban areas derived from analysis of surface form. *J. Appl. Meteor.*, **38**, pp. 1262-1292, 1999.

[78] Cionco, R., A wind-profile index for canopy flow. *Bound.-Layer Meteor.*, **3**, pp. 255-263, 1972.

[79] Pielke, R., *Mesoscale Meteorological Modeling*, Academic Press, San Diego, CA, 1994.

[80] Cionco, R. and Ellefsen, R., High resolution urban morphology data for urban wind flow modeling. *Atmos. Env.*, **32**, pp. 7-17, 1998.

[81] Wilson, N. and Shaw, R., A higher order closure model for canopy flow. *J. Appl. Meteor.*, **16**, pp. 1197-1205, 1977.

[82] Liu, J., Chen, J., Black, T. and Novak, M., E-e modeling of turbulent airflow downwind of a model forest edge. *Bound.-Layer Meteor.* **77**, 21-44, 1996.

[83] Hirt, C., Volume-fraction techniques: powerful tools for wind engineering. *1st Int. Symp. Comp. Wind Eng.*, Tokyo, 1992.

[84] Ingram, D., Batten, P., Causon, D. and Saunders, R., The representation of small scale obstructions in hydrocode calculations. from *Flow and Dispersion through Groups of Obstacles*, Clarendon Press, Oxford, 1997.

[85] Bottema, M. and Sini, J., Intercomparison and validation of different numerical approaches for simulating flow near urban roughness changes, *EACWE*, Genova, Italy, 1997.

[86] Brown, M. and Muller, C., The effect of microscale urban canyon flow on mesoscale puff dispersion. *AMS 12th Symp. Bound. Layers & Turb.*, Vancouver, BC, 1997.

[87] Rotach, M., The turbulence structure in an urban roughness

sublayer. *Flow and Dispersion through Groups of Obstacles*, Clarendon Press, Oxford, 1997.

[88] Ayotte, K, Finnigan, J. and Raupach, M., A second-order closure for neutrally stratified vegetative canopy flows. *Bound.-Layer Meteor.*, **90**, pp. 189-216, 1999.

[89] Yoshida, A. and Kunitomo, T., One-dimensional simulation of the thermal structure of urban atmospheres. *Int. J. Heat Mass Transfer,* **29**, pp. 1041-1049, 1986.

[90] Voogt, J. and Oke, T., Complete urban surface temperatures. *J. Appl. Meteor.*, **36**, pp. 1117-1132, 1997.

[91] Brown, M. and Williams, M., The effect of urban canopy parameterizations on mesoscale meteorological model simulations in the Paso del Norte area. *AWMA 90th Ann. Conf.*, Toronto, 1997.

[92] Grimmond, S., Cleugh, H. and Oke, T., An objective urban heat storage model and its comparison with other schemes. *Atmos. Env.*, **25B**, pp. 311-326, 1991.

[93] Irvine, M., Gardiner, B. and Hill, M., The evolution of turbulence across a forest edge. *Bound.-Layer Meteor*. **84**, 467-496, 1997.

[94] Shaw, R. and Schumann, U., Large-eddy simulation of turbulent flow above and within a forest. *Bound.-Layer Meteor*. **61**, pp. 47-64, 1992.

[95] Watanabe, T. and Kondo, J., The influence of canopy structure and density upon the mixing length within and above vegetation. *J. Met. Soc. Japan*, **68**, 227-235, 1990.

[96] Oke, T., The micrometeorology of the urban forest. *Phil. Trans. R. Soc. Lond. B*, **324**, pp. 335-349, 1989.

[97] Leach, M. and Chen, S., personal communication, 1999.

[98] Smith, S., personal communication, 1999.

Chapter 6

Lagrangian puff dispersion modeling and uncertainty estimation using second-order closure

Ian Sykes
ARAP Group, Titan Corporation
50 Washington Road
Princeton, NJ 08543-2229, USA

Abstract

The application of second-order turbulence closure techniques to represent the dispersion of material in the atmosphere is described. The turbulence closure model is readily adapted in a Lagrangian framework, using the Gaussian puff methodology, and is extended to treat the inhomogeneous turbulence fields encountered in the atmosphere. The specification of the mesoscale atmospheric turbulence is discussed. The role of wind shear and its representation in the puff model is also described. Second-order closure further provides a prediction of the concentration fluctuation variance. This statistical variance can be used to characterize the inherent uncertainty in a dispersion prediction. The incorporation of the fluctuation variance equation into a Lagrangian puff model is also discussed.

1 Introduction

The atmosphere contains motions on all spatial scales from sub-millimeter up to global. Mesoscale motions cover a range of scales from a few kilometers up to hundreds of kilometers, spanning the range between the planetary boundary layer and the synoptic weather scales. The mesoscale range has been subdivided by Orlanski [1] into narrower ranges associated with different dynamic processes, but our principal concern here is with the dispersive effects of the motions rather than with their dynamics. It must be noted, however, that the temporal and spatial characteristics of the velocity field are determined by the dynamic forcing and instability mechanisms and, since turbulent dispersion is a result of divergent particle trajectories, we cannot properly discuss dispersion without some understanding of the dynamical flow structures. At present, our ability to represent these structures in a statistical manner is limited, as we shall discuss in more detail, but we emphasize that improvements in dispersion modeling will come from an improved understanding of the flow dynamics.

The dispersion of material in the atmosphere is governed by the advection-diffusion equation

$$\frac{\partial c}{\partial t} + \frac{\partial}{\partial x_i}(u_i c) = \kappa \frac{\partial^2 c}{\partial x_i \partial x_i}, \tag{1}$$

which assumes that the scalar concentration, c, is transported by the velocity field, u, and is also diffused by molecular motions with diffusivity coefficient, κ. Note also that this is the simplest conservation equation for c, and assumes that the material follows the air motions exactly and is completely inert, i.e., there are no reactive or transformational source terms for c. Transport of particulate materials involves a “slip” velocity, in general, since the particles have mass and take a finite time to respond to changes in the air velocity; particles are also non-neutrally buoyant and therefore tend to fall under the gravitational acceleration. These effects can be neglected for very small particles on the order of a few microns diameter, however, and eqn. (1) is still a good approximation. The advection-diffusion equation represents the dispersion of passive, conservative scalars in a velocity field, and is therefore the starting point for our discussion of atmospheric dispersion.

Since the atmosphere is turbulent, the velocity field is generally considered to be a random process, at least on smaller scales. There are instability mechanisms on all scales in the atmosphere, and the Reynolds numbers are extremely large for all scales of interest, so the details of the velocity field are unpredictable. The "small" scales are therefore represented statistically, by applying the classical Reynolds-averaging technique to eqn. (1). Thus, using the overbar to define a mean value, and the prime to denote a fluctuation from the mean, we can write any variable in the form

$$c = \bar{c} + c', \tag{2}$$

where c' is a stochastic variable. Applying the average to the advection-diffusion equation gives

$$\frac{\partial \bar{c}}{\partial t} + \frac{\partial}{\partial x_i}\left(\bar{u}_i \bar{c}\right) = \frac{\partial}{\partial x_i}\left(\kappa \frac{\partial \bar{c}}{\partial x_i} - \overline{u_i' c'}\right), \tag{3}$$

which shows that the mean concentration, $\bar{c}$, satisfies the same type of advection-diffusion equation as the instantaneous concentration, c, with advection by the mean wind, but the diffusion term contains an additional turbulent component. The turbulent flux, $\overline{u_i' c'}$, is generally much larger than the molecular flux, and is represented as the correlation between the velocity fluctuations and the concentration fluctuations.

Equation (1) shows that molecular diffusion is the only process that is responsible for reducing the concentration, at least in a non-divergent velocity field, so the turbulent flux is not diffusive by itself. The nature of turbulent diffusion is such that the random shearing and stretching motions will tend to cascade energy to smaller scales, where molecular effects are significant, so that true mixing does ultimately take place.

The definition of "small" scale depends somewhat on the point of view. Mesoscale is "small" in a global model prediction, but would be considered as the mean flow in a mesoscale model. Similarly, the planetary boundary layer (PBL) scales are small in comparison to mesoscale phenomena, but are explicitly resolved, and therefore become the mean flow, in a Large Eddy Simulation of the PBL. In all cases, however, accurate prediction of the mean velocity field depends on a detailed specification of the initial conditions, and the reliability of the prediction always degrades with time. Thus, the velocity fields will always contain a degree of uncertainty in practice, whether they are based on observations or numerical prediction.

Our understanding of atmospheric dispersion has mostly focused on short-range phenomena, where the dominant dispersion mechanisms are associated with the PBL. The turbulence generation mechanisms, and turbulent eddy structures, are relatively well studied, and provide a sound basis for dispersion modeling (Weil [2]). It is easy to see, however, why longer-range dispersion is dependent on larger scale features. Taylor [3] showed that the spread of a tracer cloud could be directly related to the statistical properties of the velocity field for the idealized homogeneous situation, i.e., constant mean velocity and velocity statistics in both space and time. Under these assumptions, Taylor showed that the lateral spread, σ_y, is given by

$$\sigma_y^2 = \sigma_0^2 + 2\overline{v'^2} \int_0 R_L(t')dt', \tag{4}$$

where σ_0 is the initial spread, $\overline{v'^2}$ is the lateral velocity variance, and R_L is the temporal autocorrelation function, defined as

$$R_L(t') = \overline{v'(t)v'(t+t')} / \overline{v'^2}, \tag{5}$$

and the time is used in a Lagrangian sense, i.e., following a fluid particle.

For short times after release, compared with the autocorrelation timescale, it is easily seen that

$$\sigma_y^2 \cong \sigma_0^2 + \overline{v'^2} t^2, \tag{6}$$

since $R_L \sim 1$ as $t \to 0$. Also, for late times, (4) gives the well-known $t^{1/2}$ behavior for σ_y,

$$\sigma_y^2 \cong \sigma_0^2 + 2\overline{v'^2} \tau_L t, \tag{7}$$

where $\tau_L = \int_0^\infty R_L(t')dt'$ is the Lagrangian integral timescale of the velocity fluctuations.

A simple way to visualize the Taylor result is to think in terms of the spatial cloud scale in comparison with the outer scale of the turbulent eddies. When the cloud is small, it is moved bodily by the larger eddies and therefore meanders back and forth. When the cloud is large, the smaller eddies act in a similar way to molecular diffusion and produce small-scale mixing motions within the larger cloud. The important aspect to note from (7) is that the effective dispersion rate decreases as the cloud age grows larger. This is associated with the effective spatial scale of the eddies, and shows that the PBL eddies will become less and less effective as the cloud scale grows much larger than the PBL scales.

The implication of the Taylor result is that PBL eddies will be much less effective in dispersing material when the scales become larger than a few kilometers, which is the typical scale of the larger PBL eddies. Since a cloud from a small source will often reach this size after transport over a few tens of kilometers, this will be the typical range at which mesoscale processes become important for dispersion.

The Taylor theory is applicable to long averaging times, since it represents an ensemble average plume spread. Instantaneous plume spread, or relative dispersion, involves a different behavior, since eddies larger than the instantaneous cloud do not contribute to relative dispersion. Theories of relative dispersion have been proposed by Batchelor [4], Smith and Hay [5], and generally lead to a $t^{3/2}$ growth at early times, and $t^{1/2}$ at late times.

Observational studies of longer-range dispersion have consistently shown that the effective dispersion rate continues to increase long after the cloud has exceeded the planetary boundary layer scales. Angell et al. [6] released small clusters of tetroons (constant volume balloons) and tracked them with radar as they dispersed over ranges up to 1000km or travel times of about 24 hours. The tetroons were set to float at an altitude of about 2.5 km, and statistics of the separation between individual pairs or triads were reported. These statistics are a measure of relative dispersion, and Angell et al. [6] found that the effective growth rate of the r.m.s horizontal separation, σ_y, was close to linear with time over transport times of 24 hours. This led to the simple estimate, $\sigma_y = 0.5t$ (σ_y in meters and t in seconds), of Pack et al. [7], implying a constant rate of spread of 0.5 m/s. Gifford [8] also examined the existing relative dispersion data and noted that the data were consistent with a $t^{3/2}$ rate for times less than about one hour, but appeared slower for longer times. Gifford noted that the 3/2 regime might be expected at larger scales, but that the data was inconclusive. Gifford [9] later analyzed further observations of long-range plume dispersion and found a more consistent rate of growth of the cloud over transport times of the order of a day or more. This strongly implies that there is no significant mesoscale gap in the atmospheric spectrum and that larger scale motions are continually playing a role in the plume dispersion as the cloud size increases.

The random nature of the velocity fields is not only responsible for diffusive effects, which lead to dilution of the scalar concentration, but also introduces uncertainty into the prediction. Since the velocity field is not known in detail, we cannot predict the concentration field in detail. Equation (3), even with perfectly known conditions on the statistics of the velocity field, only predicts the ensemble mean concentration, where the ensemble includes all possible realizations of the velocity field. There is thus a stochastic component in the concentration and we shall discuss how to model this in Section 3. In all practical circumstances, there is also uncertainty in the mean velocity field itself, due to observational errors, interpolation errors, or numerical forecast errors, so there is also a component of uncertainty in the larger scale flow specification. One can roughly consider the two types of uncertainty as a random variability in the cloud shape due to turbulent dispersion processes and a random variability in the cloud trajectory due to wind field errors.

Discussion of uncertainty in atmospheric dispersion, especially when we extend the range from the boundary layer up to the mesoscale, requires a careful definition. In the idealized space of the wind tunnel, the definition of a "mean" quantity is relatively clear. For continuous plumes with a constant release rate, the mean can be obtained from a time average measurement. Instantaneous releases can be repeated indefinitely to construct average statistics. In this idealized case, the flow field is statistically steady, so that time averages are equivalent to ensemble averages and are therefore well defined. In the atmosphere, it is sometimes possible to perform similar averaging but only for relatively short-range experiments. If we only consider surface layer turbulence, where the time scales are very short, then the mean wind field can be considered as steady for the short duration necessary for constructing an average. For longer-range dispersion, however, a longer averaging period is needed since the concentration field is influenced by turbulent eddies with longer correlation time scales and by the diurnal variation of the boundary layer. Since the atmosphere is never homogeneous or steady, we cannot equate time averages with ensemble averages. Every concentration measurement is an individual event; repeated measurements (if possible) represent different conditions and we have no practical means for constructing a true statistical mean.

The inability to measure a mean value makes the definition of an ensemble a difficult theoretical concept. Since atmospheric events are essentially non-repeatable, the ensemble is only meaningfully defined in terms of the available input information for a model prediction (Venkatram [10]). Uncertainty can only be defined in terms of a prediction, since there is no possibility for direct measurement of any ensemble. The prediction uncertainty depends upon the ability of the input meteorology to define the actual wind field responsible for the dispersion; a more complete definition of the wind field will give a more reliable (less uncertain) dispersion prediction.

At present, our ability to characterize the uncertainty in the wind field is very limited. There are different uncertainties associated with the interpolation and analysis of observations versus the growth of uncertainty as a numerical forecast progresses. In all cases, there is a spectrum of smaller scale wind fluctuations below the resolution of the observations or forecast grid, and also a larger scale uncertainty associated with forecast or observational error. Draxler [11] investigated the sensitivity of trajectory calculations, using forecast analyses and direct wind profile observations, and found variations on the order of 20% of the travel distance associated with the different methods. There is also current interest in stochastic forecasting (e.g., Tracton and Kalnay [12]), which promises a quantitative assessment of the forecast uncertainty, but a great deal more work is needed to develop a general understanding of uncertainties.

2 Atmospheric spectra

Measurement of the velocity spectrum in the mesoscale range is made difficult by the fact that there is always energy in larger, synoptic scales. The idealized laboratory condition, where the larger scales are fixed and spectra can be obtained from long time averages, does not occur in the atmosphere. Whenever a time series of velocity measurements is collected, there will be variability on time scales comparable with the duration of the time series itself. The most reliable spectral information that we have, therefore, comes from very long period analyses that include the synoptic scales also, and characterize the complete spectrum. The best example of such a spectrum was presented by Nastrom and Gage [13], and is obtained from commercial aircraft measurements at cruise altitude over an extended period. The data are representative of the motions in the upper troposphere / lower stratosphere, but there is some evidence that the larger scale dispersion rates are comparable throughout the troposphere (see e.g., Gifford [8]). This is not altogether surprising, since the larger scale eddies are presumably driven by larger scale processes that penetrate deep into the troposphere, in contrast to the boundary layer eddies which are localized near the surface. As a starting point for a description of the mesoscale turbulence spectrum, the Nastrom and Gage data can be used to represent horizontal dispersion rates at all altitudes in the troposphere.

Gifford [14] provided a particularly useful discussion of the larger scale atmospheric spectrum. Gifford noted that theories of large-scale atmospheric turbulence, dominated by two-dimensional horizontal motions, predict a cascade of enstrophy (mean square vorticity) from smaller scales to larger scales. However, this enstrophy cascade must be accompanied by a transfer of energy toward smaller scales. The classical slopes of the two regimes are –3 for the enstrophy cascade and –5/3 for the energy cascade. These slopes are reasonably consistent with the Nastrom and Gage data, and the transition in slope occurs at a wavelength of roughly 200 km. Gifford postulates a relationship between the tropospheric energy dissipation rate, the total kinetic energy in the –5/3 range, and the Lagrangian timescale of the turbulence. The dissipation rate associated with the mesoscale spectrum is associated with the transfer of energy from the large-scale motions, so Gifford uses an independent estimate based on a planetary scale energy balance (Golitsyn [15]) for the average dissipation in the troposphere.

From the analysis of long-range plume dispersion observations together with theoretical arguments based on a linearized random-force model, Gifford [16], [17] suggested that the outer timescale should correspond to the Coriolis timescale, f^{-1}, i.e., roughly one day. Accepting these relationships, the spectral similarity description provides a quantitative estimate of the mesoscale energy spectrum, which can be used directly to determine dispersion rates. Using the tropospheric dissipation rate of 4×10^{-4} m^2/s^3, the relationships give an r.m.s. velocity of about 2.3 m/s and an outer length scale of 200 km.

We emphasize that these estimates are essentially "climatological" averages, in the sense that the spectral distributions are obtained from long-term averages covering a wide range of synoptic conditions. We expect that the mesoscale spectrum will be influenced by both local mesoscale phenomena, such as topography, land cover, convective storm activity, etc., and also synoptic conditions which can drive mesoscale features, such as fronts. At present, however, our understanding of the scale interactions and the instability and dissipation mechanisms is not sufficiently developed for us to relate mesoscale spectra to local conditions. We therefore use the "climatological" average as a reasonable estimate of the energy spectrum.

The spectral estimate can also be used to scale down the energy as appropriate for a more highly resolved wind field prediction. If we consider the use of a mesoscale meteorological model wind field in a dispersion prediction, then the model grid size determines the scale of the eddies that are explicitly resolved. An estimate of the subgrid energy is needed for a dispersion calculation, to represent the velocity fluctuations on scales smaller than the model grid size. The Gifford spectrum can be used to estimate the effective energy in the subgrid range.

3 Turbulence closure

3.1 Turbulent fluxes

The Reynolds-averaged advection-diffusion equation (3) shows that, for atmospheric dispersion situations, the turbulent flux or velocity correlation is the dominant dispersion term. The essence of any dispersion prediction is therefore an estimate of this turbulent flux term. A formal evolution equation for the turbulent flux, ignoring molecular effects, can be derived from the conservation equation, (3), as

$$\begin{aligned} \frac{\partial}{\partial t}\overline{u_i'c'} + \overline{u}_j \frac{\partial}{\partial x_j}\overline{u_i'c'} &= -\overline{u_i'u_j'}\frac{\partial \overline{c}}{\partial x_j} - \overline{u_j'c'}\frac{\partial \overline{u}_i}{\partial x_j} - \\ &\quad \frac{\partial}{\partial x_j}\overline{u_i'u_j'c'} - \overline{c'\frac{\partial p'}{\partial x_i}} + \frac{g_i}{T_0}\overline{c'\theta'} \end{aligned} \qquad (8)$$

However, this equation introduces further unknowns involving higher-order correlations, such as the pressure correlation and the third-order term. We could write equations for these higher-order correlations, but they would contain even higher order terms. This is the ubiquitous turbulence closure problem that arises from the Reynolds averaging procedure, and requires a modeling assumption, empirical or otherwise, to close the infinite hierarchy of equations. Second-order closure is a well-established method for dealing with the problem, and involves approximating all the higher-order terms in (8), i.e., evolution equations are developed for all correlations involving two fluctuating quantities.

We use the closure model of Lewellen [18] to give a closed equation for the turbulent flux,

$$\begin{aligned} &\frac{\partial}{\partial t}\overline{u_i'c'} + \bar{u}_j \frac{\partial}{\partial x_j}\overline{u_i'c'} = -\overline{u_i'u_j'}\frac{\partial \bar{c}}{\partial x_j} - \overline{u_j'c'}\frac{\partial \bar{u}_i}{\partial x_j} + \\ &\frac{\partial}{\partial x_j}\left(v_c q\Lambda \frac{\partial}{\partial x_j}\overline{u_i'c'}\right) - A\frac{q}{\Lambda}\overline{u_i'c'} + \frac{g_i}{T_0}\overline{c'\theta'} \end{aligned}, \tag{9}$$

where $A = 0.75$ and $v_C = 0.3$ are empirical model constants. The turbulent velocity scale, q, is defined as $q^2 = \overline{u_i'u_i'}$, and Λ is the turbulent length scale; these quantities will be discussed more fully below.

The buoyancy term in (9) uses the Boussinesq approximation; $g_i = (0,0,g)$ is the gravitational acceleration, T_0 is the reference temperature, and $'$ is the potential temperature fluctuation. Mesoscale motions are primarily horizontal and we will concentrate on the horizontal dispersion parameterization in our discussion, but we note that vertical diffusion can often play an important role, especially in the presence of vertical wind shear. The vertical eddy motions are generally small scale, however, and are driven by surface fluxes or small scale instabilities.

3.2 Concentration fluctuation variance

Dispersion by a turbulent velocity field is a random process and the fluctuation correlations in (9) represent an ensemble mean value for the turbulent flux. The instantaneous scalar concentration is a stochastic quantity, with a probability distribution that depends on the distribution of velocity fluctuations. The mean concentration, $\bar{c}$, is only the first moment of the probability distribution and contains no information about the statistical variability in the prediction; higher-order moments are required to give a quantitative description of the variability. Second-order closure provides a natural framework for predicting the statistical variance in the concentration, which is the second moment of the probability distribution. The mean and the variance can then be used to provide a probabilistic prediction using a parameterized probability density function, described below.

The concentration fluctuation variance equation can be obtained from the scalar mass conservation equation, (1), in the form

$$\frac{\partial \overline{c'^2}}{\partial t} + \frac{\partial}{\partial x_i}\left(\overline{u}_i \overline{c'^2}\right) = -2\overline{u'_i c'}\frac{\partial \overline{c}}{\partial x_i} - \frac{\partial}{\partial x_i}\left(\overline{u'_i c'^2}\right) - 2\kappa\overline{\left(\frac{\partial c'}{\partial x_i}\right)^2} + \kappa\nabla^2\overline{c'^2}, \tag{10}$$

where the first three terms on the right-hand side are identified as turbulent production, turbulent diffusion, and turbulent dissipation, respectively. The last term is direct molecular diffusion of the scalar variance, and is negligible for the small values of molecular diffusivity relative to atmospheric dispersion scales. However, it is important to note that the only dissipation mechanism is molecular, as is evident from consideration of the total square of the scalar concentration,

$$\frac{\partial \overline{c^2}}{\partial t} + \frac{\partial}{\partial x_i}\left(\overline{u}_i \overline{c^2}\right) = -\frac{\partial}{\partial x_i}\left(\overline{u'_i c^2}\right) - 2\kappa\overline{\left(\frac{\partial c}{\partial x_i}\right)^2} + \kappa\nabla^2\overline{c^2} \tag{11}$$

where it can be seen that all the terms except dissipation are in flux form.

For very high values of the Schmidt number, i.e., very small values of κ, the molecular terms in both equations are effectively equal to the turbulent scalar dissipation

$$\varepsilon_c = 2\kappa\overline{\left(\frac{\partial c'}{\partial x_i}\right)^2} \tag{12}$$

which is modeled using a dissipation timescale as

$$\varepsilon_c = \frac{\overline{c'^2}}{\tau_c}. \tag{13}$$

From the above, it is clear that the statistical variance in concentration due to random fluctuations in the velocity field is controlled by the turbulent flux and turbulent dissipation rate. The critical aspect of the prediction lies in the determination of the dissipation timescale, τ_c, associated with the turbulent eddies responsible for the dispersion, and the recognition that large scale eddies can simply meander the complete concentration field without producing a reduction in the variance.

The key aspect of the scalar variance prediction is the dissipation timescale. Proper characterization of the dissipation timescale allows an accurate prediction of the concentration fluctuation variance, as has been demonstrated in comparison with laboratory data (Sykes et al. [19]) and also with large-scale atmospheric dispersion observations (Sykes et al. [20]). The second-order closure model for the dissipation rate was originally developed by Sykes et al. [21] using the laboratory data of Fackrell and Robins [22]. A fundamental discovery in that study was that the scalar dissipation scales were an internal property of the scalar field itself, rather than being determined exclusively by the velocity fluctuations. Essentially, velocity eddies with scales larger than the scalar plume or cloud will meander the entire scalar distribution but not cause any nonlinear cascade of scalar fluctuation variance onto smaller scales. The turbulent cascade process is only driven by eddies with scales similar to the instantaneous plume size.

On dimensional grounds, the scalar dissipation time scale is

$$\frac{1}{\tau_c} = \frac{q_c}{\Lambda_c}, \tag{14}$$

where q_c and Λ_c represent the velocity and length scales for the scalar fluctuations. The velocity scale is obtained from a simplified description of the turbulent kinetic energy spectrum, and the length scale is predicted from a transport equation. Before discussing the details of these parameterizations, however, we introduce a generalization of the dissipation model to account for the anisotropy between horizontal and vertical directions, and the differences between plume and puff meandering.

The scalar variance dissipation model introduced by Sykes et al. [19], [21] used a single estimate of the velocity and length scales to define the dissipation timescale in (14). This is appropriate for the case of plume dispersion in two dimensions with near-isotropic turbulence. Under the more general conditions of larger scale horizontal dispersion and arbitrary sources, we require a characterization of the different dissipation rates associated with each direction.

Gifford [23] provided an idealization of the fluctuation variance calculation using the meandering plume concept. Gifford postulated a plume with instantaneous spread, σ_i, but with a randomly located plume centroid location having a Gaussian spread, σ_e. Using estimates for the instantaneous and ensemble plume widths, Fackrell and Robins [24] were able to show good agreement between Gifford's simple meandering plume model and their own wind tunnel data. We can generalize the meandering plume model and consider the concentration field to be diffusing in n dimensions, where n could be 1, 2, or 3. For example, an instantaneous puff diffuses in three dimensions initially, but once it becomes well-mixed in the vertical it only diffuses in the 2 horizontal dimensions. Similarly, a continuous plume diffuses in 2 transverse dimensions initially (the case that Gifford studied), but this is reduced to 1 dimension when it becomes well-mixed.

A simplified variance estimate can be obtained from assuming uniform distribution in place of Gaussian. In this case, if the instantaneous concentration is $\hat{c}$, then this concentration is found in a "volume" of σ_i^n for a problem in n dimensions. The probability of encountering this concentration within the ensemble "volume", however, is given by $p = \sigma_i^n / \sigma_e^n$ (assuming the ensemble centroid spread is much larger than the instantaneous spread). The mean concentration is therefore given as

$$\bar{c} = p\hat{c} \tag{15}$$

and the variance is

$$\overline{c'^2} = p(1-p)\hat{c}^2, \tag{16}$$

since there is a probability of *(1-p)* that *c* will be zero. The fluctuation intensity ration is thus

$$\frac{\overline{c'^2}}{\overline{c}^2} = \frac{(1-p)}{p} \cong \frac{1}{p} \tag{17}$$

assuming p is small.

If we assume the standard early time behavior as $t^{3/2}$ for the instantaneous spread, and t for the ensemble spread, then we obtain

$$\frac{\overline{c'^2}}{\overline{c}^2} = \frac{\sigma_e^n}{\sigma_i^n} \sim t^{-n/2}. \tag{18}$$

Sykes et al. [21] showed that the dissipation model (13) implied a power-law decay for the fluctuation intensity, with the power determined by the empirical coefficient multiplying the velocity and length scales. The implication of (18) is that the empirical coefficient must depend on the dimensionality of the diffusion process, or that the dissipation timescale is comprised of multiple components representing effective rates for each of the three spatial dimensions. As part of the concentration fluctuation prediction, therefore, we require three length scales to determine the fluctuation dissipation rate. The internal scales are denoted as Λ_{cx}, Λ_{cy}, and Λ_{cz} for the two horizontal and vertical directions, respectively. Note that the x and y subscripts refer to streamwise and cross-stream scales respectively. The concentration length scales are used to estimate the dissipation velocity scales, using a simplified Kolmogorov spectrum assumption. Thus, the horizontal velocity scale for the concentration fluctuations is modeled as

$$q_{cH}^2 = \left(\overline{u'^2} + \overline{v'^2}\right) \min\left(\left(\frac{\Lambda_H}{\Lambda_y}\right)^2, \left(\frac{\Lambda_y}{\Lambda_H}\right)^{2/3}\right), \tag{19}$$

where Λ_H is the length scale associated with the mesoscale horizontal velocity fluctuations. Note that we use the cross-stream scale in (19); the streamwise scale is used to distinguish between plume and puff dispersion, but does not determine the velocity scale.

The vertical dissipation velocity scale, q_{cV}, is defined similarly using the vertical velocity variance and the vertical scale. However, one important feature of the vertical dispersion is the finite limit under daytime conditions with an inversion-capped boundary layer. When the cloud becomes well-mixed through the layer, both the vertical diffusion and the effective vertical velocity scale are zero, since there is no further mixing associated with the vertical direction.

Using the two velocity scales defined above, the scalar dissipation timescale is then modeled as

$$\frac{1}{\tau_c} = bs\left(\frac{q_{cH}}{\Lambda_{cx}} + \frac{q_{cH}}{\Lambda_{cy}} + \frac{q_{cV}}{\Lambda_{cz}}\right). \tag{20}$$

The internal fluctuation scales grow with the turbulent diffusion, and the internal velocity scale gives a measure of the growth rate. The model for the two horizontal internal length scales is

$$\frac{d\Lambda_{cx}}{dt} = \frac{d\Lambda_{cy}}{dt} = \alpha q_{cH}, \tag{21}$$

where $\alpha_c = 0.25$ for $\Lambda_{cy} < \Lambda_H$, otherwise $\alpha_c = 0.098$, and similarly for the other two components. The empirical coefficients in (21) were determined by Sykes et al. [21] from comparison with the wind tunnel data of Fackrell and Robins [22], and the direct application to the mesoscale range of atmospheric motions is questionable. At the present time, however, there is no real information on the dissipative effects of mesoscale motions; further research is definitely needed in this area.

4 Lagrangian puff model

4.1 Gaussian moment definition

The Lagrangian puff model is a practical methodology for atmospheric dispersion applications, in which a collection of Gaussian distributions is used to describe the concentration field of a dispersing contaminant (Bass [25]). The Gaussian distributions are Lagrangian in the sense that their centroids move with the wind field, and spread due to turbulent diffusion. The principal advantage of the Lagrangian approach is the absence of any numerical grid, so that the enormous range of scales involved in dispersion from a localized source onto the mesoscale or larger can be accurately represented. The Lagrangian methods also facilitate a proper description of the effective turbulent diffusivity coefficients, which has been shown to depend on travel time from the source.

Lagrangian methods include both particle and puff representations. Particle methods use stochastic techniques to represent the effects of turbulent eddies, and generally require a large number of particles to provide accurate representation of a continuous distribution. Many applications are described in the literature (e.g., Diehl et al. [26]; Wilson and Flesch [27]; Hernandez et al. [28]) and many complex features such as non-Gaussian turbulence statistics and complicated wind shear effects can be accurately represented by stochastic particle models. The puff approach attempts to reduce the required number of particles by using a Gaussian distribution to represent a continuous concentration field instead of a large number of individual particles, and we shall restrict our discussion to the puff approach in the following.

A three-dimensional Gaussian distribution is completely described by its spatial integral moments up to second-order, and can be written in the form

$$c(x) = \frac{Q}{(2\pi)^{3/2}\,(Det(\sigma))^{1/2}} \exp\left[-\frac{1}{2}\sigma_{ij}^{-1}(x_i - \bar{x}_i)(x_j - \bar{x}_j) \right]. \tag{22}$$

For general atmospheric dispersion problems, we must also consider the effects of the ground surface and the capping inversion at the top of the planetary boundary layer, which are usually represented as reflective surfaces. For our present discussion, we restrict attention to the moment equations, which are independent of the shape assumption.

Using an angle bracket to denote an integral over all space, the spatial moments in (22) are defined as

Zeroth moment – mass

$$Q = \langle c \rangle , \tag{23}$$

First moment – centroid

$$Q\bar{x}_i = \langle c x_i \rangle , \tag{24}$$

Second moment – spread

$$Q\sigma_{ij} = \left\langle c\left(x_i - \bar{x}_i\right)\left(x_j - \bar{x}_j\right)\right\rangle \ . \tag{25}$$

Note that this generalized tensor definition of the moments is completely independent of any specific coordinate system, and can therefore describe an arbitrarily oriented Gaussian shape. The full moment description requires ten numbers to represent the mass, the three centroid coordinates, and six spread moments since σ_{ij} is a symmetric second-rank tensor.

Note that the specific Gaussian variation (22) applies to an individual puff, but in general the local concentration field will be composed of a sum of contributions from a number of such puffs. We next describe the transport and diffusion model equations for the individual puff moments.

4.2 Moment transport equations

The species conservation equation, (3), is readily integrated over the spatial dimensions to give conservation of mass

$$\frac{dQ}{dt} = 0 \, , \tag{26}$$

where Q now represents the integrated mean concentration, $\bar{c}$. Note that (26) applies to an inert conserved tracer, with no loss or transformation processes.

Equations for the higher spatial moments of $\bar{c}$ are obtained similarly, but require some assumption about the spatial variation of the velocity field. The simplest assumption is a constant velocity based on the value at the puff centroid, but this neglects any effect of shear. We therefore use a linear representation for the local velocity field

$$\bar{u}_i(x) = \bar{u}_i(x) + \left(x_j - \bar{x}_j\right)\frac{\partial \bar{u}_i}{\partial x_j} \, , \tag{27}$$

where the velocity gradient is also evaluated at the centroid location. Multiplying (3) by x_i and integrating by parts gives

$$\frac{d\bar{x}_i}{dt} = \frac{\left\langle \bar{u}_i \bar{c}\right\rangle}{Q} + \frac{\left\langle \overline{u_i' c'}\right\rangle}{Q} \, , \tag{28}$$

where the first term on the right-hand side represents transport by the mean wind and the second term is the turbulent drift. The latter arises from correlation between the velocity fluctuations and the concentration fluctuations. Substituting for $\overline{u}$ from (27) and assuming any symmetric spatial distribution for c, the velocity gradient term vanishes to give

$$\frac{d\overline{x}_i}{dt} = \overline{u}_i(\overline{x}) + \frac{\left\langle \overline{u_i' c'} \right\rangle}{Q}. \tag{29}$$

The second-moment equation is obtained by multiplying (3) by $x_i' x_j'$, where $x_i' = x_i - \overline{x}_i$, and integrating by parts to give

$$\frac{d\sigma_{ij}}{dt} = \sigma_{ik}\frac{\partial \overline{u}_j}{\partial x_k} + \sigma_{jk}\frac{\partial \overline{u}_i}{\partial x_k} + \frac{\left\langle x_i' \overline{u_j' c'} \right\rangle}{Q} + \frac{\left\langle x_j' \overline{u_i' c'} \right\rangle}{Q}, \tag{30}$$

where the first two terms represent the shear distortion of the puff and the second two terms represent the turbulent diffusion. The pairs of terms are required for the ij-symmetry of the second moment tensor.

This completes the specification of the concentration moment equations, but we have introduced turbulent flux moments in (29) and (30). The closure model described in the previous section can now be employed to represent the velocity correlation terms.

4.3 Turbulence closure diffusion model

The general equation (9) for the turbulent fluxes can be integrated spatially to provide transport equations for the flux moments in (29) and (30). In order to simplify the discussion, we restrict attention to the horizontal components and further assume that the turbulence is isotropic in the horizontal. There is evidence that the zonal component is more energetic than the meridional component at larger scales, but over most of the mesoscale range there is no directional preference. We therefore only need consider one of the diagonal correlation terms for the horizontal diffusivity, and the evolution equations is derived directly from (9) as

$$\frac{d}{dt}\left\langle y' \overline{v' c'} \right\rangle = Q\overline{v'^2} - A\frac{q}{\Lambda_H}\left\langle y' \overline{v' c'} \right\rangle, \tag{31}$$

where we have used the component values for the velocity and position vectors, i.e., $x_i' = (x', y', z')$ and $u_i' = (u', v', w')$.

Under steady homogeneous conditions, (31) gives a very simple prediction for the effective horizontal diffusivity and lateral spread. The diffusivity in the y-direction can be defined as

$$K_y = \frac{\langle y'\overline{v'c'} \rangle}{Q} . \tag{32}$$

Then, if $K_y = 0$ at $t = 0$, the solution for K_y is

$$K_y = \overline{v'^2}\tau(1 - \exp(-t/\tau)), \tag{33}$$

where the horizontal timescale $\tau = \Lambda_H/(Aq)$. The solution for the lateral spread, $\sigma_y^2 = \sigma_{22}$, is then

$$\sigma_y^2 = \sigma_0^2 + 2\overline{v'^2}\tau(1 - \tau + \tau \exp(-t/\tau)), \tag{34}$$

where σ_0 is the initial spread. This result is consistent with Taylor's [3] exact analytic expression, (4), using an exponential form for the Lagrangian autocorrelation function and an integral timescale of τ.

The result (34) represents an ensemble or absolute dispersion, but relative dispersion can also be described using the closure framework. Sykes and Gabruk [29] show that instantaneous spread can be predicted by restricting the energy spectrum to the instantaneous plume or cloud size. The spectral shape is used to reduce the effective velocity scale, q, and the effective turbulence scale, Λ_H, is replaced by the local cloud scale, Λ_{cy}. The model was compared with short-range observations by Sykes and Gabruk [29], and can be shown to give the expected $t^{3/2}$ growth for cloud sizes within the Kolmogorov inertial range.

4.4 Concentration fluctuation variance

As discussed above, we also predict an uncertainty in the mean concentration due to the random wind fluctuations. This requires that the puffs include an equation for the concentration fluctuation variance. This is a nonlinear quantity and therefore requires a treatment of puff overlap interactions. We have found that it is easier to deal with the total square concentration, $\overline{c^2}$, since the conservation equation is simpler than the fluctuation variance, $\overline{c'^2}$, cf. (10) and (11). Since the total square concentration equation only involves a dissipation term when integrated over the spatial volume, the puff transport equation is simply

$$\frac{d}{dt}\left\langle \overline{c^2} \right\rangle = -\frac{\left\langle \overline{c'^2} \right\rangle}{\tau_c}, \tag{35}$$

where the dissipation timescale, τ_c, has been defined in the previous section.

Note that the puff integral on the right-hand side of (35) is the fluctuation variance, which must be estimated from the relation

$$\left\langle \overline{c^2} \right\rangle = \left\langle \overline{c'^2} \right\rangle + \left\langle \bar{c}^2 \right\rangle. \tag{36}$$

If we represent the mean concentration as a sum of Gaussian puffs, then the expression for the local concentration is

$$\bar{c}(\mathrm{x}) = \sum_{\alpha} Q^{(\alpha)} G^{(\alpha)}(\mathrm{x}), \tag{37}$$

where the sum is taken over all the contributing puffs, denoted by a superscript α, and $G^{(\alpha)}$ represents the spatial Gaussian distribution of puff-α. The square of the mean concentration is therefore

$$\bar{c}^2(\mathrm{x}) = \sum_{\alpha,\beta} Q^{(\alpha)} Q^{(\beta)} G^{(\alpha)}(\mathrm{x}) G^{(\beta)}(\mathrm{x}), \tag{38}$$

and requires a sum over all puff pairs. The distribution of $\bar{c}^2$ amongst the individual puffs for substitution in (36) is not unique, but an obvious partition is given by

$$\left\langle \bar{c}^2 \right\rangle^{(\alpha)} = Q^{(\alpha)} \sum_{\beta} Q^{(\beta)} I_{\alpha\beta}, \tag{39}$$

where the interaction integral is defined as

$$I_{\alpha\beta} = \left\langle G^{(\alpha)}(\mathrm{x})G^{(\beta)}(x)\right\rangle, \tag{40}$$

i.e., the overlap integral between the two Gaussian puffs.

The overlap integrals can be represented analytically for the Gaussian shapes, but must be calculated efficiently to avoid having to compute N^2 pairs when there are N puffs. Sykes and Henn [30] describe a numerical scheme for limiting the number of overlap pair computations to those with significant interaction.

4.5 Wind shear and inhomogeneity

The mesoscale spectrum discussed above contains many scales of motion that can be resolved by modern mesoscale weather prediction models. When the wind fields from these models are used for the basis of the dispersion prediction, we only need to parameterize the sub-grid motions since larger scales are represented explicitly. As model grid sizes get smaller, the resolved portion of the spectrum expands and the sub-grid motions become less important. In this situation, however, it is essential that the dispersion model be able to represent the effects of the resolved scale velocity fields.

A traditional shortcoming of Lagrangian puff models has been the inability to describe complex flow situations, due to use of the local velocity at the puff centroid and the assumption of a simplified Gaussian shape with principal axes aligned with the coordinate directions. This is partially remedied by the inclusion of the shear distortion terms in the generalized Gaussian moment equation (30). This enables the puff model to represent the effects of local wind shear, which includes both horizontal and vertical gradients. The assumption of linear velocity variation is still a limitation, however, and the puff must be split when it grows too large for the local approximation to be valid. Sykes and Henn [30] describe a scheme for representing the flow distortion effects in a Lagrangian puff model, and compare numerical predictions with an analytical solution for a severely distorted distribution.

5 Calculation examples

5.1 Calculations with the mesoscale spectrum

The first result from the modeling discussed above is the spread prediction from the Gifford mesoscale spectrum. In the idealized situation with no mean wind, the instantaneous spread prediction is shown in Figure 1, for a transport period of 96 hours. The data compiled by Gifford [8, 9] are also shown in the figure. Several points are evident in the figure. First, there is significant variability in the observations and the model prediction lies generally in the lower half of the data scatter.

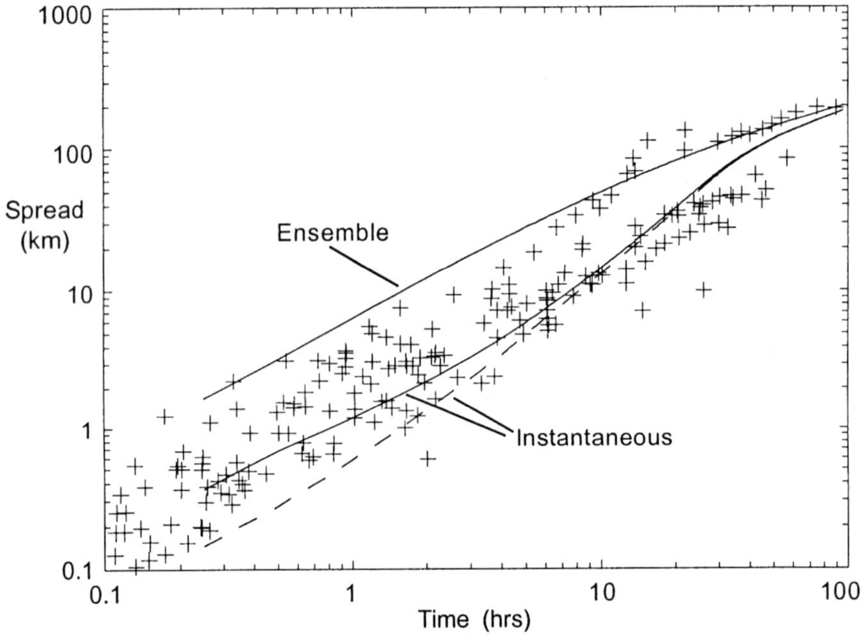

Figure 1: Closure model predictions for instantaneous cloud spread compared with observational data from Gifford [8], [9] (shown as symbols). The lower lines represent the closure model results; the dashed line shows the result for stable conditions, the solid line is for convective conditions. The upper line shows the ensemble average spread prediction.

This is not unreasonable, since the observations generally include systematic wind shear effects in addition to the random spectrum of horizontal velocity fluctuations. We shall examine some of the shear effects below, but we note here that vertical wind shear will generally lead to increased spread and the effect is not included in the mesoscale spectrum.

A second aspect of the mesoscale dispersion observations is the sensitivity to release conditions at early time. The initial size of the cloud, and the specific meteorological conditions in the first hour or two plays a large role in the spread during this early period. If the source is released into a convective boundary layer, one might expect a different initial diffusion rate from a nocturnal release. Two release conditions are shown in Figure 1 for the model calculations, illustrating the early time dependence but also showing that the longer-range prediction becomes independent of the initial conditions. The observations contain many different types of release, with variation in initial spread, initial elevation from near surface up to a few kilometers, and initial time of release. The large variability in the early time is most likely due to this variability in the initial conditions.

Figure 1 also shows the ensemble spread prediction using the same mesoscale spectrum. As expected, the ensemble spread is much larger than the instantaneous value, and the differences are reduced after about one day. This is consistent with the outer scale of the spectrum as the inverse Coriolis timescale; after several time scales the cloud spread is larger than the largest eddies in the spectrum and the instantaneous spread converges toward the ensemble spread. In this regime we expect the synoptic scales to become the dominant mechanism, but these eddies are not included in the calculations for Figure 1.

One measure of the uncertainty in a dispersion prediction, as exemplified in Gifford's [22] meandering plume model, is the discrepancy between the instantaneous plume or cloud size and the ensemble average size. Figure 1 shows that the ratio between these two can be very large, especially at relatively early times when the mesoscale motions meander the small plume dramatically. As we saw in Section 2, the r.m.s. velocity associated with the mesoscale spectrum is about 2.5 m/s, while the effective energy level at the scale of the initial cloud, 50 m in this case, is much smaller. Under stable conditions, the small-scale turbulence energy is even smaller than the convective conditions; hence the uncertainty is larger in this situation.

The fluctuation intensity for the ensemble calculation is illustrated in Figure 2. The development shows a very rapid rise at early time and slow decay as the cloud spreads onto the larger scales. The figure shows the uncertainty associated with the mesoscale spectrum for an instantaneous release, which can meander in both horizontal dimensions. This uncertainty is not particularly realistic, since the mesoscale spectrum is not typically a good measure of the overall wind uncertainty in a real case. Also, the four-day steady state is far from realistic, so we next consider some of the variability that plays an important role in the longer-range dispersion prediction.

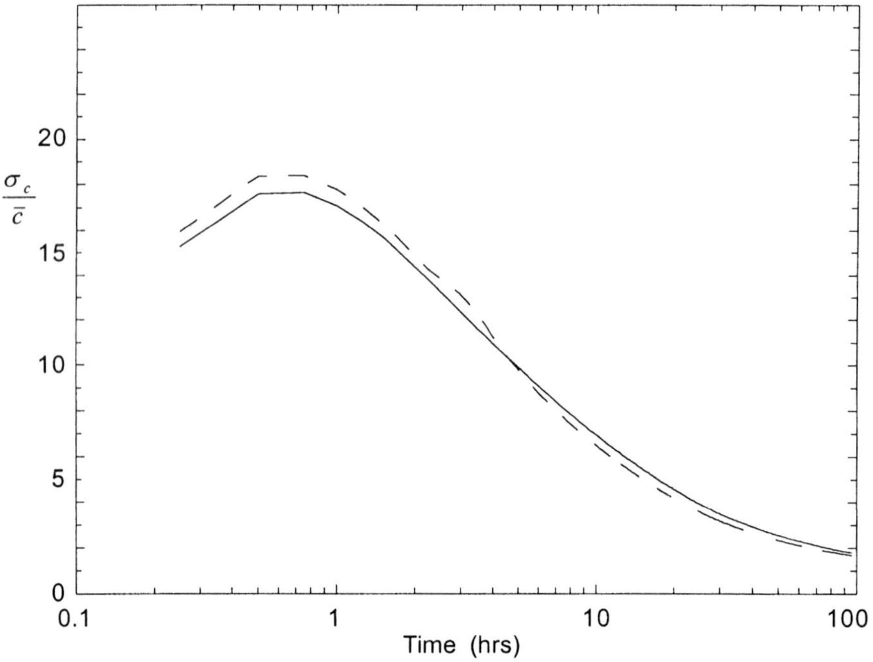

Figure 2: Closure model predictions for the concentration fluctuation intensity, ratio of the r.m.s. fluctuation to the mean concentration, for the ensemble prediction shown in Figure 1. Dashed line is for stable conditions, solid line is for convective conditions.

5.2 Effects of wind shear

Vertical wind shear can have a dramatic effect on horizontal dispersion rates. Vertical gradients can be relatively large, especially under stable conditions, so that two material particles in close vertical proximity can be transported at significantly different velocities. Vertical wind shear effects are often associated with large-scale flow features and are therefore persistent over long times, enhancing the dispersive effect of the relative motions.

Wind shear effects are subtle, because wind shear by itself does not cause any dispersion. If the flow field is incompressible, then the velocity field cannot change the concentration since concentration is exactly conserved along a Lagrangian trajectory. Wind shear can combine with vertical diffusion, however, enhancing its effectiveness and generating diffusion in the horizontal.

The maximum horizontal dispersion occurs for the lowest vertical diffusivity; larger vertical diffusivity tends to mix material faster in the vertical so that a material particle does not sample the velocity field at any particular level for very long. Small vertical diffusivities, on the other hand, allow the cloud to shear strongly but do not mix material very much in the vertical.

The atmosphere typically provides a strong shear-driven dispersion effect due to the diurnal variation in the planetary boundary layer. The typical diurnal variation is illustrated in Figure 3, which shows a daytime mixed-layer cloud becoming sheared as the nocturnal inversion quenches the vertical diffusivity and establishes a wind shear profile. The release was made in the late morning and becomes mixed throughout the daytime convective layer before the onset of the nocturnal inversion in the first frame. The cloud is then stretched by the shear all through the night, and is mixed in the vertical as the convective boundary layer grows in the next day. The shear displacement in the early morning frame is seen to be roughly equal to the cloud width at the surface, so that the subsequent fumigation virtually doubles the horizontal width. This simplified example shows the importance of describing the shear distortion and also the vertical splitting of puffs to describe the morning fumigation as the mixed-layer grows.

McNider et al. [31] have also studied the role of the nocturnal boundary layer in lateral plume dispersion. Their work shows that

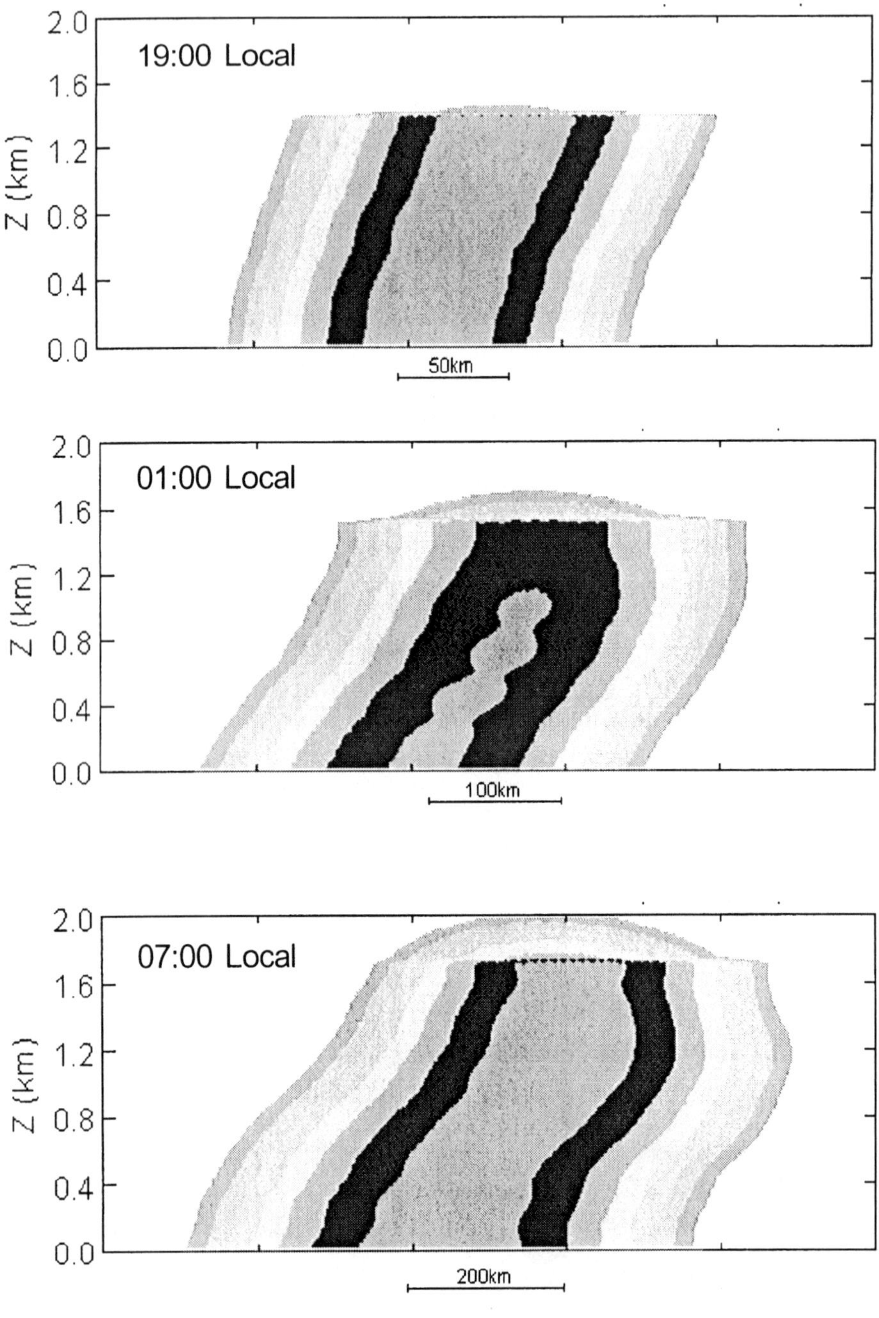
2.0
1.6
1.2
0.8
0.4
0.0
Z (km)
19:00 Local
50km
01:00 Local
100km
07:00 Local
200km

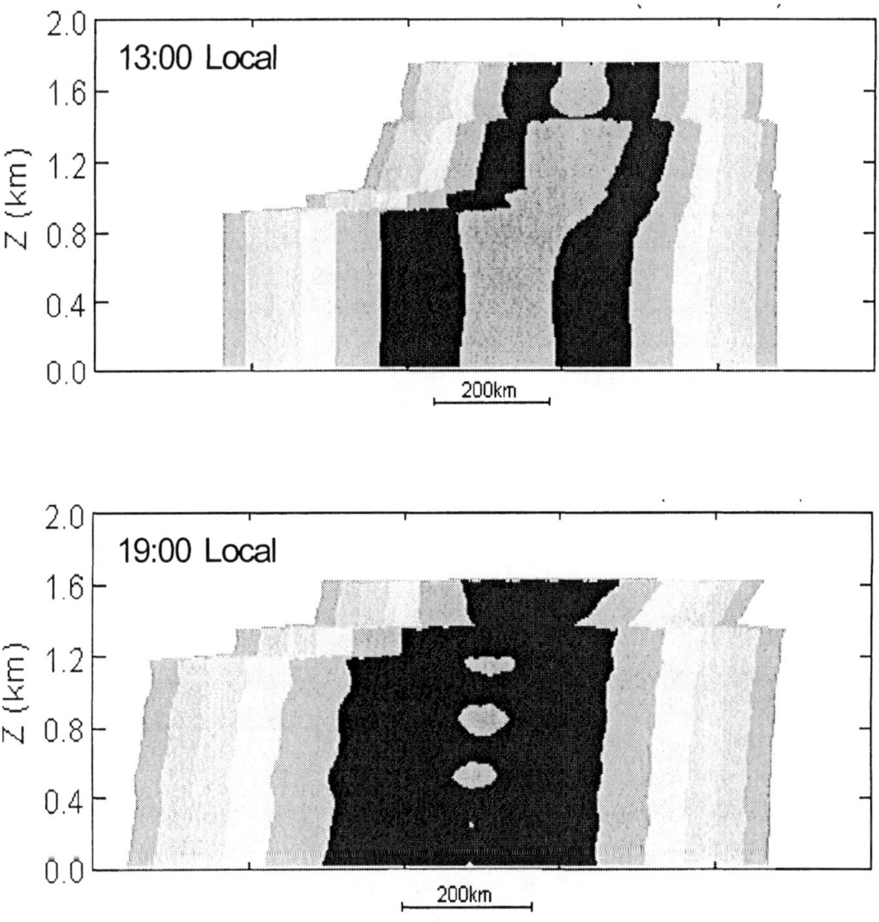

Figure 3: Sample puff calculation for a 24-hour period showing nocturnal shear and daytime fumigation. Note the change in the horizontal scale as cloud evolves.

the inertial oscillation, driven by the Coriolis force after the mixed-layer has collapsed, can greatly enhance lateral dispersion since the oscillation involves direction shear as well as speed shear. Again, the long persistence of these motions is responsible for the large effective dispersion during the nocturnal period.

6 Summary

Mesoscale dispersion involves many complex flow effects due to both spatial and temporal inhomogeneity. Some of these features can be represented explicitly in mesoscale weather prediction models, but are often unresolved by routine observation networks or larger scale forecast models. In most cases, there is a gap between the boundary layer eddy scales and the resolution of the wind field; so this part of the spectrum must be represented statistically. The larger scales are important because the dispersion of a cloud is dominated by the largest scale eddies, and dispersion on time scales longer than a few hours will typically involve scales larger than the boundary layer.

Gifford [14] presents a quantitative model for the mesoscale spectrum, based on observational data, and this can be used as the basis for a second-order closure model for mesoscale dispersion. The statistical predictions from this model are in reasonable agreement with observations of relative dispersion over time periods of a few days, although there is a significant scatter in the observations, probably due to the variability in source conditions and specific meteorological conditions during the various releases.

Mesoscale dispersion predictions generally need to account for resolved scale variations in the wind field. The diurnal cycle of the planetary boundary layer can contribute significant dispersion effects through the persistent nocturnal shear and subsequent daytime vertical mixing. Many other mesoscale phenomena can produce persistent gradients of velocity and enhance the dispersion in a similar manner. Any dispersion prediction for the longer range needs to account for these shear and time-dependent effects.

Second-order closure turbulence closure provides a natural framework for representing the concentration fluctuation variance, which can be used as a measure of the uncertainty due to random velocity fluctuations in a concentration prediction. The sources of uncertainty in a mesoscale dispersion prediction, however, are difficult to describe since they are not the traditional "turbulence" as experienced in a wind tunnel. In the wind tunnel, and to some extent for short-range boundary later dispersion, the notion of a mean wind and a fluctuation is well defined. For mesoscale applications, where the flow is inherently unsteady, the "mean" wind is more difficult to define. In any practical situation, our definition of the mean wind is based on either the interpolation of observations, or a numerical forecast model prediction. In both cases, there are significant uncertainties associated with the numerical process of interpolation or forecasting.

The mesoscale spectrum observed by Nastrom and Gage [13] is a long-term average spectrum of actual wind fields, and does not include any of the interpolation/forecast uncertainties. There are thus two pressing needs for a proper assessment of the wind field uncertainties in longer-range dispersion predictions. First, we need to understand how the mesoscale energy spectrum is influenced by the synoptic conditions and local terrain variations. We expect that the energy cascade from larger scales and the energy production within the mesoscale range will depend on the larger scale instabilities and also on the local terrain forcing. It is hoped future work with higher resolution mesoscale meteorological models will provide insight into these phenomena. Similarly, the introduction of ensemble forecasting techniques will lead to a quantitative assessment of forecast uncertainties appropriate for specific situations.

References

[1] Orlanski, I., A rational subdivision of scales for atmospheric processes. *Bull. Am. Met. Soc.,* **56**, pp. 527-530, 1975.

[2] Weil, J. C., Updating applied diffusion models. *J. Climate Appl. Meteor.,* **24**, pp. 1111-1130, 1985.

[3] Taylor, G. I., Diffusion by continuous movements. *Proc. Lond. Math. Soc.(Ser. 2),* **20**, pp. 196-211, 1921.

[4] Batchelor, G. K., Diffusion in a field of homogeneous turbulence II. The relative motion of particles. *Proc. Camb. Phil. Soc.,* **48**, pp. 345-362, 1952.

[5] Smith, F. B. and Hay, J. S., The expansion of clusters of particles in

the atmosphere. *Quart. J. Roy. Met. Soc.,* **87**, pp. 82-101, 1961.

[6] Angell, J. K., Allen, P. W. and Jessup, E. A., Mesoscale relative diffusion estimates from tetroon flights. *J. Appl. Met.,* **10**, pp. 43-46, 1971.

[7] Pack, D. H., Ferber, G. J., Heffter, J. L., Telegadas, K., Angell, J. K., Hoecker, W. H. and Machta, L., Meteorology of long-range transport. *Atmos. Env.,* **12**, pp. 425-455, 1978.

[8] Gifford, F. A., Tropospheric relative diffusion observations. *J. Appl. Met.,* **16**, pp. 311-313, 1977.

[9] Gifford, F. A., Atmospheric diffusion in the range 20 to 2000 kilometers. *Air Pollution Modeling and its Applications*, ed. F. A. S. C. D. Wespelaere and N. V. Gillani, Plenum Press, pp. 247-265, 1985.

[10] Venkatram, A. K., A framework for evaluating air quality models. *Boundary-Layer Met.,* **24**, pp. 371-385, 1982.

[11] Draxler, R. R., The accuracy of trajectories during ANATEX calculated using dynamic model analyses versus Rawinsonde observations. *J. Appl. Met.,* **30**, pp. 1446-1467, 1991.

[12] Tracton, M. S. and Kalnay, E., Operational ensemble prediction at the National Meteorological Center: Practical aspects. *Wea. Forecasting,* **8**, pp. 379-398, 1993.

[13] Nastrom, G. D. and Gage, K. S., A climatology of atmospheric wavenumber spectra of wind and temperature observed by commercial aircraft. *J. Atmos. Sci.,* **42**, pp. 950-960, 1985.

[14] Gifford, F. A., A similarity theory of the tropospheric turbulence energy spectrum. *J. Atmos. Sci.,* **45**, pp. 1370-1379, 1985.

[15] Golitsyn, G. S., A similarity approach to the general circulation of planetary atmospheres. *Icarus,* **13**, pp. 1-24, 1970.

[16] Gifford, F. A., Horizontal diffusion in the atmosphere: a Lagrangian, dynamical theory. *Atmos. Env.,* **16**, pp. 505-512, 1982.

[17] Gifford, F. A., The random-force theory: application to meso- and large-scale atmospheric diffusion. *Boundary-Layer Met.,* **30**, pp.

159-175, 1984.

[18] Lewellen, W. S., Use of invariant modeling. *Handbook of Turbulence*, ed. W. Frost and T. H. Moulden, Plenum Press, pp. 237-280, 1977.

[19] Sykes, R. I., Lewellen, W. S. and Parker, S. F., A Gaussian plume model of atmospheric dispersion based on second-order closure. *J. Clim. & Appl. Met.,* **25**, pp. 322-331, 1986.

[20] Sykes, R. I., Parker, S. F., Henn, D. S. and Lewellen, W. S., Numerical simulation of ANATEX tracer data using a turbulence closure model for long-range dispersion. *J. Appl. Met.,* **32**, pp. 929-947, 1993.

[21] Sykes, R. I., Lewellen, W. S. and Parker, S. F., A turbulent-transport model for concentration fluctuations and fluxes. *J. Fluid Mech.,* **139**, pp. 193-218, 1984.

[22] Fackrell, J. E. and Robins, A. G., Concentration fluctuations and fluxes in plumes from a point source in a turbulent boundary layer. *J. Fluid Mech.,* **117**, pp. 1-26, 1982.

[23] Gifford, F. A., Statistical properties of a fluctuating plume dispersal model. *Adv. Geophys.,* **6**, pp. 117-137, 1959.

[24] Fackrell, J. E. and Robins, A. G., The effects of source size on concentration fluctuations in plumes. *Boundary-Layer Met.,* **22**, pp. 335-350, 1982.

[25] Bass, A., Modeling long range transport and diffusion, Second Joint Conf. on Appl. of Air Poll. Met., AMS & APCA, New Orleans, LA, pp. 193-215, 1980.

[26] Diehl, S. R., Smith, D. T. and Sydor, M., Random-walk simulation of gradient-transfer processes applied to dispersion of stack emission from coal-fired power plants. *J. Appl. Met.,* **21**, pp. 69-82, 1982.

[27] Wilson, J. D. and Flesch, T. K., Flow boundaries in random-flight dispersion models: enforcing the well-mixed condition. *J. Appl. Met.,* **32**, pp. 1695-1707, 1993.

[28] Hernandez, J. F., Cremades, L. and Baldasano, J. M., Dispersion modeling of a tall stack plume in the Spanish Mediterranean coast by a particle model. *Atmos. Env.,* **29**, pp. 1331-1341, 1995.

[29] Sykes, R. I. and Gabruk, R. S., A second-order closure model for the effect of averaging time on turbulent plume dispersion. *J. Appl. Met,* **36**, pp. 165-184, 1997.

[30] Sykes, R. I. and Henn, D. S., Representation of velocity gradient effects in a Gaussian puff model. *J. Appl. Met.,* **34**, pp. 2715-2723, 1995.

[31] McNider, R. T., Singh, M. P. and Lin, J. T., Diurnal wind-structure variations and dispersion of pollutants in the boundary layer. *Atmos. Env.,* **27A**, pp. 2199-2214, 1993.

Chapter 7

Description of atmospheric transport processes in Eulerian air quality models

Daewon W. Byun*
Atmospheric Sciences Modeling Division,
Air Resources Laboratory, NOAA,
Research Triangle Park, NC 27711, USA
**On assignment to the National Exposure Research Laboratory, U.S. Environmental Protection Agency*

Abstract

Many types of air quality models treat atmospheric advection and turbulent diffusion processes differently. Gaussian models use analytical solutions of the advection-diffusion equations. Lagrangian models use a hypothetical air parcel concept effectively removing the need for explicit treatment of transport processes. Eulerian models require solution of the advection diffusion equation using some numerical techniques. Presented here are modeling techniques for atmospheric transport processes (except for the subgrid scale cloud mixing) in Eulerian air quality models. First, the presentation describes computational coordinates for atmospheric studies, diagnostic and prognostic meteorological modeling techniques, and assumptions regarding atmospheric dynamics. Then, it emphasizes the use of the flux-form formulations that ensure conservation of mass during the transport processes. Instead of dwelling on specific numerical algorithms or comparisons among them, it provides general descriptions of numerical techniques for the transport processes used in Eulerian air quality models.

1 Introduction

Pollutants injected into the atmosphere undergo various physical and chemical transformations including transport processes. The transport processes primarily consist of advection and diffusion, and the mixing of pollutants by convective clouds. Accurate treatments of transport processes in air quality models are very critical for simulating air quality. Therefore, the air quality modeling techniques are typically classified by the methods by which the transport processes are treated.

Gaussian models express transport of inert pollutants with analytical solutions obtained through the use of simplifying assumptions to the advection-diffusion equations. Validity of the Gaussian models deteriorates due to such factors as vertical wind shear, diurnal evolutions of wind and atmospheric stability, topography, and land use variations. Furthermore, they are strongly affected by temporal and spatial scales of applications. Because of the limiting assumptions of the Gaussian formula, various types of special purpose models have been built to satisfy specific transport conditions. Many of these models were distributed by the U.S. EPA (Environmental Protection Agency) as a part of the UNAMAP (User's Network for Applied Modeling of Air Pollution) models starting in the late 1960s. Several of them, in upgraded forms, are now available through SCRAM (the Support Center for Regulatory Air Models). These models have been used as the workhorse air pollution models for regulatory purposes. However, EPA has advertised that these models are, in principle, guideline models and their applications must be compatible with the model formulations.

Box models assume that pollutants are uniformly distributed in an arbitrary imaginary box or column bounded by the ground and the inversion base. The transport processes are treated as simple boundary flux problems. For example, horizontal advection is represented as the inflow and outflow fluxes across the walls, and the entrainment of the background pollutants due to the growth of the inversion layer is expressed as a downward flux from the top. Because the pollutants are assumed to mix instantaneously within the box, there is no separate treatment of the diffusion process. These box models solve the complicated sets of nonlinear equations representing photochemical reactions of the primary and secondary pollutants. They predict temporal variations of concentrations, and estimate mass balances of multiple pollutants, spatially averaged over a domain represented by the box. Because of the simplifications in representing the transport processes, these box models are incapable of predicting air quality for regions with significantly inhomogeneous (spatially) emissions, or where the characteristic turbulent mixing time scale is larger than the chemical reaction time scales. In the mid-1970s, EPA proposed the Empirical Kinetic Modeling Approach (EKMA), a chemistry-intensive box model for estimating the effects of organic hydrocarbon and NO_x (NO and NO_2) emission changes on peak ozone concentrations. During the 1970s, development work resulted in several models, but except for EKMA, their applications by regulatory agencies were limited.

Lagrangian models are based on species conservation equations for atmospheric diffusion and chemical reactions, stated in terms of a moving coordinate system. The observer follows sets of hypothetical columns of air similar to the box models. The advection is replaced with movement of each air column along the prevailing winds, e.g., trajectory. Like box models, the turbulence diffusion processes are not explicitly simulated in single-layer Lagrangian models, except for the entrainment of the background pollutants through the top when the boundary layer grows. Primary pollutant emissions are injected into the columns when they pass over source regions. Similar to box models, trajectory models simulate chemical reactions in each column. While trajectory models go one step beyond box models, their assumptions break down when complex topographic features cause complicated wind fields or vertical wind shears within air columns. Although pollutant masses are well conserved, accuracy of the movement is relatively limited because advection is not explicitly simulated. Trajectory modeling is only as good as the validity of the back-trajectory concept if it can adequately describe the motion of pollutants. In earlier Lagrangian models, the description of the source functions and sink processes were rather simplistic.

Eulerian air quality models (AQMs) have been used to meet various scientific challenges and to help in formulating environmental policy decisions. Eulerian models utilize a grid system defined in appropriate coordinates to describe atmospheric transport (advection and diffusion), emission sources, and chemical production and destruction. The observer's reference frame is fixed in the Eulerian model, usually at the earth's surface. Most numerical weather prediction models rely on this paradigm. Eulerian models generate four-dimensional (space and time) trace species concentrations. Unlike the other modeling techniques, the atmospheric transport processes are explicitly solved in Eulerian models. Incorporating different physical and chemical processes is relatively straightforward. Terms in the conservation equation can be implemented as independent science process modules. Numerical diffusion in the advection process, and difficulties in representing atmospheric mixing processes, are some of the drawbacks of Eulerian models. Model predictions are limited to the grid resolution, and sub-grid processes must be approximated through parameterizations.

To compensate for this deficiency, some Eulerian models include either trajectory or Gaussian dispersion models to treat initial transport and chemical transformations of pollutants coming from large point source emissions within a grid cell. These hybrid ("plume-in-grid") models attempt to minimize instantaneous dilution of pollutants over the entire grid cell. Once the plumes reach a certain size, their concentrations are added to the existing concentrations in the appropriate grid cells, and subsequently go through transformation and transport processes within the grid model.

Eulerian air quality models also depend on wind-field models or weather prediction models to provide detailed meteorological descriptions. Mass conservation characteristics of Eulerian models are heavily dependent on the quality of the meteorology data used.

In this chapter, atmospheric transport modeling techniques used in Eulerian AQMs are discussed, except for subgrid scale cloud mixing processes. It includes descriptions of the atmospheric dynamics (computational coordinates, meteorological modeling techniques, and assumptions on dynamics), necessary meteorological parameters for air quality simulations, properties of numerical transport algorithms, and general approaches for numerical advection solvers and diffusion algorithms.

2 Meteorological data and atmospheric dynamics description

Air quality models need detailed information regarding wind and atmospheric turbulence structure. Even simple Gaussian models require mean wind, atmospheric stability criteria, and other planetary boundary layer (PBL) parameters that define dispersion characteristics. The EKMA box models require meteorology data averaged over the box. Such data include PBL height, entrainment rate, and wind speed as well as photolytic rates that determine the efficiency of photochemical reactions. Lagrangian models require similar meteorological data along the trajectories. Trajectories are often determined with the gridded data either from analyses of observed meteorological data or direct output of weather forecasting models. Compared with other types of air quality models, Eulerian models must be used with comprehensive meteorological data.

Eulerian air quality modeling systems use meteorological data across the science-processing components. Emissions processing requires various meteorological parameters: To compute plume rise, temperature, PBL height, atmospheric stability and winds are used; to estimate biogenic emissions, temperature, radiation, humidity, cloudiness and rain amounts are needed; and to estimate mobile source emissions, temperature and wind speed are used. In the chemical transport model (CTM), atmospheric advection is characterized by mean wind components, whereas diffusion is determined by the PBL height and eddy diffusivity. State variables such as temperature, pressure, and air density influence thermal reaction rates. Actinic flux determines photolytic reaction rates. Humidity, cloud, and precipitation information are used to characterize aqueous reactions, cloud induced mixing, and wet deposition. Cloudiness also affects the actinic fluxes. Additionally, post processing of CTM output (such as diagnostic evaluation, model result interpretation, data extraction and analysis, and trajectory estimation) requires detailed meteorological information.

The current observational networks provide basic meteorological parameters at locations designed to detect synoptic weather systems. However, air quality modeling usually requires much finer scale information than the routine measurements can provide. Remote sensing techniques based on satellite and radar can improve the quality and quantity of the meteorological data. Yet the critical data, such as PBL height, moisture, cloud cover, and precipitation processes, are available only at intermittent times. Therefore, meteorological observations are either processed by diagnostic methods (objective analysis schemes using actual observations) or assimilated into prognostic meteorological models to provide essential meteorological information.

Some requirements for the meteorological data to perform Eulerian air quality modeling are as follows:

• **Faithful representation of the atmosphere** Meteorological data should provide a realistic description of the past atmosphere, including evolution of flow fields and key meteorological events.
• **Accurate estimation of derived PBL and cloud parameters** PBL height, surface heat and moisture fluxes, turbulent diffusivity, and cloud characteristics should be available as precisely as possible.
• **Sufficient resolution of data** The meteorological data should resolve the topographic features at the grid resolution suitable for handling the scales and domains of air quality problems.

2.1 Meteorological coordinates

Eulerian modeling systems compute with discretized representations of atmospheric state variables. Choice of the grid system influences accuracy and efficiency of the computation. Often, the Cartesian coordinates are not appropriate in mesoscale atmospheric simulations. To better represent the atmospheric volume above the ground, curvilinear coordinates that include a terrain-influenced vertical coordinate and a conformal map projection, such as Lambert Conformal, Polar Stereographic, or Mercator, are often used. One advantage of conformal mapping is the ability to maintain certain differential invariants in governing equations after transformation to the map coordinates. For example, it preserves the angle between two intersecting curves (i.e., the orthogonality of latitude circles and meridians is preserved on the map). Although the distances and areas are changed in conformal mapping, they can be accounted for by the map scale factor.

With a conformal map projection, we relate generalized curvilinear coordinates $(\hat{x}^1, \hat{x}^2, \hat{x}^3, \hat{t})$ to the reference rotated earth-tangential Cartesian coordinates (x, y, z, t), on which most atmospheric observations are based:

$$\begin{cases} \hat{x}^1 = mx \\ \hat{x}^2 = my \\ \hat{x}^3 = s \\ \hat{t} = t \end{cases} \qquad \begin{cases} x = m^{-1}\hat{x}^1 \\ y = m^{-1}\hat{x}^2 \\ z = h(\hat{x}^1, \hat{x}^2, \hat{x}^3, \hat{t}) = h_{AGL}(\hat{x}^1, \hat{x}^2, \hat{x}^3, \hat{t}) + z_{sfc}(\hat{x}^1, \hat{x}^2) \\ t = \hat{t} \end{cases} \tag{1}$$

where m is the map scale factor, s is the generalized vertical coordinate, z_{sfc} is the topographic height, and h is the geometric height, and h_{AGL} represents height above ground level (AGL). In the derivation of relations in eqn. (1), we neglected the first-order variations of the map scale factor in x- and y-directions. The approximation establishes a quasi-orthogonality of the vertical coordinate to the horizontal plane on the conformal map. The differential element of distance ds can be expressed in terms of the curvilinear coordinates as:

$$(ds)^2 = dx_i dx^i = \frac{\partial x^i}{\partial \hat{x}^j}\frac{\partial x^i}{\partial \hat{x}^k} d\hat{x}^j d\hat{x}^k = \hat{\gamma}_{jk} d\hat{x}^j d\hat{x}^k , \tag{2}$$

where the subcripts (i,j,k) represent the co-variant components (i.e., following the tangential basis vectors in coordinate axes) and the superscripts represent the contra-variant components (i.e., following the basis vectors normal to the surface defined by the intersecting coordinates). The covariant metric tensor relating the measurement of distance in the coordinate transformation is defined as:

$$\hat{\gamma}_{jk} = \frac{\partial x^i}{\partial \hat{x}^j}\frac{\partial x^i}{\partial \hat{x}^k} . \tag{3}$$

Note that the system $(\hat{x}^1, \hat{x}^2, \hat{x}^3, \hat{t}) = (\hat{x}^1, \hat{x}^2, s, \hat{t})$ is not orthogonal because not all the off-diagonal components of the covariant metric tensor of the system vanish;

$$\hat{\gamma}_{jk} = \begin{bmatrix} m^{-2} + (\frac{\partial h}{\partial \hat{x}^1})^2 & (\frac{\partial h}{\partial \hat{x}^1})(\frac{\partial h}{\partial \hat{x}^2}) & (\frac{\partial h}{\partial \hat{x}^1})(\frac{\partial h}{\partial s}) \\ (\frac{\partial h}{\partial \hat{x}^1})(\frac{\partial h}{\partial \hat{x}^2}) & m^{-2} + (\frac{\partial h}{\partial \hat{x}^2})^2 & (\frac{\partial h}{\partial \hat{x}^2})(\frac{\partial h}{\partial s}) \\ (\frac{\partial h}{\partial \hat{x}^1})(\frac{\partial h}{\partial s}) & (\frac{\partial h}{\partial \hat{x}^2})(\frac{\partial h}{\partial s}) & (\frac{\partial h}{\partial s})^2 \end{bmatrix} \tag{4}$$

Its determinant is given as:

$$\hat{\gamma} = \left|\hat{\gamma}_{jk}\right| = \frac{1}{m^4}(\partial h / \partial s)^2 \, . \tag{5}$$

The Jacobian for vertical coordinate transformation is related to the determinant of the covariant tensor as

$$J_s = \left|\frac{\partial h}{\partial s}\right| = m^2 \sqrt{\hat{\gamma}} \, . \tag{6}$$

The components of horizontal wind vector in the Cartesian coordinate system (U and V) are related to the contravariant wind components in the conformal map projection ($\hat{v}^1$ and $\hat{v}^2$) as

$$\mathbf{V}_z = U\mathbf{i} + V\mathbf{j} = (\hat{v}^1 / m)\mathbf{i} + (\hat{v}^2 / m)\mathbf{j} \, , \tag{7}$$

where **i** and **j** represent the unit vectors in the eastward and northward directions, respectively. In the generalized coordinate system, the continuity equation is given as

$$\frac{\partial(\rho J_s)}{\partial t} + m^2 \nabla_s \bullet \left(\frac{\rho J_s \hat{\mathbf{V}}_s}{m^2}\right) + \frac{\partial(\rho J_s \hat{v}^3)}{\partial s} = J_s Q_\rho \, , \tag{8}$$

where ρ is the air density, $\hat{V}_s = \hat{v}^1 i + \hat{v}^2 j$, $\hat{v}^3$ is the contravariant vertical component of the velocity. Although the source term for air density, Q_ρ, should be zero in an ideal case, it is retained in eqn. (8) to capture the possible density error originating from numerical procedures in meteorological models and its impact on the transport process in air quality models. Q_ρ is a measure of mass consistency that quantifies how well the density and wind fields satisfy the continuity equation.

Many different types of vertical coordinates have been used for various meteorological simulations. For example, the geometric height is used to study boundary layer phenomena because of its obvious advantage of relating near surface measurements to modeled results. Pressure coordinates are natural choices for atmospheric studies because many upper atmospheric measurements are made on pressure surfaces. Because most radiosonde measurements are based on hydrostatic pressure, one may prefer to use the pressure coordinate for studying cloud dynamics. Several different generic vertical coordinates can be used in a CTM for describing different atmospheric processes while the underlying model structure is based on a specific coordinate that is consistent with the meteorological model.

To account for the effects of topography on the meteorology, various types of terrain-influenced coordinates have been applied to mesoscale models. Phillips [1] introduced the so-called sigma-coordinate system that maps the earth's surface onto a normalized pressure coordinate surface to accommodate the spatially and temporally varying surface pressure whilst at the same time maintaining the advantage of the pressure coordinate system. Several variations of the sigma coordinate based on hydrostatic pressure have since been developed. As an alternative, a terrain-influenced height coordinate has been applied especially for studying nonhydrostatic atmosphere.

Characteristics of a few terrain-influenced coordinates are summarized in Table 1. Hierarchy of the classification is; (1) temporal dependency of coordinates, (2) base physical characteristics of coordinate variables, and (3) method of treating the topography. Application assumptions, such as hydrostatic/nonhydrostatic and compressible/incompressible atmosphere approximations, are not part of the classification criteria. Isentropic coordinates are not included here because they are not suitable for the regional and urban scale air quality simulation due to their inherent difficulties representing the PBL structure. For larger-scale simulations, an isentropic coordinate system can serve as an interesting alternative (Arakawa et al. [2]). Also, there are new developments of hybrid coordinates that combine isentropic coordinates with other coordinates to mitigate the problem of representing the PBL.

Table 1: Summary of terrain-influenced height and reference hydrostatic pressure coordinate systems. [Note '$\neq f(\,)$' represents that the parameter is independent of the argument.] (Modified from Byun [3]).

Coordinate	Application	Vertical Jacobian	Geopotential height
normalized geometric height (σ_z) $\sigma_z = \dfrac{z - z_{sfc}}{H - z_{sfc}}$	hydrostatic	$J_{\sigma_z} = H - z_{sfc}$ $\neq f(\hat{x}^3, t)$	$\Phi = gz = g[z_{sfc} + \sigma_z (H - z_{sfc})]$
	generalized hydrostatic		
	non-hydrostatic		
terrain-influenced reference pressure (σ_{p_o}) $\sigma_{po} = \dfrac{p_o - p_T}{p_o(z_{sfc}) - p_T}$	hydrostatic	$J_{\sigma_{p_o}} = \dfrac{p_o^*}{\rho_o(\hat{x}^3) g}$ $\neq f(t)$, $p_o^* = p_o(z_{sfc}) - p_T$	$\Phi = \Phi_{sfc} - \int_{\sigma_{posfc}}^{\sigma_{po}} \dfrac{p_o^*}{\rho_o} d\sigma_{po}'$
	non-hydrostatic		
terrain-influenced hydrostatic pressure (σ_π) $\sigma_\pi = \dfrac{\pi - \pi_T}{\pi_{sfc} - \pi_T}$	hydrostatic	$\tilde{\rho} J_{\sigma_{\tilde{p}}} = \dfrac{\tilde{p}^*}{g}$ $\neq f(\hat{x}^3)$	$\Phi = \Phi_{sfc} - \int_{\sigma_{\tilde{p}sfc}}^{\sigma_{\tilde{p}}} \dfrac{\tilde{p}^*}{\tilde{\rho}} d\sigma_{\tilde{p}}'$
	non-hydrostatic	$\pi^* = \pi_{sfc} - \pi_T$ $\rho J_{\sigma_\pi} = \dfrac{\pi^*}{g}$ $\neq f(\hat{x}^3)$	$\Phi = \Phi_{sfc} - \int_{\sigma_{\pi sfc}}^{\sigma_\pi} \dfrac{\pi^*}{\tilde{\rho}} d\sigma_\pi'$
terrain-influenced large-scale hydrostatic pressure ($\sigma_{\tilde{p}}$)	non-hydrostatic	$\tilde{\rho} J_{\sigma_{\tilde{p}}} = \dfrac{\tilde{p}^*}{g}$ $\neq f(\hat{x}^3)$	$\Phi = \Phi_{sfc} - \int_{\sigma_{\tilde{p}sfc}}^{\sigma_{\tilde{p}}} \dfrac{\tilde{p}^*}{\tilde{\rho}} d\sigma_{\tilde{p}}'$
step-mountain ETA (η) with σ_π, $\eta = \sigma_\pi \eta_{sfc}$, $\eta_{sfc} = \dfrac{p_o(z_{sfc}) - p_T}{p_o(0) - p_T}$	hydrostatic	$J_\eta = \dfrac{\tilde{p}^*}{\tilde{\rho} g \eta_{sfc}}$ $\tilde{\rho} J_\eta \neq f(\hat{x}^3)$	$\Phi = \Phi_{sfc} - \int_{\eta_{sfc}}^{\eta} \dfrac{\tilde{p}^*}{\tilde{\rho}\eta_{sfc}} d\eta'$
	non-hydrostatic	$J_\eta = \dfrac{\pi^*}{\rho g \eta_{sfc}}$ $\rho J_\eta \neq f(\hat{x}^3)$	$\Phi = \Phi_{sfc} - \int_{\eta_{sfc}}^{\eta} \dfrac{\pi^*}{\rho \eta_{sfc}} d\eta'$

In a time dependent vertical coordinate, the vertical Jacobian is also time dependent. Therefore, it is more difficult to derive a diagnostic relation from the continuity equation. For a coordinate whose Jacobian-weighted air density is independent of height, a diagnostic equation for the vertical velocity is available when appropriate top and bottom boundary conditions are used. Vertical layers defined with this type of vertical coordinate are considered as material surfaces because mass continuity can be satisfied in a diagnostic fashion. Air particles are not expected to cross material surfaces during the advection process. An atmospheric model based on this type of coordinate may not have a mass consistency problem except for numerical reasons such as finite differencing, treatment of top and bottom boundary conditions, and computer precision. Appropriate choices of coordinates and assumptions on atmospheric dynamics yield relationships that can be used to estimate the mass-consistency errors. With this information, air density or wind field can be reconstructed to ensure mass conservation of pollutants.

2.2 Diagnostic wind field modeling

Models providing wind flow fields can be divided into two types: prognostic primitive equation models and diagnostic or kinetic models. Prognostic models are based on the solution of time-dependent primitive equations. Solving these equations still remains a laborious and expensive task. The effort requires more reliable input data as the governing equations become more elaborate in order to exploit the advantages offered by the completeness of the model. Often such data are not available.

Diagnostic models generate wind fields by satisfying physical constraints. The wind model is defined as mass consistent only if the continuity equation is imposed as a constraint. Such fields are not necessarily divergence-free. This fact must be kept in mind when building numerical solutions. The numerical transport formulations must be in the conservative forms. Otherwise the divergence in wind fields may lead to mass conservation errors. The diagnostic wind field modeling approach can be used for nonsteady cases as well. Assuming there exists a modified wind field that eliminates the source (error) term in the continuity equation for air, one can find the relationship between the original (either observed or modeled) wind field and modified wind field (with superscript M) as

$$m^2\nabla_s \bullet \left[\frac{\rho J_s}{m^2}(\hat{\mathbf{V}}_s - \hat{\mathbf{V}}_s^{M})\right] + \frac{\partial}{\partial s}[\rho J_s(\hat{v}^3 - \hat{v}^{3M})] = J_s Q_\rho . \tag{9}$$

Note that in deriving eqn. (9), the steady-state assumption to eqn. (8) was not applied. A modified velocity field that has changed minimally in a least square sense can be found by the application of the variational calculus technique (Sasaki [4], [5]). The relevant Euler-Lagrangian equations for wind components are given as

$$\hat{\mathbf{V}}_s^{M} = \hat{\mathbf{V}}_s + \frac{1}{2\alpha_1^2}\nabla_s\lambda \tag{10a}$$

$$\hat{v}^{3M} = \hat{v}^3 + \frac{1}{2\alpha_3^2}\frac{\partial\lambda}{\partial s}, \tag{10b}$$

where λ is the Lagrangian multiplier to be determined and α_1 and α_3 are the weights for the horizontal and vertical wind components. One solves for λ in the Poisson equation

$$m^2\nabla_s \bullet \left[\frac{\rho J_s}{2\alpha_1^2 m^2}\nabla_s\lambda\right] + \frac{\partial}{\partial s}\left[\frac{\rho J_s}{2\alpha_3^2}\frac{\partial\lambda}{\partial s}\right] = J_s Q_\rho , \tag{11}$$

with the associated boundary conditions that $\lambda = 0$ for flow-through boundaries and $\partial\lambda/\partial s = 0$ for impenetrable boundaries (i.e., at the topographic surface).

The modified wind components are subject to the same top and bottom boundary conditions imposed by the given coordinate system and dynamic assumptions. One obvious limitation of the wind field modeling is that its fidelity is heavily dependent on the quality of input data. It works well where surface measurements are dense enough to represent the underlying topographic and mesoscale features. However, unfortunately, this is not the case for usual situations. Also, because the scheme does not assume any functional shape of the vertical wind profile, the result does not follow the surface-layer similarity theory inherently. There are a few wind field models that utilize additional constraints like the surface layer similarity theory. In such a case, the formula for the Lagrangian multiplier must be modified accordingly. In addition, the diagnostic method can be applied to enhance the mass-consistent characteristics of the prognostic meteorological model output. However, it cannot totally eliminate the large mass inconsistency originating from the input data because it merely tries to minimize the error in a least-square sense.

2.3 Mesoscale meteorological models

Most meteorological models currently used for air quality modeling were originally developed for the simulation of various mesoscale phenomena and the prediction of weather. They usually predict winds, horizontal and vertical distributions of thermal and moisture structures, and cloud and precipitation predictions. Their performance has mainly been judged by their ability to predict stormy and active weather situations. However, the strong interaction of forces during stormy situations tends to obscure a few deficiencies in the models with respect to their use in air quality systems, where relatively stable and calm weather situations may be critical.

Dynamic assumptions used in meteorological models have profound impacts on the description of atmospheric transport processes. Eulerian air quality models must follow the dynamic descriptions as implied in the meteorological model. For example, the hydrostatic approximation has been applied widely to air quality modeling even for domains with considerable terrain features. The hydrostatic equilibrium assumption states that gravity and the vertical pressure gradient force are in balance. When a hydrostatic equilibrium is established, the atmospheric pressure at a given height is simply related to the weight of the air above. In the real world where the atmosphere is constantly moving, the hydrostatic assumption relates not only the static components of pressure and density but also the perturbation pressure field (which is responsible for the horizontal velocity field) to the perturbation density field. The result is that the vertical acceleration due to gravity is much less than the horizontal acceleration due to the perturbation pressure gradient force. The hydrostatic approximation fails to be valid for small-scale motions such as convection where the vertical acceleration has a magnitude similar to gravity. For mesoscale atmospheric models with small horizontal grid resolutions (e.g., less than 2 km), the hydrostatic pressure calculation may lead to less realistic solutions.

To simulate small-scale atmospheric features, the nonhydrostatic primitive sets of equations are applied. One obvious difference between a nonhydrostatic model and its hydrostatic counterpart is that the nonhydrostatic model requires explicit integration of the vertical velocity component. There are two methods for nonhydrostatic pressure calculation. One uses a fully compressible continuity equation and the other employs an elastic continuity equation. For the latter, a Poisson partial differential equation for pressure has to be solved at each time step. Furthermore, solution of the Poisson equation becomes very costly when terrain-influenced coordinates are used. However, researchers have found that the integration time step for an elastic nonhydrostatic model can be much larger than that for a hydrostatic model, especially when the acoustic waves are solved separately from other meteorological waves (e.g., Klemp and Wilhelmson [6]). This can speed up the nonhydrostatic models.

Recently, a few air quality models using nonhydrostatic meteorological model output have been developed. There are two major difficulties in applying the nonhydrostatic meteorological output to air quality simulations. For example, it is customary that the transformations for horizontal and vertical components are performed separately even for nonhydrostatic models, where the computational coordinates become significantly nonorthogonal. The problem is manifested in the form of abnormal convergence and divergence of air mass in a limited area around complex topography. Another serious problem is the representativeness of the vertical transport velocity where strong convective activity exists. Because the nonhydrostatic meteorological model relies on this vertical velocity to trigger cloud formation and precipitation, strong scale-dependent interactions between the thermodynamic variables and airflow exist. Depending on the cloud parameterizations used, the detailed descriptions of these interactions differ and they could have significant impact on the flow field near large convective cells. These abnormal convergent/divergent conditions cause serious problems in calculating chemical reactions in air quality models.

It must be noted that the hydrostatic or nonhydrostatic atmospheric description, which is a characterization of the vertical motion, is rather independent from either the compressible/incompressible atmosphere or the anelastic atmosphere assumption, which is an approximation of the mass continuity equation. Choices of the assumptions from the two distinct groups have been used to simplify atmospheric motions, although some of the combinations, such as incompressible but nonhydrostatic atmosphere, are rarely used in atmospheric studies. An incompressible atmosphere assumption leads to the nondivergent flow approximation. There is no *a priori* connection between this assumption and atmospheric thermodynamics. Therefore, we cannot infer relationships among the state variables, such as density, temperature and pressure fields. Basically, an incompressible atmosphere assumption is a shallow-water approximation for an adiabatic atmosphere. With the incompressibility assumption, distinction between the continuity equation and the advective form equation becomes blurred. Consequently, concentrations in the form of either density or mixing ratio are used indiscriminately in some atmospheric models. References such as Byun [3], [7], [8] provide detailed information on the implications of the atmospheric dynamic assumptions to air quality modeling.

There are many other meteorological modeling features that can influence simulation of transport processes in air quality models. To use AQMs for evaluating past high air pollution episodes and assessing effectiveness of emissions control strategies, past meteorological data are needed. In such applications, one can use a four-dimensional data assimilation (FDDA) technique to enhance the fidelity of prognostic model simulation (Seaman et al. [9]). Because of the diverse weather and topographical conditions and the wide range of quality (representativeness, density, and frequency) of available meteorological observations, application of the FDDA, however useful, is more of an art than an exact universal method. For example, when a Newtonian nudging is used for the FDDA, weighting coefficients for the nudging should be determined differently for different state variables empirically to obtain useful results.

2.4 Meteorology-chemistry interface processor

As described earlier, simulation with an Eulerian air quality modeling system requires a comprehensive set of meteorological data. Diagnostic models usually provide wind fields alone unless other objective analysis packages are used together. Even detailed prognostic meteorological models may not provide all the parameters necessary for simulating transport processes in CTMs. Because of this, a secondary processor is generally used to calculate the additional data needed. The processor is run following execution of the primary meteorological model and before execution of the CTM. In some air quality modeling systems it is called the (meteorological) preprocessor, although it is a postprocessor to a meteorological model. To avoid the relativism in the naming, it is preferable to call it a meteorology-chemistry interface processor (MCIP). Some functions of the processor are:

- No degradation of the key meteorological data
- Consistent interpolation of meteorological parameters
- Diagnostic estimation of additional required parameters, such as species dry deposition velocities, and actinic fluxes
- Flexible transformations among different grids
- Linkage of meteorological data with emissions processing
- Generation data in an efficient portable file format.

3 Desired properties of numerical transport algorithms

Mass transport is a fundamental process in an air quality model. Two transport processes, convection (or advection) and (turbulent) diffusion, must be accounted for in atmospheric models. Advection can only transport a disturbance in the direction of wind velocity. Turbulent diffusion, on the other hand, can spread a disturbance in every direction. The atmospheric continuity equation that governs pollutant transport describes these transport mechanisms mathematically. Most numerical algorithms used for the advection processes are known to possess unwanted numerical diffusion. Any disturbances being advected along the line of flow are also spread in a direction normal to the line of flow. This diffusion is referred to as artificial diffusion because it does not describe any physical diffusion (molecular or turbulent). Numerical transport algorithms for air quality models should satisfy several computational requirements. The following is a collection of the desirable properties:

Consistency: A numerical approximation is said to be consistent when the equations on the discretized computational domain approach the partial differential equations of the physical domain when the mesh is refined (i.e., when the grid spacing and time step size approach zero). Thus, consistency measures how well the numerical form approximates the governing equations in some limiting cases.

Stability: Stability in a numerical algorithm defines the behavior of a functional value as the number of iterations or time steps increases. The stability infers that the value remains bounded, although it may increase or decrease depending on the nonsteady feature of the system. An algorithm is stable if a small error at any stage in the calculation produces a smaller cumulative error. In an unstable algorithm, errors in algorithm parameters may grow in an unbounded fashion.

Convergence: Convergence means that the solution to a numerically approximated equation approaches the true solution to the partial differential equation with the same boundary and initial conditions as the mesh is refined. Lax's equivalence theorem states that consistency and stability are necessary and sufficient conditions of convergence for numerical solutions. Most computational work assumes that this theorem applies to both the linear and nonlinear problems.

Conservation: Conservation laws state that various quantities remain constant, regardless of the fluctuations and flow phenomena occurring. Conserved quantities typically express integrals of continuous fluid properties over the region of interest. Conservation laws exist for mass, momentum, energy, species number, densities, and various other ("conservative") quantities. Conservation laws can also be used to formulate discretized finite-difference algorithms in which finite sums of simulated physical quantities over the grid cells are conserved. These sums are discrete analogs of the conservative integrals. When conservation laws are not enforced, truncation and round-off errors can grow without bound, making even very simple systems behave bizarrely. If the strict conservation of a property in an algorithm is given up, something else should be demanded as a trade-off. For example, asymptotic algorithms do not conserve strictly but produce accurate answers for very long time steps. When a conservation law for a quantity is built into an algorithm, the conservative characters of the quantity becomes a measure of its accuracy.

Causality: Causality states that effects do not happen before causes, that they have finite propagation speeds, or require finite times to transmit information. It has important implications for the choice of the solution method.

Reversibility: Reversibility means that the equation is invariant under the transformation $t \rightarrow -t$. Computationally, reversibility means that if at time t_n the calculation is stopped and the time step is made negative, the solution retraces its evolution back to time t_o. It is important to preserve reversibility for waves and convection, but not really relevant to diffusion or chemical kinetics. Reversibility may not be easy to achieve in practice even when an algorithm is designed to be reversible, because of nonlinear terms in the equations or other complicating effects.

Positivity: Positivity means that a quantity (such as the mass density) which is initially positive everywhere cannot become negative. This property holds for any quantity described by the convection or diffusion equations. This property is important for all the transport processes except for waves.

Monotonicity: Monotonicity implies that the algorithm does not generate new extreme. A monotonic algorithm maintains the positivity, but a simple positive algorithm is not necessarily monotonic. A nonmonotonic algorithm can maintain the positivity.

Accuracy: Accuracy involves computer precision, numerical convergence, and algorithm stability. A method need not always be convergent for it to be useful. Asymptotic methods, used near equilibrium or for the solution of stiff equations, are examples of this. Asymptotic numerical approximations do not necessarily get better as the step size is made smaller.

Robustness: Robustness represents the extent to which an algorithm can handle a full spectrum of situations, continuing to perform despite some violation of the assumptions in its specification.

Adaptability: Adaptability states the applicability of an algorithm to different physical environments (e.g. air-water) and hardware platforms (e.g. parallel machines).

Speed and memory: It is important to characterize the efficiency of an algorithm's use of the computer resources (memory, for example) available to perform its function in the minimum amount of time (speed).

Numerical algorithms cannot satisfy all the requirements listed above and they are imperfect in varying degrees of accuracy. Although many new methodologies have been suggested for the treatment of advection and diffusion, there have been few systematic comparative studies on the merits and the problems of combining different transport algorithms. Useful algorithms must give quantitatively correct answers, must converge, must be stable and preferably fast. In addition, the memory requirements of algorithms must be considered. For many problems, users' requirements are so diverse that it is seldom possible to meet all of these criteria with one piece of software. As a result, a suite of different routines is needed to meet the different requirements of each user.

4 Numerical solvers for advection

As mentioned above, non-Eulerian air quality models do not explicitly solve the advection equation. There have been many studies of the numerical advection algorithms for use in Eulerian air quality models. The reason why this subject has attracted so much attention is that each numerical advection scheme has certain undesirable computational characteristics. The advection equation is a hyperbolic partial differential equation that requires both initial and boundary conditions. Spatial discretization of the solution generates a finite number of Fourier modes that travel at different speeds and lead to constructive and destructive interference. If the high wavenumber Fourier modes are damped significantly, then numerical diffusion becomes prevalent. Various errors introduced by advection algorithms have long been a major source of inaccuracy in air quality models. The presence of highly nonlinear chemical interactions between concentration fields of different species further complicates the problem.

In air quality modeling, it is essential that the advection scheme be mass conservative and positive-definite. Accuracy must not be compromised to achieve these goals. Recent algorithmic developments and rapid progress in computer architectures have led to improved simulation capabilities. However, because of the critical role of AQMs in determining source-receptor relations and establishing emissions control strategies, more stringent requirements are imposed on advection algorithms. Examples include resolution of signals of the order of grid size (i.e., $2\Delta x$ wave) and conservation of mixing ratio or mass at grid-cell level under realistic meteorological conditions. Depending on the characteristics of advection algorithms, they can meet some requirements better while failing to satisfy others. A perfect advection algorithm does not exist and users must decide what algorithm will work best for the problem at hand. This section presents the basic concepts behind the two advection algorithm types, semi-Lagrangian and flux-forms.

4.1 Equations for advection

The advection equation can be written in two different forms depending on the choice of the prognostic variables. When concentration in density units is used, the conservation equation for trace species follows the continuity equation for air density:

$$\frac{\partial(\varphi_i J_s)}{\partial t} + m^2 \nabla_s \bullet \left(\frac{\varphi_i J_s \hat{\mathbf{V}}_s}{m^2} \right) + \frac{\partial(\varphi_i J_s \hat{v}^3)}{\partial s} = J_s Q_{\varphi_i} , \tag{12}$$

where Q_{φ_i} is the sources or sinks of trace species. For simplicity, eqn. (12) can be rewritten in terms of the scaled density $\varphi_i^* = \varphi_i J_s / m^2$. When there is no source term, we have

$$\frac{\partial \varphi_i^*}{\partial t} + \nabla_s \bullet \left(\varphi_i^* \hat{\mathbf{V}}_s\right) + \frac{\partial(\varphi_i^* \hat{v}^3)}{\partial s} = 0 . \tag{13}$$

Because eqn. (13) expresses the transport process in terms of flux divergence and assures mass conservation, it is called either a flux form or conservation form advection equation. It represents the instantaneous continuity equation. Readers should be reminded that the following descriptions apply to the transport by the mean wind components. However, actual decomposition of the equation into mean and turbulent components is deferred to a later section where turbulent diffusion is discussed.

For convenience, the advection equation is divided into horizontal and vertical equations by applying the fractional time-splitting method,

$$\left.\frac{\partial \varphi_i^*}{\partial t}\right|_{hadv} = -\nabla_s \bullet \left(\varphi_i^* \hat{\mathbf{V}}_s\right) = -\frac{\partial(\varphi_i^* \hat{v}^1)}{\partial \hat{x}^1} - \frac{\partial(\varphi_i^* \hat{v}^2)}{\partial \hat{x}^2}, \tag{14a}$$

$$\left.\frac{\partial \varphi_i^*}{\partial t}\right|_{vadv} = -\frac{\partial(\varphi_i^* \hat{v}^3)}{\partial s}. \tag{14b}$$

Although the coordinate system is not exactly orthogonal, separation of vertical and horizontal components is justifiable because the determinant of the covariant tensor is a simple multiplication of the vertical and horizontal components of the Jacobian matrix (i.e., J_s and m^{-2}) as shown in eqn. (6).

On the other hand, one can rewrite eqn. (12) in terms of mixing ratio q_i,

$$\frac{dq_i}{dt} = \frac{d(\varphi_i / \rho)}{dt} = \frac{\varphi_i}{\rho}\left(\frac{Q_{\varphi_i}}{\varphi_i} - \frac{Q_\rho}{\rho}\right). \tag{15}$$

For pure advection processes with ideal meteorological input having no mass-consistency error, the right-hand side (r.h.s) of eqn. (15) vanishes and we have a Lagrangian form advection equation

$$\frac{dq_i}{dt} = \frac{\partial q_i}{\partial t} + \hat{\mathbf{V}}_s \bullet \nabla_s q_i + \hat{v}^3 \frac{\partial q_i}{\partial s} = 0. \tag{16}$$

This is the basis of the Lagrangian model. This concept can be used in Eulerian models as well. However, because the information must be discretized on the Eulerian grid, a numerical advection scheme based on eqn. (16) is often called a Semi-Lagrangian Transport (SLT) method. Usually the SLT method is applied in computing horizontal two-dimensional advection where the last term in eqn. (16) drops out. Vertical advection is computed separately, or sometimes solved together with vertical diffusion. In the following we discuss a few advection algorithms representative of these two distinct types of advection equations.

4.2 Semi-Lagrangian transport (SLT) algorithms

There are several semi-Lagrangian schemes listed in the literature (e.g., Williamson and Rasch [12], Smolarkiewicz and Grell, [13]). In the SLT method, one estimates a set of backward trajectories of particles that arrive exactly at grid points at the end of an advection time step. The following two-step process accomplishes this:

Step 1: Determine the trajectory and origin of a particle arriving at the grid points. The Lagrangian solution to eqn. (16) determines the departure point (x_D, y_D) of a particle currently at (x_A, y_A) as:

$$\begin{aligned} x_D &= x_A - \Delta t\, u \\ y_D &= y_A - \Delta t\, v \end{aligned}, \qquad (17)$$

where we used notations $x = \hat{x}^1$, $y = \hat{x}^2$, $u = \hat{v}^1$, and $v = \hat{v}^2$ for simplicity. For a nonuniform and time dependent wind field, the trajectory must be first determined through the use of an iterative scheme. For example, the mid-point of the trajectory is determined with an iterative process such as:

$$\begin{aligned} x_M^{k+1} &= x_A - \frac{\Delta t}{2} u(x_M^k, y_M^k) \\ y_M^{k+1} &= y_A - \frac{\Delta t}{2} v(x_M^k, y_M^k) \end{aligned}. \qquad (18)$$

Step 2: Assign concentration at the origin to the destination. Once the starting point of the trajectory is estimated, the concentration from the origin is assigned to the grid point of arrival. Since the origin does not always coincide with a grid point, an interpolation scheme is necessary to estimate the concentration at the starting point. Depending on the algorithms used to accomplish the two distinct steps, characteristics of SLT schemes can differ greatly. The interpolation scheme used determines the damping characteristics. For example, linear interpolation has unacceptably large damping whereas cubic interpolation gives very little damping. However, the higher order interpolation schemes introduce nonmonotonic and sometimes negative results. The method is not mass conservative, and spurious oscillations may be generated.

Assuring mass conservation and obtaining positive or monotonic results require special treatments. Noninterpolating schemes can be monotonic (i.e., maintain the peaks and valleys without generating spurious ones). However, because schemes force displacement of a field from one mesh to another, they are very dispersive (i.e., shifting of relative locations of local peaks and valleys).

When a SLT scheme is applied in AQMs, the predicted mixing ratio represents the value at the local position, which is different from the grid cell-averaged value. This is the source of an apparent mass conservation problem with the SLT methods that sums of the mixing ratios are not conserved. An interpolation scheme that assures conservation of mass can be devised. The noninterpolating scheme assumes the mixing ratio to be uniform in a cell, and thus the sum of mixing ratios is conserved for a nondivergent flow. One advantage of the SLT scheme is that a large time step can be used without being subject to the Courant stability condition. This feature does not benefit current Eulerian AQMs because the time-dependent precursor emissions are injected into the cell at each time step (limited by the Courant condition) before the computation of chemical reactions. Additional investigation must be pursued to take advantage of the SLT's large time steps such as by studying relative time scales of different atmospheric processes and rearranging the computational sequence among them.

Even with the exact solution, the total mass apparently fails to be conserved depending on the relative location of the grid points and origins of the trajectories. When the Lagrangian advection equation is solved analytically, the solution propagates along the characteristic curves called trajectories. Figure 1 shows a cyclic fluctuation of domain average mixing ratios when three cones are advected by a rotational flow with the angular speed of one revolution per day. The cyclic fluctuation of mixing ratio sums, although small in magnitude, is caused by the sampling of mixing ratios from the initial condition field that are sometimes not exactly on the grid points. The relative grid positions to the signal coincide at every six hours and the time series repeats after 24 hours, when the signal returns to the original position. The symptom is analogous to the mass conservation problem of the SLT algorithm with interpolation. The difference is that the analytical solution can reconstruct the initial field when the origins coincide with the grid points used at the start of simulation. Unlike the analytical example, an actual SLT scheme does not remember the original field and thus the recovery is not possible. The analytical solution can be implemented by means of a discretized initial condition field as well. The latter case corresponds to the non-interpolating SLT scheme and the test example shows that the sum of mixing ratio is not conserved as well.

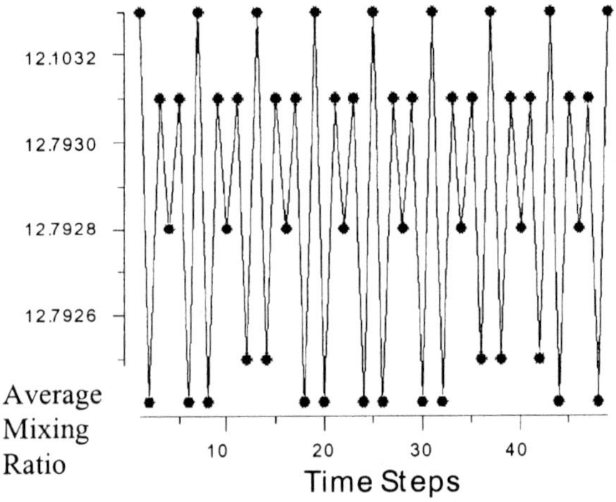

Figure 1: Change of domain average mixing ratio under a rotational flow computed from the analytical solution. The signal (three cones) advected is the same as shown in Figure 2.

4.3 Flux-form advection algorithms

A flux-form advection algorithm asserts that the rate of change of the total mass contained in a fixed cell equals the flux of the mass across the boundaries of the cell. Most Eulerian AQMs use the flux form algorithms to conserve properties during advection. The flux form is a natural choice because it is based on the continuity equation without any assumptions regarding the atmospheric dynamics.

Many AQMs split the horizontal advection equation further in two directions. They solve for the one-dimensional (1-D) equation in sequence for each direction, using the solution of one as the initial condition of the other. This is called a time-splitting method. Errors associated with the time-splitting method are discussed below. When using appropriately interpolated contravariant wind components, the 1-D advection in the generalized coordinate system is equivalent to the 1-D equation in the Cartesian coordinate system. Therefore, it is sufficient to discuss advection algorithms in Cartesian coordinates:

$$\frac{\partial \varphi}{\partial t} + \frac{\partial (u\varphi)}{\partial x} = 0 \quad . \tag{19}$$

The differential equation can be discretized in finite difference form in such a way (Odman [10]):

$$\varphi_j^{n+1} = \varphi_j^n - \frac{\Delta t}{\Delta x}\left(F_{j+1/2}^n - F_{j-1/2}^n\right). \tag{20}$$

where $F_{j+1/2}^n$ and $F_{j-1/2}^n$ denote the advective fluxes ($F = u\varphi$) through the faces of cell j, Δt is the time step, and Δx is the horizontal grid spacing. A uniform spaced and staggered grid system is used here for simplicity (the grid cell average of the concentration, φ_j, is represented at the center of the cell, while the advection velocity, $u_{j+1/2}$, is defined at grid cell interfaces). To maintain the numerical stability and to accommodate other physical changes such as emissions in a synchronized way, the time step of 1-D advection must satisfy the Courant-Friedrich-Lewy (CFL) condition for hyperbolic equations,

$$\max_j (\beta_{j+1/2}) \leq 1 \quad , \tag{21}$$

where $\beta_{j+1/2} = |u_{j+1/2}| \Delta t / \Delta x$ is the Courant number for advection. We define a nondimensional coordinate as $\eta = (x - x_{j-1/2})/\Delta x$, $0 \leq \eta \leq 1$, in a grid cell j. Depending on the direction of the velocity, the flux $F_{j+1/2}$ can be defined as

$$F_{j+1/2} = \begin{cases} \dfrac{\Delta x}{\Delta t} \displaystyle\int_{1-\beta_{j+1/2}}^{1} \varphi_j(\eta)d\eta, & u_{j+1/2} \geq 0 \\ \dfrac{\Delta x}{\Delta t} \displaystyle\int_{0}^{\beta_{j+1/2}} \varphi_{j+1}(\eta)d\eta, & u_{j+1/2} < 0 \end{cases} . \tag{22}$$

Equation (22) provides a general expression of the flux computation in several finite-volume methods, such as the piecewise parabolic method (Colella and Woodward [14]), the Bott scheme (Bott [15]), and the Yamartino-Blackman Cubic scheme (Yamartino [16]) and others.

As with other numerical methods, the flux-form advection algorithms suffer from inaccuracies inherent in the difference between a numerical algorithm and the analytical form of the equation. With a finite difference scheme defined by eqns. (20) and (22), one can attempt to reconstruct the original differential equation. Application of a Taylor series expansion of the concentration field near the grid point to the finite difference form reveals that an algorithm is a different approximation to eqn. (19). Often, the difference is in the form of diffusion and thus we call this a numerical diffusion. This reduces peak values more than would happen in a physical diffusion process. Similarly, the finite difference form can be replaced with a spectral expression

$$\varphi(x_j,t) = \varphi(j\Delta x, \omega n\Delta t) = \hat{\varphi}(k,\omega)e^{i(kj\Delta_x + \omega? \Delta_t)} , \tag{23}$$

where κ is the wavenumber, ω is the frequency, and Δt is the advection time step. Depending on the wavelengths of signals, the phase speeds can be different from the advection velocity, u. In such a case, we say that the approximation has numerical dispersion. A numerical algorithm with a large dispersion causes erroneous estimation of trajectories of signals with different wavelengths.

In addition, an attempt to maintain a high-order accuracy can cause spurious oscillations associated with the scheme. To remove the spurious oscillation, one may introduce a (monotonic) correction mechanism. Typically, a nonlinear flux-correction, or nonlinear filtering procedure, provides this mechanism. In advection schemes, such adjustments are either applied implicitly through the solution or explicitly as a subsequent step to the linear solution. The flux correction step is an important source of nonlinearity.

4.4 Time-splitting error

Time-splitting errors occur when a one-dimensional advection algorithm is applied consecutively to simulate the two- or three-dimensional advection. To understand the nature of the problem, we use a two-dimensional Taylor series expansion of concentration $\varphi(x,y,t)$ about the reference point:

$$\varphi_{lm}^{n+1} = \varphi_{lm}^{n} - u\Delta t \left.\frac{\partial \varphi}{\partial x}\right|_{lm}^{n} - v\Delta t \left.\frac{\partial \varphi}{\partial y}\right|_{lm}^{n} + \frac{u^2\Delta t^2}{2}\left.\frac{\partial^2 \varphi}{\partial x^2}\right|_{lm}^{n} + \frac{v^2\Delta t^2}{2}\left.\frac{\partial^2 \varphi}{\partial y^2}\right|_{lm}^{n}$$

$$+ \frac{uv\Delta t^2}{2}\left.\frac{\partial^2 \varphi}{\partial x \partial y}\right|_{lm}^{n} + HOT, \qquad (24)$$

where index n stands for the current time step, l and m are indices for the reference position, and HOT represents higher order terms. Neglecting the high order term, eqn. (24) can be written in a symbolic operator form

$$\varphi_{lm}^{n+1} = (I - A_x - B_y + C_{xy})\varphi_{lm}^{n}, \qquad (25)$$

where I is the unit matrix, and differential operators A_x, B_y and C_{xy} are defined as

$$A_x = u\Delta t \left.\frac{\partial}{\partial x}\right|_{lm}^{n} - \frac{u^2\Delta t^2}{2}\left.\frac{\partial^2}{\partial x^2}\right|_{lm}^{n},$$

$$B_y = v\Delta t \left.\frac{\partial}{\partial y}\right|_{lm}^{n} - \frac{v^2\Delta t^2}{2}\left.\frac{\partial^2}{\partial y^2}\right|_{lm}^{n},$$

$$C_{xy} = \frac{uv\Delta t^2}{2}\left.\frac{\partial^2}{\partial x \partial y}\right|_{lm}^{n}.$$

A_x and B_y are proportional to the net fluxes in x- and y-directions, respectively. C_{xy} is the operator responsible for the cross-term. The time splitting application of a second-order 1-D numerical scheme can be expressed as

$$c_{lm}^{n+1} = (I - A_x - B_y + A_x B_y)c_{lm}^{n}. \qquad (25')$$

The so-called cross-term error for the time splitting method is the difference between eqns. (25) and (25'):

$$A_x B_y - C_{xy} = -\frac{uv^2\Delta t^3}{2}\frac{\partial}{\partial x}\frac{\partial^2}{\partial y^2}\bigg|_{lm}^{n} - \frac{u^2 v\Delta t^3}{2}\frac{\partial}{\partial y}\frac{\partial^2}{\partial x^2}\bigg|_{lm}^{n} + \frac{u^2v^2\Delta t^4}{4}\frac{\partial^4}{\partial x^2\partial y^2}\bigg|_{lm}^{n}. \quad (26)$$

The definitions of elementary streamline patterns, i.e., divergence, vorticity, stretching and shearing flows, respectively, are given as

$$S_\Delta = \frac{1}{2}(\partial u/\partial x + \partial v/\partial y), \quad (27a)$$

$$S_\Omega = \frac{1}{2}(\partial v/\partial x - \partial u/\partial y), \quad (27b)$$

$$S_\Gamma = \frac{1}{2}(\partial u/\partial x - \partial v/\partial y), \quad (27c)$$

and

$$S_\Lambda = \frac{1}{2}(\partial v/\partial x + \partial u/\partial y). \quad (27d)$$

When we neglect the higher order term $O(\Delta t^4)$ for the moment, and substitute u and v with eqns. (27a-d), eqn. (26) can be rewritten as

$$A_x B_y - C_{xy} = -\frac{uv\Delta t^3}{2}\left\{S_\Lambda\left(\delta_x\frac{\partial^2}{\partial y^2}\bigg|_{lm}^{n} + \delta_y\frac{\partial^2}{\partial x^2}\bigg|_{lm}^{n}\right) + S_\Omega\left(\delta_x\frac{\partial^2}{\partial y^2}\bigg|_{lm}^{n} - \delta_y\frac{\partial^2}{\partial x^2}\bigg|_{lm}^{n}\right)\right.$$

$$\left. + S_\Gamma\left(-\delta_y\frac{\partial^2}{\partial x\partial y}\bigg|_{lm}^{n} + \delta_x\frac{\partial^2}{\partial x\partial y}\bigg|_{lm}^{n}\right)\right\} + O(\Delta t^4), \quad (28)$$

where the standard difference notation $\delta_x f = f(x+\Delta x/2) - f(x-\Delta x/2)$ and $\delta_y f = f(y+\Delta y/2) - f(y-\Delta y/2)$ are used. Inspection of eqn. (28) reveals that the part of the operator related to the stretching deformation (S_Γ) is nonzero for the linear or high-order variation of concentrations in each direction. On the other hand, the operators involved with the shearing deformation (S_Λ) and rotation (S_Ω) are the second order quasi-linear operators (when the finite differencing approximation is applied) in each direction. These terms vanish for linearly varying signals. Therefore, in most testing conditions, the stretching deformation is often responsible for the cross-term effects.

Several 1-D numerical algorithms have been modified to minimize the cross-term effects when applied in multi-dimensional situations. They mostly utilize the error analysis, eqn. (26), to maintain the order of accuracy of the 1-D schemes. Also, the mass-consistency adjustment algorithm described below can effectively remove the time splitting error.

4.5 Adjustment of mass consistency error

A fundamental requirement for the numerical transport algorithms is the conservation of trace species in the domain. Ideally, meteorological data for air quality simulations should be mass consistent. However, meteorological data are often not mass consistent. When simulating meteorological conditions for a limited area such as on an urban or regional scale, the total air mass in the modeling domain is subject to the inflow, outflow, top and bottom boundary conditions imposed by large synoptic scale weather systems and surface exchanges of heat and moisture. It is also difficult to make $J_s Q_\rho$ in eqn. (8) vanish everywhere in the computational cells.

In such a case, an algorithm that theoretically conserves mass may fail to conserve trace species mass in the application. Byun [7] and others proposed to handle the mass inconsistency by forcing conservation of mixing ratio during the advection process. This forces the r.h.s of eqn. (16) to vanish. The undesirable effects of the mass-inconsistent meteorology data can be corrected with

$$(\varphi_i^*)^{cor} = (\varphi_i^*)^T \exp\left[\int \frac{Q_\rho}{\rho} dt\right], \tag{29}$$

where superscripts '*T*' and '*cor*' represent values after transport (advection) and after correction, respectively.

The mixing-ratio conservation scheme is demonstrated to be useful for photochemical air quality models where chemical production and loss terms are computed using molar mixing ratio. This type of correction method should be used only for improving mass conservation characteristics of the numerical advection process after the mass inconsistency in the meteorology data is minimized using a variational wind field adjustment scheme discussed in the section above. Characteristics of the correction procedures can be summarized as follows:

• The correction scheme fixes mass conservation errors due to the time splitting, numerical algorithms, and the mass inconsistent meteorological data input.
• It does not, however, improve the inherent properties of a numerical advection scheme such as monotonicity, or numerical diffusivity.
• In a budget study, such as in process analysis, the adjustment process must be considered as an integral part of the three-dimensional advection. Horizontal advection, vertical advection, and the mass correction step all together simulate the atmospheric advection process.

Table 2 summarizes three correction methods used in some AQMs. A correction scheme should conserve the total tracer mass under dynamically consistent wind and density fields. Further, it should at least conserve mixing ratio even in the event that wind and density fields are not mass consistent.

Table 2: Mass correction schemes tested. Superscripts *cor*, *T* and *int* represent corrected, transported and interpolated quantities, respectively.

Correction method	Air quality model using the correction scheme	Mass correction algorithm
Advection of calibration gas	Urban Airshed Model (UAM) (Schieffe and Morris[17])	$\varphi_i^{cor} = \frac{\varphi_i^T}{\varphi_r^T}\varphi_r,\ \varphi_r = 1.0$
Advection of air density	SARMAP Air Quality Model (SAQM) (Chang et al.[18])	$\varphi_i^{cor} = \frac{\varphi_i^T}{\rho^T}\rho^{int}$
Advection of ρJ_s with two-step time splitting	Community Multi-scale Air Quality (CMAQ) model (Byun and Ching[19])	$(\frac{\varphi_i J_s}{m^2})^{cor} = \frac{(\varphi_i J_s / m^2)^T}{(\rho J_s / m^2)^T}(\frac{\rho J_s}{m^2})^{int}$

The CMAQ correction scheme is expected to maintain conservation of mixing ratio to the machine precision. The other two correction schemes ignore effects of variation in density field or coordinate/grid structures during the correction process. For Cartesian coordinates where the Jacobian is uniform and constant with time, the SAQM scheme is equivalent to the CMAQ scheme. All of the correction schemes become identical for the atmosphere with uniform (spatially) and constant (temporally) density distribution and nondivergent flows in a Cartesian coordinate system.

Here, we demonstrate performance of the CMAQ correction scheme with two-dimensional rotational flow. A more complete description of the study will be presented in another paper, Byun and Lee [20]. The density field is non-uniform and has an additional mass-inconsistent component. The test result provided in Figure 2 is for a rotational flow with ~10% mass-inconsistent density distribution in the diagonal direction for the 90x90 cell 4 km resolution domain. Plate (a) of Figure 2 reflects the effect of this mass-inconsistency on the advection, demonstrating problems associated with certain meteorology inputs used in air quality modeling. Plate (b) shows that transport with the CMAQ adjustment removes the problem. Here, not only the mixing ratio, but also the total mass of the trace species is conserved. For a limited-area air quality simulation, operational meteorological input and pollutant distributions cannot guarantee the boundary condition for maintaining mass conservation and thus, in all practical purposes, only conservation of the mixing ratio should be expected.

For the time-dependent flows, the mass continuity, eqn. (8), should be satisfied all of the time. In the following, an advection test with a divergent nonsteady but mass consistent flow is presented. Figure 3 shows the density and wind fields with a combination of rotational, stretching, shearing and divergent components. The small divergence in the flow field is compensated by the decrease in the total mass according to the continuity equation. The divergent rate is set to a small value; that is, the change in the parameter for the 24-hour simulation is around 1%. Even with this little divergence, the non-corrected results lose mass by about 25% (figure not present here). Most of the error results from the time-splitting error described in Section 4.4. Figure 4 shows that the three correction schemes effectively remove the time splitting errors and that the CMAQ scheme maintains the constancy of mixing ratio throughout the simulation.

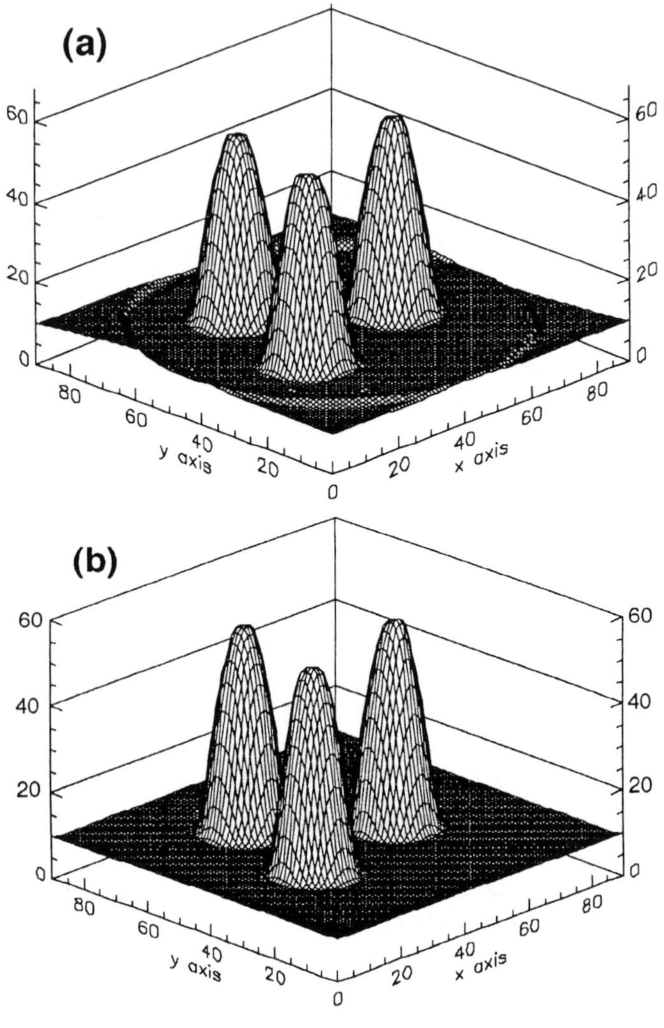

Figure 2: Advection of three cones with the piece-wise parabolic method under a mass-inconsistent rotational flow. Plate (a) shows result without the mixing ratio adjustment scheme and (b) with the CMAQ adjustment scheme.

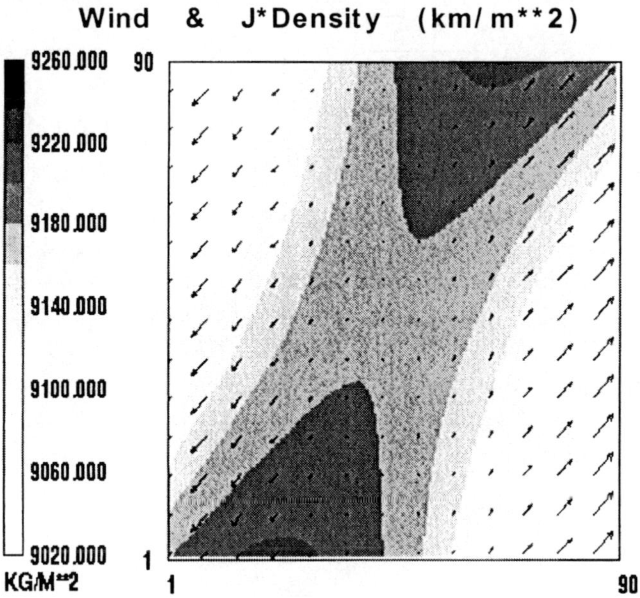

Figure 3: A weakly divergent two-dimensional flow with mass consistent nonhomogeneous density field (multiplied with Jacobian). The wind field remains the same throughout the simulation period (24 hours) while the density field is modified by the amount of divergence to maintain the mass consistency.

5 Atmospheric diffusion algorithms

Because of the stochastic nature of atmospheric motions, the trace species continuity equations must be averaged to form a set of deterministic equations before they can be solved numerically. First, velocity components and concentrations in eqn. (10) are decomposed into mean ($\bar{\hat{v}}^j$) and turbulent ($\hat{v}^{j\prime}$) terms, that is $\hat{v}^j = \bar{\hat{v}}^j + \hat{v}^{j\prime}$. The map scale factor and Jacobian are not decomposed because they are the fundamental parameters that define the structure of the model. Then application of ensemble averaging leads to the governing diffusion equation used in Eulerian air quality models:

$$\frac{\partial \varphi_i^*}{\partial t} + \hat{\nabla}_s \bullet \left[\varphi_i^* \overline{\hat{\mathbf{V}}}_s \right] + \frac{\partial (\varphi_i^* \overline{\hat{v}^3})}{\partial s} + \hat{\nabla}_s \bullet \left[\overline{\rho} \sqrt{\hat{\gamma}} \hat{\mathbf{F}}_{q_i} \right] + \frac{\partial (\overline{\rho} \sqrt{\hat{\gamma}} \hat{F}_{q_i}^3)}{\partial s} = Q_{\varphi_i}, \qquad (30)$$

where $\hat{\mathbf{F}}_{q_i} = \hat{\mathbf{i}}\overline{q_i' \hat{v}^{1\prime}} + \hat{\mathbf{j}}\overline{q_i' \hat{v}^{2\prime}}$ represents the horizontal turbulent diffusion flux vector, $\hat{F}_{q_i}^3 = \overline{q_i' \hat{v}^{3\prime}}$ is the vertical diffusion flux, and Q_{φ_i} stands for all other sources and sinks, including emissions and chemical reactions of trace species. Resulting covariance terms from the linear decomposition of the state variables and ensemble averaging are called Reynolds fluxes. In the following, general numerical methods for simulating the turbulence diffusion processes in Eulerian AQMs are presented. Refer to Byun et al. [11] for the details.

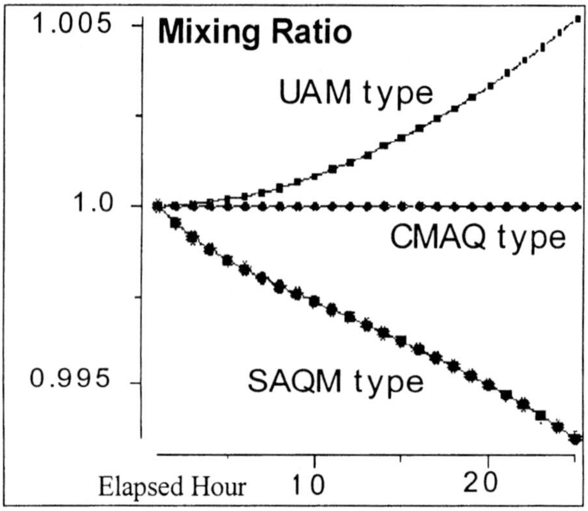

Figure 4: While the three correction schemes effectively remove the time splitting errors well, the CMAQ correction scheme performs the best.

5.1 Closure techniques

Introducing the turbulent diffusion in terms of the Reynolds fluxes leads to a new problem set in which the number of unknowns is larger than the number of equations. This so-called closure problem is caused by the attempt to represent nonlinear processes such as momentum advection using a linear decomposition such as the Reynolds decomposition. Closure techniques provide methods of parameterizing the unknown Reynolds fluxes in terms of known mean quantities. When the parameterization results in the flux exchange only between adjacent cells (or layers), the scheme is called local. When the parameterization allows turbulence flux exchange among the grid cells (layers) that are not necessarily adjacent, the scheme is called nonlocal. In the following, general characteristics of local and nonlocal closure schemes are discussed.

Local closure: Local closure assumes that turbulence is analogous to molecular diffusion, i.e., that an unknown turbulence flux at any point in space is parameterized by values of known quantities at the same point. An example of a local closure scheme is the first-order eddy diffusion method that approximates Reynolds fluxes with a gradient transport theory, or a mixing length theory. One of the problems with this is finding a rational basis for parameterizing the eddy diffusivity. Also, the theory fails when eddies larger than the grid size are present, as in the convective boundary layer.

The one-and-a-half order closure in a meteorological model retains the prognostic equations for the mean state variables and adds equations for the variances of those variables. The set of one-and-a-half order equations is obtained by simplifying the full second-order turbulence equations. Instead of the velocity component variance equations, the turbulent kinetic energy (TKE) equation is often used. By including the variance equations, the number of unknowns to be parameterized has been increased compared to the first-order closure approach. However, the benefit is that the eddy diffusivity can be parameterized not only with the mean quantities but also with the TKE and the temperature variances that characterize turbulence intensity.

On the other hand, if an air quality model is based on one-and-a-half order closure in a true sense, the prognostic equations of the variances for the tracer concentrations should be included explicitly. In practical Eulerian photochemical AQMs, solving additional prognostic variance equations for the trace species is computationally very expensive. Also, the higher order terms related with the trace species fluctuations must be parameterized. Therefore, the one-and-a-half order closure for an air quality model often actually means that the diffusion equations for the tracer species are formulated with the first order closure, while eddy diffusivities are estimated with the TKE information from a meteorological model with one-and-a-half order schemes for wind, temperature and humidity.

The set of second-order turbulence equations includes all the second moment terms. To derive these terms, parameterizations of the full set of third-order moments are required. Similar to the first-order case, second-order closure approximates terms involving third-order moments. Several basic closure assumptions such as down-gradient diffusion, return to isotropy, and turbulent dissipation in the inertial subrange are used in the parameterization of the third moment terms. These parameterizations should be valid, especially, for the scales of the energy containing eddies that are sensitive to atmospheric stability. Measurements of high-order moments in the real atmosphere are difficult because of large scatter in the direct flux measurements. A long averaging time or a very large sample size of data is required to estimate higher-order moments using eddy-correlation methods.

For air quality applications, especially for a complex chemical reaction system, the second-order closure technique requires too many ad-hoc assumptions that cannot be confirmed by observations or other theoretical reasoning. In addition, it incurs a prohibitively high computational (CPU and memory) cost. We can solve the first-order tracer diffusion equations with the variances and covariances for wind components, temperature and humidity provided from the second-order meteorological models. For air quality applications, the true second-order closure formulation must solve for the cross-species covariances explicitly. Some researchers have attempted second-order closure for simple chemical mechanisms with a limited number of photochemically reactive species. The introduction of additional parameterizations for the third-order moment terms among the tracer species themselves and wind components (about which we lack sufficient knowledge), and the added cost of solving for a large number of covariance terms, make this scheme impractical and prohibitively costly for operational Eulerian photochemical models.

Nonlocal closure: Non-local closure recognizes that larger-size eddies can transport fluid across finite distances before the smaller eddies have a chance to cause mixing. This advection-like concept is supported by observations of thermals rising with undiluted cores, finite size swirls of leaves or snow, and the organized circulation patterns sometimes visible from cloud photographs.

Two main approaches of non-local closure methods are the transilient turbulence theory and the spectral diffusion theory. Both allow a range of eddy sizes to contribute to the turbulent mixing process. The spectral diffusion theory attempts to simulate the mixing process by transforming signals into a spectral space. For example, a spectral diffusion model of Otte and Wyngaard [21] represents the mean variables within the PBL by a truncated series of Legendre polynomials. The first Legendre mode represents the layer average, and additional modes add structure to the vertical profiles. Only a few modes are necessary to resolve vertical profiles comparable to high-resolution diffusion models. However, the need to fit a different number of spectral modes for each trace species makes the scheme less attractive for air quality applications and thus it has not been applied to an operational AQM. The transilient turbulence theory (e.g., Stull [22]) provides a general representation of the turbulent flux exchange process. The Latin word transilient, meaning to jump over, is used because turbulent eddies can transport mass and momentum directly across several grid layers. A variety of mixing processes can be modeled with the transilient scheme depending on the form of the transilient matrix. Examples include complete mixing, top-down/bottom-up mixing, asymmetric convection mixing, small-eddy mixing, cloud top entrainment, a detraining updraft core, patchy turbulence, no turbulence, or eddies triggered by the surface layer. Non-local closure is most suitable for describing vertical turbulence mixing process, which should represent turbulent diffusion and atmospheric transport by eddies of different sizes simultaneously.

5.2 General description of vertical mixing algorithms

The numerical solver kernel for the flux form vertical diffusion should solve for:

$$\frac{\partial q}{\partial t} = -\frac{\partial F}{\partial s} \cdot \qquad (31)$$

To account for the loss process due to deposition in the lowest model layer (j=1), dry deposition flux is considered as the flux boundary condition at the surface, that is:

$$\left.\frac{\partial q_1}{\partial t}\right|_{dep} = -\frac{v_d}{h_{dep}} q_1 , \tag{32}$$

where the geometric thickness of the lowest model layer is used for h_{dep}. The flux form representation is extremely useful in describing the algorithms based on both local and non-local closure methods. The transilient matrix formulation provides a powerful general method describing the flux exchanges among vertical layers.

To represent turbulent mass exchange processes with the transilient parameterization, the boundary layer height must coincide with the height of layer interfaces of the vertical grid. For most situations, the index for the boundary layer top (L_p) is less than the total number of model layers (i.e., $L_p < N$). With the transilient turbulence formulation, the new values of the trace species mixing ratio q due to turbulent mixing for a layer j at a future time $(t + \Delta t)$ can be written as:

$$q_j(t+\Delta t) = \sum_{k=1}^{L_p} c_{jk}(t,\Delta t) q_k(t) , \tag{33}$$

where c_{jk} are the components of a transilient matrix and subscripts j and k are indices of two different grid boxes (vertical layers) below boundary layer top in a column of atmosphere. If we consider turbulent mixing between grid boxes j and k, c_{jk} represents fraction of air mass ending in the grid box j that came from grid box k. The grid box j is considered as the "destination" box while grid box k is considered as the "source" box. Thus, the change in the tracer concentration due to turbulent mixing for grid box j at a time interval Δt is a simple matrix multiplication with concentration from the source cell.

The conservation of air mass requires that the sum over k of all mixing fractions be unity:

$$\sum_{k=1}^{L_p} c_{jk}(t,\Delta t) = 1 , \tag{34}$$

and the conservation of trace gas amount requires that the sum over j of all mass-ratio weighted transilient coefficients be unity as well:

$$\sum_{k=1}^{L_p} \frac{\Delta s_j}{\Delta s_k} c_{jk}(t,\Delta t) = 1 , \tag{35}$$

where $\Delta s_j = s_j - s_{j-1}$, and $\Delta s_j / \Delta s_k$ represents the mass ratio (i.e., ratio of layer thickness for mixing ratio q) between the source and destination boxes. In order for the transilient turbulence theory to be useful, the coefficients should be determined with appropriate turbulent flux parameterizations.

Because the transilient matrix describes the exchange of mass between grid boxes, the kinematic turbulent fluxes across the j-th level can be expressed, for a vertical layering with non-uniform grid spacing (Stull [23]), for $2 \le j \le L_p$ as:

$$F_j(t,\Delta t) = F_{j-1}(t,\Delta t) + \frac{\Delta s_j}{\Delta t} \sum_{k=1}^{L_p} c_{jk}(t,\Delta t)[q_j(t) - q_k(t)] , \tag{36a}$$

and for j=1 as:

$$F_1(t,\Delta t) = \frac{\Delta s_1}{\Delta t} \sum_{k=1}^{L_p} c_{1k}(t)[q_1(t) - q_k(t)] . \tag{36b}$$

The transilient turbulence representation presented above is an explicit expression in which the magnitudes of mixing coefficients depend on the numerical integration time-step. To use the transilient matrix representation as a general expression for the turbulent closure methods, we need to relate eqns. (36a) and (36b) with semi-implicit representations. Using Blackadar's [24] definition of m_{jk}, which is the fraction of mass exchange between two levels j and k per unit time, the turbulent fluxes at a given time t can be rewritten as:

$$F_j(t) = F_{j-1}(t) + \Delta s_j \sum_{k=1}^{L_p} m_{jk}(t)[q_j(t) - q_k(t)] . \tag{37}$$

Note that $m_{jk} = c_{jk} / \Delta t$ for an explicit method. The concentration prediction at the new time step is obtained from

$$q_j^{n+1} = q_j^n - \frac{\Delta t}{\Delta s_j}\left[\vartheta(F_j^{n+1} - F_{j-1}^{n+1}) + (1-\vartheta)(F_j^n - F_{j-1}^n)\right]. \tag{38}$$

Depending on ϑ, the finite difference scheme is explicit ($\vartheta = 0$), semi-implicit ($\vartheta = 1/2$), or fully implicit ($\vartheta = 1$). It can be rewritten in a general matrix equation:

$$\mathbf{Pq}^{n+1} = \mathbf{Rq}^n, \tag{39}$$

where the coefficients of **P** and **R** are given as:

$$p_{jj} = 1 + \vartheta\Delta t\sum_{\substack{k=1\\k\neq j}}^{L_p} m_{jk}\,;\ p_{\substack{jk\\j\neq k}} = -\vartheta m_{jk}\Delta t,$$

$$r_{jj} = 1 - (1-\vartheta)\Delta t\sum_{\substack{k=1\\k\neq j}}^{L_p} m_{jk}\,;\ r_{\substack{jk\\j\neq k}} = (1-\vartheta)m_{jk}\Delta t.$$

If the matrix **P** is nonsingular, we have a general expression for the transilient turbulence:

$$\mathbf{q}^{n+1} = \mathbf{P}^{-1}\mathbf{Rq}^n = \mathbf{Cq}^n, \tag{40}$$

where $\mathbf{C}=\mathbf{P}^{-1}\mathbf{R}$. The relationship between the coefficients of the transilient matrix and the mass exchange ratio among the grid boxes are somewhat complicated for the semi-implicit scheme. However, the semi-implicit scheme becomes attractive for the closure algorithms with sparse **P** matrix (i.e., when the matrix inversion is not so expensive) because it allows longer integration time-steps than the explicit method. In the following, we describe local and non-local flux-form atmospheric turbulence algorithms.

Eddy-diffusion algorithm (local): The turbulence flux is parameterized with eddy diffusivity (K).

$$\hat{F}_{q_i}^3 = \overline{q_i'\hat{v}^{3\prime}} = -K\frac{\partial q}{\partial s}. \tag{41}$$

Either parametrizations based on the PBL height or the TKE information provided by the meteorological model can prescribe the eddy diffusivity.

The diffusion equation can be discretized for level j $(1 < j < N)$ in a N-layer model and time-steps t^n and t^{n+1} as:

$$q_j^{n+1} = q_j^n + \frac{\Delta t}{\Delta s_j}\left[\frac{K_{j+1/2}}{\Delta s_{j+1/2}}\left(\vartheta(q_{j+1}^{n+1} - q_j^{n+1}) + (1-\vartheta)(q_{j+1}^n - q_j^n)\right)\right.$$

$$\left. - \frac{K_{j-1/2}}{\Delta s_{j-1/2}}\left(\vartheta(q_j^{n+1} - q_{j-1}^{n+1}) + (1-\vartheta)(q_j^n - q_{j-1}^n)\right)\right], \tag{42}$$

where $s_{j+1/2} = (s_j + s_{j+1})/2$ and $\Delta s_{j+1/2} = s_{j+1/2} - s_{j-1/2}$. For the lowest model layer, we need to account for the loss due to deposition:

$$q_1^{n+1} = q_1^n + \frac{\Delta t}{\Delta s_1}\left[\frac{K_{1+\frac{1}{2}}}{\Delta s_{1+\frac{1}{2}}}\left(\vartheta(q_2^{n+1} - q_1^{n+1}) + (1-\vartheta)(q_2^n - q_1^n)\right)\right]$$

$$-\Delta t\frac{v_d}{h_{dep}}\left[\vartheta q_1^{n+1} + (1-\vartheta)q_1^n\right], \tag{43}$$

and for the top layer we have zero flux through the top boundary:

$$q_N^{n+1} = q_N^n - \frac{\Delta t}{\Delta s_N}\frac{K_{N-1/2}}{\Delta \xi_{N-1/2}}\left(\vartheta(q_N^{n+1} - q_{N-1}^{n+1}) + (1-\vartheta)(q_N^n - q_{N-1}^n)\right). \tag{44}$$

The above relations can be rewritten in a tridiagonal matrix equation, which can be solved with an efficient numerical solver

$$\begin{pmatrix} d_1 & c_1 & \cdots & 0 & \cdots & 0 \\ a_2 & d_2 & c_2 & 0 & \cdots & 0 \\ \vdots & a_2 & \ddots & \ddots & \cdots & 0 \\ 0 & \vdots & \vdots & d_j & \ddots & \vdots \\ \vdots & 0 & 0 & \ddots & \ddots & c_{Lp-1} \\ 0 & \cdots & 0 & 0 & a_{Lp} & d_{Lp} \end{pmatrix}\begin{pmatrix} q_1^{n+1} \\ q_2^{n+1} \\ \vdots \\ q_j^{n+1} \\ \vdots \\ q_{Lp}^{n+1} \end{pmatrix} = \begin{pmatrix} b_1 \\ b_2 \\ \vdots \\ b_j \\ \vdots \\ b_{Lp} \end{pmatrix}. \tag{45}$$

Refer to Byun et al. [11] for the detailed expressions of the coefficients, (a_j, b_j, c_j, d_j).

Although the numerical solver algorithm for the semi-implicit scheme is stable in the sense that the amplitude of the signal either decays or stays the same, the positivity condition may not be satisfied if we choose a large time-step for the integration, especially for signals with small wavelengths. For example, with equal grid spacing, we can use the Von Neuman stability analysis technique to demonstrate that some parts of the short-wavelength spectrum show negative amplitude when the time-step is too long. The semi-implicit scheme is positive definite given the rather stringent CFL condition.

Blackadar algorithm (nonlocal): This scheme, first introduced by Blackadar [24], has long been used as one of the PBL schemes in the Mesoscale Meteorology model generation 4 (MM4) and generation 5 (MM5). The Blackadar model is a simple non-local closure scheme designed to simulate vertical transport by large convective eddies during conditions of free convection. Therefore, this scheme is used only in the convective boundary layer and must be coupled with another scheme for non-convective conditions and above the boundary layer, such as K-theory.

With the Blackadar scheme, the convective mixing is assumed to be dominated by eddies of varying sizes but all having roots in the surface layer, each eddy exchanging a certain amount of its mass with the air around it as it ascends. Here, the upward mass exchange ratio is estimated from a relation that is based on a balance between the mixing and surface heat flux;

$$M_u = H_{sfc} \,/\, \int_{s_{sfc}}^{s_h} C_{pd}\rho(\Theta_{sfc} - \Theta)w(s)J_s ds \,, \tag{46}$$

where H_{sfc} is the sensible heat flux leaving the surface layer, $w(s)$ is a weight function that accounts for the variation of exchange rate, Θ_{sfc} and Θ are potential temperatures at the surface and at height s, and C_{pd} is the specific heat at constant pressure. Then the mixing algorithm is represented as follows: for the lowest model layer

$$\frac{\partial q_1}{\partial t} = -\frac{v_d}{h_{dep}} q_1 - \sum_{k=1}^{L_p} M_u \frac{\Delta s_k}{\Delta s_1}[q_1(t) - q_k(t)] \,, \tag{47a}$$

and for $2 \le j \le L_p$,

$$\frac{\partial q_j}{\partial t} = -M_u \frac{\Delta s_1}{\Delta s_j}[q_j(t) - q_1(t)] \quad . \tag{47b}$$

Finite differencing approximation of the above leads to the following arrow-shaped sparse matrix equation:

$$\begin{pmatrix} d_1 & f_2 & \cdots & f_k & \cdots & f_{L_p} \\ e_2 & d_2 & 0 & 0 & \cdots & 0 \\ \vdots & 0 & \ddots & \cdots & \cdots & 0 \\ e_j & \vdots & \vdots & d_j & \vdots & \vdots \\ \vdots & 0 & 0 & 0 & \ddots & 0 \\ e_{L_p} & \cdots & 0 & 0 & 0 & d_{L_p} \end{pmatrix} \begin{pmatrix} q_1^{n+1} \\ q_2^{n+1} \\ \vdots \\ q_j^{n+1} \\ \vdots \\ q_{L_p}^{n+1} \end{pmatrix} = \begin{pmatrix} b_1 \\ b_2 \\ \vdots \\ b_j \\ \vdots \\ b_{L_p} \end{pmatrix} . \tag{48}$$

Refer to Byun et al. [11] for the detailed expressions of the coefficients, (b_j, d_j, e_j, f_j). A numerical algorithm similar to the Thomas algorithm for the tridiagonal system can be used to solve the sparse matrix equation.

Asymmetric convective mixing algorithm (nonlocal): The Asymmetric Convective Model (ACM), developed by Pleim and Chang [25], is based on Blackadar's non-local closure scheme but with a different scheme for downward mixing in the convective boundary layer (CBL). Observational evidence and large eddy simulation modeling studies indicate that mixing processes in a convective boundary layer are essentially asymmetric with fast upward buoyant plumes and slow broad compensatory subsidence. Therefore the direct, non-local downward transport of the Blackadar scheme is replaced with layer by layer subsidence which increases in mass flux as it descends, like a cascading waterfall (Figure 5). As with the Blackadar model, the ACM can only be used during convective conditions in the PBL. For other stability regimes, one needs to rely on other schemes such as K-theory.

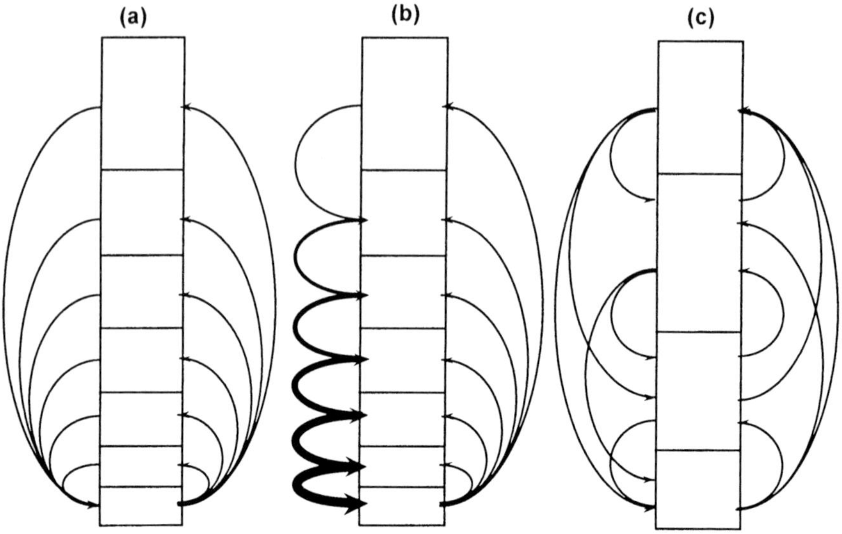

Figure 5: Schematics of (a) the Blackadar Scheme (b) the Asymmetric Convective Model, and (c) the transilient turbulence model.

Because the mass influx to the lowest model layer is from the second layer only in ACM, we can write the time rate change of mixing ratio as follows: for the lowest layer,

$$\frac{\partial q_1}{\partial t} = -\frac{v_d}{h_{dep}} q_1 - \frac{1}{\Delta s_1}[M_u(s_h - s_1)q_1(t) - M_{d2}\Delta s_2 q_2(t)], \tag{49a}$$

and for $2 \le j \le L_p$:

$$\frac{\partial q_j}{\partial t} = M_u q_1 - M_{dj} q_j + M_{dj+1}\frac{\Delta s_{j+1}}{\Delta s_j} q_{j+1}, \tag{49b}$$

where M_u represents upward mixing rate. M_{dj} represents downward mixing rate at layer j and is defined as:

$$M_{dj} = M_u \frac{s_h - s_{j-1}}{\Delta s_j} \quad . \tag{50}$$

Finally, we rewrite the prediction equation in terms of the upward mixing rate:

$$\frac{\partial q_1}{\partial t} = -M_u \frac{s_h - s_1}{\Delta s_1}(q_1 - q_2) - \frac{v_d}{h_{dep}} q_1 , \tag{51a}$$

and for $2 \le j \le L_p$:

$$\frac{\partial q_j}{\partial t} = M_u \left[q_1 - \left(\frac{s_h - s_{j-1}}{\Delta s_j} \right) q_j + \left(\frac{s_h - s_j}{\Delta s_{j+1}} \right) q_{j+1} \right] . \tag{51b}$$

Note that this scheme does not include the effects of vertical wind shear in generating turbulent mixing. The magnitude of the mixing rates of the transilient matrix is based on the conservation of sensible heat flux in the vertical direction:

$$M_u = H_{sfc} / \left[c_p \rho (\bar{\theta}_{sfc} - \theta_2) J_s (s_1 - s_h) \right] . \tag{52}$$

The semi-implicit finite differencing of the time derivative results in a sparse matrix equation of the form:

$$\begin{pmatrix} d_1 & c_1 & \cdots & 0 & \cdots & 0 \\ e_2 & d_2 & c_2 & 0 & \cdots & 0 \\ \vdots & 0 & \ddots & \ddots & \cdots & 0 \\ e_j & \vdots & \vdots & d_j & c_j & \vdots \\ \vdots & 0 & 0 & 0 & \ddots & c_{L_p-1} \\ e_{L_p} & \cdots & 0 & 0 & 0 & d_{L_p} \end{pmatrix} \begin{pmatrix} q_1^{n+1} \\ q_2^{n+1} \\ \vdots \\ q_j^{n+1} \\ \vdots \\ q_{L_p}^{n+1} \end{pmatrix} = \begin{pmatrix} b_1 \\ b_2 \\ \vdots \\ b_j \\ \vdots \\ b_{L_p} \end{pmatrix} , \tag{53}$$

Refer to Byun et al. [11] for the detailed expressions of the coefficients, (b_j, c_j, d_j, e_j). Again, a numerical algorithm based on a backward or forward substitution can efficiently solve the sparse matrix system.

<u>Transilient parameterization with TKE (nonlocal):</u> In order to use the transilient turbulence concept for mixing trace gases, the mass exchange coefficient matrix must be known. This is the closure problem with this parameterization. A couple of methods have been presented in the literature. One method is based on the TKE equation (Stull and Driedonks [26] and Raymond and Stull [27]) and the other is the one

based on non-local Richardson number (Zhang and Stull [28]). In the following we describe briefly the TKE based scheme and discuss its associated difficulties. The horizontally homogeneous form of the TKE equation in the Cartesian coordinate is given as:

$$\frac{\partial E}{\partial t} = -\overline{u''w''}\frac{\partial u}{\partial z} - \overline{v''w''}\frac{\partial v}{\partial z} + \frac{g}{\Theta_o}\overline{\theta''w''} - \varepsilon \quad . \tag{54}$$

Note that the pressure and turbulent transport terms have been ignored. After normalizing with E, the non-local analogy of the finite difference form of eqn. (54) can be written as:

$$\frac{\Delta_t E_{jk}}{E_{jk}} = \left[\frac{\left(-\overline{u''w''}\right)_{jk}}{E_{jk}}\left(\frac{\Delta u}{\Delta z}\right)_{jk} + \frac{\left(-\overline{v''w''}\right)_{jk}}{E_{jk}}\left(\frac{\Delta v}{\Delta z}\right)_{jk} + \frac{g}{\Theta_j}\frac{\left(\overline{\theta''w''}\right)_{jk}}{E_{jk}} - \frac{\varepsilon_{jk}}{E_{jk}}\right]\Delta_t t\,, \tag{55}$$

where the symbol Δ_t represents temporal change while $\varDelta$ represents spatial gradient. To close the system, the unknown parameters are written in terms of known parameters by introducing three scaling parameters T_O, Ri_C, and $D = \varepsilon_{jk} / E_{jk}$, which are the time scale of turbulence, the critical Richardson number, and dissipation factor, respectively. Thus, weighted kinematic fluxes can be written as:

$$\frac{\left(-\overline{u''w''}\right)_{jk}}{E_{jk}} = T_o\left(\frac{\Delta u}{\Delta z}\right)_{jk} ; \quad \frac{\left(-\overline{v''w''}\right)_{jk}}{E_{jk}} = T_o\left(\frac{\Delta v}{\Delta z}\right)_{jk} ; \quad \frac{\left(-\overline{\theta''w''}\right)_{jk}}{E_{jk}} = \frac{T_o}{Ri_c}\left(\frac{\Delta\Theta}{\Delta z}\right)_{jk} \tag{56}$$

Then eqn. (55) can be rewritten as:

$$Y_{jk} = \frac{\Delta_t E_{jk}}{E_{jk}} = \frac{T_o \Delta_t t}{(\Delta z)^2_{jk}}\left[(\Delta u)^2_{jk} + (\Delta v)^2_{jk} - \frac{g}{Ri_c \Theta_j}(\Delta\Theta)_{jk}(\Delta z)_{jk}\right] - D\Delta_t t\,. \tag{57}$$

Since we are dealing with fraction of masses that are coming from and going to different layers (i.e., $j \neq k$), Y_{jk} in the above equation is for off-diagonal elements only. Diagonal elements in eqn. (57), Y_{jj}, represent a mass of air that remains in each layer without interaction with other layers. Observations during convective conditions (Ebert et al. [29]) indicated that turbulent eddies cause a well-mixed rather than convectively overturned boundary layer. This requires that the values of the mixing potential elements (Y_{jk}) should increase monotonically from the upper right-most element towards the main diagonal. Further, to account for the subgrid scale mixing in each layer, an independent parameter Y_{ref} was introduced. Thus, the diagonal elements can be written as:

$$Y_{jj} = MAX(Y_{j,j-1}, Y_{j,j+1}) + Y_{ref}, \tag{58}$$

given values for T_O, Ri_C, and D. Usually, Y_{ref} is estimated based on observations. Finally, the off-diagonal elements of the transilient matrix are estimated with:

$$c_{jk} = Y_{jk} / \|\mathbf{Y}\|_{\infty}, \tag{59}$$

where $\|\mathbf{Y}\|_{\infty}$ is the infinite norm of matrix **Y**. Once the transilient matrix is determined, the concentration due to turbulent processes can be computed with the explicit form eqn. (33), instead of the implicit form eqn. (38). However, the time-steps should be such that the mixing ratio cannot be negative. Also, the parameterization introduced above depends on many free parameters (T_O, Ri_C, D, and Y_{ref}) that must be verified through observations.

5.3 Horizontal diffusion algorithms

The horizontal diffusion in the curvilinear coordinate system (see eqn. (30)) is given as:

$$\left.\frac{\partial \varphi_i^*}{\partial t}\right|_{hdiff} = \left.\frac{\partial\left(\sqrt{\hat{\gamma}}\,\bar{\rho}\bar{q}_i\right)}{\partial t}\right|_{hdiff} = -\hat{\nabla}_s \bullet \left[\sqrt{\hat{\gamma}}\,\bar{\rho}\hat{\mathbf{F}}_{q_i}\right]. \tag{60}$$

In ideal cases, the process must represent the effects of physical diffusion on pollutant dispersion. However, our understanding of horizontal turbulence is rather limited due to lack of adequate turbulence measurements as well as the scale dependency of the problem. In general, the simulation of the advective horizontal transport introduces numerical diffusion. In earlier days of atmospheric modeling, the horizontal diffusion process is often omitted because the numerical diffusion associated with the advection algorithm was large. With the advent of less diffusive numerical advection schemes and the emerging need for simulating air quality at high resolutions, representation of horizontal diffusion in atmospheric modeling has become important.

To provide a horizontal diffusion algorithm that accounts for the physical diffusion, the amount of numerical diffusion induced by the advection schemes must be considered. It is difficult, though not impossible, to accomplish this because the numerical diffusion characteristics of a specific advection algorithm are not only dependent on the wave number of the signal but also the Courant number of the transport flow. Refer to Odman [30] for a method that quantifies numerical diffusion errors associated with advection algorithms.

In practice, there are not many choices for the horizontal diffusion algorithms. Frequently, the horizontal turbulent fluxes are parameterized with the eddy-diffusion theory. When off-diagonal terms are neglected, the diffusion equation is given as

$$\left.\frac{\partial \varphi_i^*}{\partial t}\right|_{hdiff} = \frac{\partial}{\partial \hat{x}^1}\left[\sqrt{\hat{\gamma}}\,\bar{\rho}(\hat{K}^{11}\frac{\partial \bar{q}_i}{\partial \hat{x}^1})\right] + \frac{\partial}{\partial \hat{x}^2}\left[\sqrt{\hat{\gamma}}\,\bar{\rho}(\hat{K}^{22}\frac{\partial \bar{q}_i}{\partial \hat{x}^2})\right]. \tag{61}$$

The horizontal diffusion is often solved with an explicit finite difference method to minimize memory requirements in Eulerian AQMs, with a sufficiently small time step to ensure the positivity of solution.

The contravariant eddy diffusivity components are related to the Cartesian counterparts as $\hat{K}^{11} = mK_{xx}$ and $\hat{K}^{22} = mK_{yy}$. In Eulerian air quality modeling, we do not usually distinguish between eddy diffusivities in two different horizontal directions (i.e., $K_{xx} = K_{yy} = K_H$). Then, the problem is reduced to parameterization of the eddy diffusivity to reflect the sub-grid scale diffusion. The simplest approach of estimating K_H is to assume a space independent diffusivity (e.g., $K_H = 50$ m^2/s). Smagorinsky [31] formulated a horizontal diffusivity that accounts for diffusion due to distortion or stress in the horizontal wind field:

$$K_H = 2\alpha_o^{\,2}(S_\Gamma^{\,2} + S_\Lambda^{\,2})^{1/2}(\Delta x)^2, \tag{62}$$

where $\alpha_o \cong 0.28$. It characterizes the impact of stress in the wind field. Yamartino [32] suggests a method of determining appropriate horizontal diffusion coefficients by subtracting the numerical diffusivity already accompanying the advection scheme. He highlighted the fact that the form of numerical diffusivity has a similar form to eqn. (62). However, eqn. (62) relies on the grid-scale wind components and therefore is not suitable for estimating the subgrid scale diffusion not resolved by the input winds. Hanna [33] summarized that the procedure for estimating horizontal diffusion at the subgrid scale has not yet been resolved in a consistent manner in three-dimensional Eulerian models. He concluded that the Eulerian models must employ wind fields that include the full spectrum of mesoscale and regional fluctuations in space and time. Overly smooth wind fields provided by the diagnostic wind field modeling or mesoscale meteorological models can not yield sufficient horizontal diffusion in the model. McNider et al. [34] stated that in current meteorological models the energy spectrum corresponding to 120 km (or four times the horizontal grid spacing used) to 1 km (below which boundary layer turbulence is parameterized) is not well represented. Furthermore, the parameterizations of sub-grid scale horizontal diffusivities incorporated in commonly used regional models produce K_H values that range over several orders of magnitude.

6 Conclusions

In this chapter, a few issues involved with describing atmospheric transport processes in Eulerian air quality models have been discussed. It has discussed computational coordinates, contrasted diagnostic and prognostic meteorological modeling techniques, highlighted impacts of atmospheric dynamics on air quality simulations, and has presented a set of desirable properties of numerical transport algorithms. The primary objective was to suggest numerical transport algorithms that assure key numerical and physical characteristics for successful air quality modeling. Use of the flux-form formulations that has been emphasized ensures conservation of mass during the transport processes. This chapter has neither dwelt on specific numerical algorithms nor compared them. Instead, it has provided a consistent description of atmospheric dynamics and generalized numerical techniques of transport processes recommended for Eulerian air quality modeling.

Disclaimer

This document has been reviewed in accordance with the U.S. Environmental Protection Agency's peer and administrative review policies and approved for presentation and publication. Mention of trade names or commercial products does not constitute endorsement or recommendation for use.

References

[1] Phillips, N. A., A coordinate system having some special advantages for numerical forecasting. *J. Meteor.*, **14**, pp. 184-185, 1957.

[2] Arakawa, A., Mechso, C. R. and Konor, C. S., An isentropic vertical coordinate model: Design and application to atmospheric frontogenesis studies. *Meteor. Atmos. Phys.*, **50**, pp. 31-45, 1992.

[3] Byun, D. W., Dynamically consistent formulations in meteorological and air quality models for multi-scale atmospheric applications: Part I. Governing Equations in Generalized Coordinate System. *J. Atmos. Sci.*, **56**, pp. 3789-3807, 1999.

[4] Sasaki, Y., An objective analysis based on the variational method. *J. Meteor. Soc. Japan* , **36**, pp. 77-88, 1958.

[5] Sasaki, Y., Some basic formalisms in numerical variational analysis. *Mon. Wea. Rev.,* **98**, 875-883, 1970.

[6] Klemp, J. B. and Wilhelmson, R. B., The simulation of three-dimensional convective storm dynamics. *J. Atmos. Sci.*, **35**, pp. 1070-1096, 1978.

[7] Byun, D. W., Dynamically consistent formulations in meteorological and air quality models for multi-scale atmospheric applications: Part II. Mass conservation issues. *J. Atmos. Sci.*, **56**, pp. 3808-3820, 1999.

[8] Byun, D. W., Fundamentals of one-atmosphere dynamics for multiscale air quality modeling. pp 38, Chapter 5 in *Science*

Algorithms of the EPA Models-3 Community Multiscale Air Quality (CMAQ) Modeling System. Byun, D.W. and J.K.S. Ching, ed., National Exposure Research Laboratory, U.S. Environmental Protection Agency, Research Triangle Park, NC., 1999.

[9] Seaman, N. L., Stauffer, D. R. and Lario-Gibbs, A. M., A multiscale four-dimensional data assimilation system applied in the San Joaquin Valley during SARMAP. Part I: Modeling design and basic performance characteristics. *J. Appl. Met.*, **34**, pp.1739-1761, 1995.

[10] Odman, M. T., *Research on Numerical Transport Algorithms for Air Quality Simulation Models*. EPA Report. EPA/660/R-97/142, 1998. [Available from National Exposure Research Laboratory, U.S. Environmental Protection Agency, Research Triangle Park, NC 27711]

[11] Byun, D. W., Young., J., Pleim, J., Odman, M. T. and Alapaty, K., Numerical transport algorithms for the community multiscale air quality (CMAQ) chemical transport model in generalized coordinates. pp 58, Chapter 7 in *Science Algorithms of the EPA Models-3 Community Multiscale Air Quality (CMAQ) Modeling System.* Byun, D.W. and J.K.S. Ching, ed., National Exposure Research Laboratory, U.S. Environmental Protection Agency, Research Triangle Park, NC., 1999.

[12] Williamson, D. L. and Rasch, P. J., Two-dimensional semi-Lagrangian transport with shape-preserving interpolation. *Mon. Wea. Rev.,* **117**, 102-117, 1989.

[13] Smolarkiewicz, P. K. and Grell, G. A., A class of monotone interpolation schemes. *J. Comp. Phys.,* **84**, 355-375, 1992.

[14] Colella, P. and Woodward, P. R., The piecewise parabolic method (PPM) for gas-dynamical simulations. *J. Comp. Phys.,* **54**, pp. 174-201, 1984.

[15] Bott, A., A positive definite advection scheme obtained by nonlinear renormalization of the advective fluxes. *Mon. Wea. Rev.,* **117**, pp. 1006-1015, 1989.

[16] Yamartino, R. J., Nonnegative, conserved scalar transport using

grid-cell-centered, spectrally constrained Blackman cubics for applications on a variable-thickness mesh. *Mon. Wea. Rev.,* **121**, pp. 753-763, 1993.

[17] Schieffe, R. D. and Morris, R. E., A review of the development and application of the Urban Airshed Model. *Atmos. Environ.*, **27B**, pp. 23-39, 1993.

[18] Chang J. S., Jin, S., Li, Y., Beauharnois, M., Lu, C. –H., Huang, H. – C, Tanrikulu, S. and DaMassa, J., *The SARMAP air quality model. Final Report*, SJVAQS/AUSPEX Regional Modeling Adaptation Project, 53 pp, 1997. [Available from California Air Resources Board, 2020 L Street, Sacramento, California 95814]

[19] Byun, D. W. and Ching, J. K. S., ed., *Science Algorithms of the EPA Models-3 Community Multiscale Air Quality (CMAQ) Modeling System.* EPA Report. EPA/600/R-99/030, 1999. [Available from National Exposure Research Laboratory, U.S. Environmental Protection Agency, Research Triangle Park, NC 27711.]

[20] Byun, D. W. and Lee, S.–M., Mass conservative numerical integration of trace species conservation equation: Experiment with idealized two-dimensional flows. (in preparation)

[21] Otte, M. J. and Wyngaard, J. C., A general framework for an "unmixed layer" PBL model. *J. Atmos. Sci.*, **53**, pp. 2652-2670, 1996.

[22] Stull, R. B., *An Introduction to Boundary Layer Meteorology*. Kluwer Academic, 666 pp, 1988.

[23] Stull, R. B., Review of non-local mixing in turbulent atmosphere: Transilient turbulence theory. *Bound.-Layer Meteor.,* **62**, pp. 21-96, 1993.

[24] Blackadar, A. K., Modeling pollutant transfer during daytime convection, *Preprints, Fourth Symposium on Atmospheric Turbulence, Diffusion, and Air Quality*, Reno, Am. Meteor. Soc., pp. 443-447, 1978.

[25] Pleim, J.E. and J. Chang, A non-local closure model for vertical

mixing in the convective boundary layer. *Atmos. Environ.*, **26A**, pp. 965-981, 1992.

[26] Stull, R. B. and Driedonks, A. G. M., Applications of the transilient turbulence parameterization to atmospheric boundary layer simulations. *Bound.-Layer Meteor.,* **40**, pp. 209-239, 1987.

[27] Raymond, W. H. and Stull, R. B., Application of transilient turbulence theory to mesoscale numerical weather forecasting. *Mon. Wea. Rev.*, **118**, pp. 2471-2499, 1990.

[28] Zhang, Q. and Stull, R. B., Alternative nonlocal descriptions of boundary-layer evolution. *J. Atmos. Sci.*, **49**, pp. 2267-2281, 1992.

[29] Ebert, E. E., Schumann, U. and Stull, R. B., Nonlocal turbulence mixing in the convective boundary layer evaluated from large-eddy simulation. *J. Atmos. Sci.*, **46**, pp. 2178-2207, 1989.

[30] Odman, M. T., A quantitative analysis of numerical diffusion introduced by advection algorithms in air quality models. *Atmos. Environ,*. **31**, pp. 1933-1940, 1997.

[31] Smagorinsky, J., General circulation experiments with the primitive equations: The basic experiment. *Mon. Wea. Rev.,* **91**, pp. 99-164, 1963.

[32] Yamartino, R. J., Improvements to horizontal transport in grid models. Preprints, 23rd International Technical Meeting on Air Pollution Modelling and its Application, Varna, Bulgaria, NATO/CCMS, pp. 51-56, 1998.

[33] Hanna, S.R., Random variability in mesoscale wind observations and implications for diffusion models. Preprints. *Eighth Joint Conference on Applications of Air Pollution Meteorology with A&WMA*. Nashville, Tennessee, AMS, pp. 1-5, 1994.

[34] McNider, R. T., Singh, M. P. and Gupta., S., Nocturnal wind structure and plume growth rates due to inertial oscillation. Preprints. *Ninth Joint Conf. on App. of Air Pollution Meteorology with A&WMA,* Atlanta, Georgia, AMS, pp. 455-458, 1996.

Chapter 8

Atmospheric chemistry and chemical transport models

Ananthakrishna Sarma[1] & Jeremy M. Hales[2]
[1] *Science Applications International Corporation,*
1710 SAIC Dr
McLean, VA 22102, USA
[2] *Envair,*
60 Eagle Reach
Pasco, WA 99301, USA

Abstract

In the recent past, researchers have started to study the various aspects of the mesoscale atmosphere. Advances in computer technology have facilitated the investigation of several aspects of the atmosphere in much greater detail than was possible before. One such aspect is the mesoscale atmospheric chemistry. In this chapter we explore some of the impacts of the mesoscale circulation on atmospheric chemistry and air quality. It is important to understand and characterize chemical reactions that are significant in the mesoscale. Such understanding will enable the formulation of parameterizations and schemes to treat atmospheric chemistry in conjunction with atmospheric physics and dynamics in mesoscale models. This is especially important to quantify plume chemistry and urban air quality. Some recent advances in the fields of atmospheric chemistry as well as high-resolution atmospheric modeling are introduced in this chapter.

1 Introduction

Like many topics in atmospheric science, atmospheric chemistry is scale spanning. Physical and chemical processes occur at all scales of the atmosphere, and mechanistic interactions between the different scales can have profound influences on overall atmospheric behavior. Atmospheric dynamics scales reflect the dominant wavelengths of atmospheric motions. On the synoptic scale, for example, the Rossby waves and frontal systems are dominant; on the mesoscale, convective complexes as well as terrain and thermally driven flows become significant. Atmospheric chemistry processes, on the other hand, are influenced strongly by the rates of individual reactions; slowly reacting species are often transported over large spatial domains, while those reacting quickly tend to induce variability on very fine scales. As we focus our attention on smaller and smaller length scales, the faster reactions become increasingly significant.

To understand fully the impact of mesoscale circulations on atmospheric chemistry leading to ambient air quality, one must characterize the chemical reactions occurring in the troposphere from a mesoscale perspective. Towards this end, it is appropriate to view the atmospheric chemistry system as consisting of three parts: 1) sources of pollutants, 2) a reaction volume in which the chemical reactions occur, and 3) a set of removal mechanisms.

Most anthropogenic pollutant emissions vary significantly on meso and smaller scales. For example, industrial emissions, for the most part, occur as continuous point sources. Sometimes a collection of point sources can be considered to form an area source such as in a refinery complex, or automobile emissions from an urban area. Even in volcanic eruptions which can impact a distance of several thousand kilometers, significant chemical reactions occur in the first few hundred kilometers of the plume which are dominated by strong cooling and aerosol fallout.

The reaction volumes from a mesoscale perspective are $O(10^{10})$ m^3 which corresponds to 10 km × 10 km × 100 m. Plume transport in such environments depend strongly on the vertical scale and hence on the vertical mixing due to mesoscale processes such as Planetary Boundary Layer (PBL) physics. Time scales in such volumes are therefore of the order of tens of seconds or less. Such time scales will dictate which chemical reactions can be parameterized and which have to be treated more explicitly.

Atmospheric removal mechanisms are strongly tied to mesoscale meteorological processes. For example, precipitation scavenging plays a significant role in the removal of aerosol and water-soluble gases. The PBL properties can alter the dry deposition of aerosol and gases by changing the near surface wind fields and thermal stability. Vertical mixing in the convective boundary layer also impacts the deposition by lowering the concentration of the species near the ground. In this discussion we will concentrate on the chemical processes themselves rather than on the sources or removal processes.

2 Tropospheric reaction chemistry

The task of characterizing tropospheric reaction-chemistry processes, and their resulting influence on ambient pollutant concentrations, can be viewed in terms of two rather broad and overlapping issues. The first of these concerns understanding the chemistry itself, that is, determining unambiguously the mechanistic pathways of the important reactions in the pollution complex, and quantifying these in terms of mathematical rate expressions. Owing to the huge number of potentially important reaction pathways in typical urban pollution environments, this is a demanding task. It is far from complete at the present time.

The second issue concerns the need to represent this reaction chemistry for use in computer codes describing pollution behavior. Referred to generally as "chemical-transport models", these codes simulate the emission, advection, diffusion, deposition, and reaction of pollution, producing as a product computed pollutant concentrations c_i as functions of space and time by approximating solutions to families of chemical conservation equations of the form

$$\underbrace{\frac{\partial c_i}{\partial t}}_{\text{time rate of change}} = \underbrace{-\nabla\cdot(c_i\mathbf{u})}_{\text{advective transport}} + \underbrace{\nabla\cdot(\mathbf{K}\nabla c_i)}_{\text{diffusive transport}} + \underbrace{R_i}_{\text{generation and decay}} . \tag{1}$$

These codes are typically quite complex and require extensive computer resources, with chemical-reaction calculations (included within the generation and decay term, R), accounting for a large fraction of this demand. Because of this, it is necessary to economize the models' chemistry calculations by condensing the mathematical description of reaction behavior — a procedure that invariably requires estimation and approximation. Combined with our incomplete knowledge of reaction chemistry, this imposes substantial levels of uncertainty on model calculations and the corresponding computed pollutant concentration fields.

The following sections examine these two issues in sequence. Because any comprehensive review of these subjects is well beyond the scope of this chapter, we will confine this text to provide a somewhat broad picture and itemize key areas where further information is required. Readers wishing to pursue these subjects in greater detail may do so by consulting the several comprehensive reviews listed in the references to this chapter.

2.1 Homogeneous gas-phase processes

From a tropospheric mesoscale perspective, especially in continental environments, gas-phase pollutant reactions occur primarily as a consequence of photochemical activation. At the expense of some oversimplification, one can view the core of this reaction system — or at least our current wisdom regarding it — in terms of the schematic shown in Figure 1. Here the primary photochemical steps include photolysis of ozone, nitrogen dioxide, and nitrous acid. These reactions produce activated products, which in turn react with pollutant species in a manner that generally moves the pollution system toward a higher oxidation state.

Several points should be noted when viewing this relatively simple scheme. First, the figure indicates profound differences between daytime chemistry and that occurring at night, when photolysis is absent. This is particularly important in the context of nitrogen trioxide, often referred to as "nitrate radical", whose formation is favored under night time conditions. Although not shown explicitly on this schematic, nitrate radical voraciously attacks a wide variety of organic pollutants, resulting in an important nocturnal sink for these compounds. Atkinson [1] gives a comprehensive review of our current state of knowledge pertaining to nitrate radical's atmospheric chemistry.

In addition to nitrate, other radicals produced by this core chemistry go on to participate in important reactions that are not featured here. The hydroxyl radical, HO$^{\bullet}$, is of particular importance in this regard owing to its high reactivity with a number of pollutant species such as, for example, sulfur dioxide:

$$SO_2 + HO^{\bullet} \longrightarrow HOSO_2 \xrightarrow{O_2} HO_2^{\bullet} + SO_3 \xrightarrow{H_2O} H_2SO_4. \qquad (2)$$

The general schematic layout of Figure 1 has been accepted for several years and is considered generally to be "textbook chemistry". This, however, does not mean that all elements of this scheme are completely understood or that other, competing reactions do not exist. The rate coefficient for the reaction between nitrogen dioxide and hydroxyl radical, for example, has been re-measured recently and revised accordingly (Dransfield et al. [2]). In addition, there is a growing body of evidence (DeHaan et al. [3]) suggesting that free chlorine and its radical products may participate in this general core chemistry to a significant extent, at least under some circumstances.

The reversible linkage between nitrogen dioxide, and peroxyacetyl nitrate (PAN) and its higher homologues illustrates nitrogen dioxide's ability to form "reservoir products", which temporarily remove it from the photochemical sequence. PAN formation is favored at lower temperatures, and under such conditions it may sequester nitrogen dioxide, thus allowing it to be transported non-reactively and then re-emitted over extended downwind distances for subsequent photochemical activity. Although not depicted on the core scheme of Figure 1, additional types of nitrogen-dioxide reservoir species can be generated — particularly through interaction with nitrate radical. The chemistry associated with such reservoir species is understood incompletely at the present time.

Figure 1 also implies that different photochemical regimes should dominate, depending on whether nitric oxide and nitrogen dioxide concentrations ($NO + NO_2 = NO_x$) are large or small, kinetically speaking, compared with those of carbon monoxide and volatile organic compounds (VOC). In the high-NO_x regime, nitric oxide titrates out ozone, while both nitric oxide and nitrogen dioxide scavenge the free radicals necessary for VOC oxidation. The result is a depression of VOC

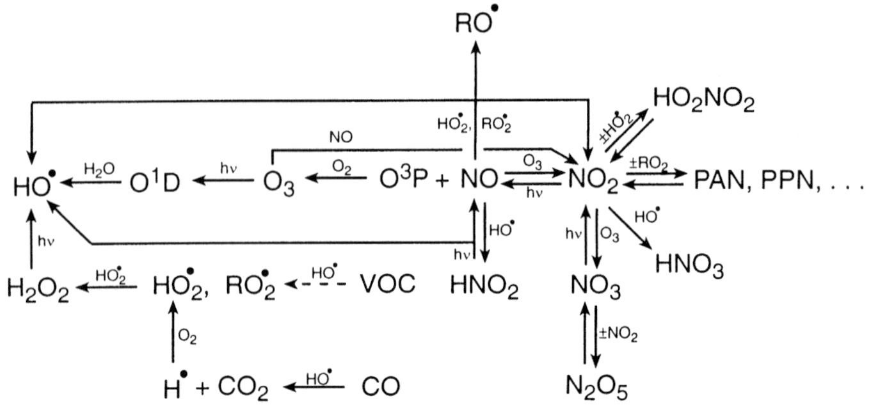

Figure 1: Simplified schematic of "core" tropospheric reaction chemistry.

reactivity and ozone concentration, in direct contrast to the usual presumption that NO_x enhances ozone formation. This phenomenon has been referred to widely as the "NO_x disbenefit". The opposite, low-NO_x situation precludes nitrogen oxides as a source of radicals, and whatever photochemistry that occurs must rely on alternative sources, such as natural ozone photolysis. Quite obviously, optimum radical activity, and maximum ozone production, must occur at some intermediate NO_x concentration. Such conditions have been described as "maximum incremental reactivity" by Carter and his co-workers [4].

It is important also to observe that the dashed arrow representing the reaction of VOC to form peroxy radicals is a grossly incomplete representation. Multitudes of different hydrocarbon species exist, which are attacked by the hydroxyl radical in a variety of ways to form a plethora of radicals and partially oxidized compounds such as carbonyls, acids, alcohols, and esters, which in turn react in a variety of fashions. While initial hydroxyl-radical attack on straight and branched-chain hydrocarbons is relatively well quantified, our understanding of the subsequent product reactions is incomplete. Some of these partially oxidized products, such as formaldehyde, can be photolyzed directly in the troposphere to produce new radicals for subsequent reaction:

$$H_2CO \xrightarrow{h\nu} HCO^{\bullet} + H^{\bullet}. \qquad (3)$$

Finally, the potential for linkage of Figure 1's homogeneous reaction scheme to various elements of heterogeneous reaction behavior should be noted. This is important for the formation of aerosol particles through condensation of VOC products, creation of sulfate particles through radical interactions with reduced sulfur species, interactions of nitric acid with ammonia to form nitrate aerosols, and similar phenomena. In addition, the radicals and intermediates produced through this scheme can migrate to particle and droplet surfaces and subsequently participate in, or at least influence, aqueous-phase and surface chemistry. Some aspects of these heterogeneous processes will be discussed in Section 3, below.

Atkinson [5] provides a rather comprehensive, modern review of many of the aspects of the extended chemistry alluded to here. In addition, and from a textbook perspective, the book by Finlayson-Pitts and Pitts [6], and its recent expanded update, Finlayson-Pitts and Pitts [7] provide particularly useful references in this regard. Based on these and additional reviews, one may conclude that substantial additional quantitative information is required regarding the reaction rates and product yields of organic reaction intermediates, in order to reduce satisfactorily uncertainty levels associated with homogeneous chemistry calculations. Moreover, the recent results of DeHaan, et al. [3] and Spicer et al. [8] suggest that significant additional information pertaining to the interactions of organics with chlorine and its radicals may be required for full chemical characterization, at least in coastal areas containing significant sodium chloride aerosol levels. In somewhat more specific terms, these authors recommend the need for further quantitative efforts in the following areas, in order to advance significantly our understanding and our ability to quantify reactive aspects of pollution behavior in a satisfactory manner:

- Determination of rate constants, mechanisms, and product yields for the reactions of organic peroxy radicals with other peroxy radicals, nitrate radicals, and nitric oxide.
- Determination of reaction rates of alkoxy radicals for decomposition, and isomerization reactions as well as for reaction with oxygen. Behavior of radicals other than those formed from alkanes and alkenes are especially important, in view of the current state of knowledge.

- Determination of detailed mechanisms for ozone reactions with unsaturated organic compounds. This involves understanding the reactions of the biradical intermediates formed in these reactions.
- Determination of thermal-decomposition rates, as well as rates and mechanisms of the other atmospherically important reactions of the higher PAN homologues.
- Determination of the products and mechanisms of the reactions of monoterpenes and oxygenated VOCs emitted from vegetation, resulting from reactions with hydroxyl radical, nitrate radical, and O_3.
- Determination of the mechanisms and products of the reactions of HO• – aromatic adducts with O_2 and NO_2.
- Determination of the tropospheric chemistry of many oxygenated VOCs formed as first-generation products of VOC photo-oxidations.
- Development of an understanding of the reaction sequences leading to products which gas/particle partition and lead to secondary organic aerosol formation.

2.2 Aerosol dynamics and heterogeneous processes

The term "heterogeneous processes", in the lexicon of this chapter, denotes those chemical and microphysical phenomena that involve a solid or non-aqueous liquid particle phase. This demarcation is somewhat problematic in that it excludes homogeneous aqueous-phase chemistry. In the cloud-droplet nucleation process, where aerosol particles accrete water-vapor molecules to grow progressively larger, it is difficult to define a clear demarcation between a "wet aerosol particle" and a "dirty cloud droplet". This notwithstanding, we will proceed with our ad hoc classification, with the understanding that many of the chemical processes discussed here will have considerable overlap between what we define operationally as "homogeneous" and "heterogeneous" chemistry.

Even given the above-noted restriction on our current purview, this subject encompasses several areas, including homogeneous aerosol nucleation, condensational-growth and agglomeration of aerosol particles, and chemical reaction on the surfaces and interiors of aerosol particles. As with the homogeneous rate processes discussed in Section

2.2, all of these features are included, typically, with the "generation and decay" terms of eqn. (1). We discuss these topics roughly in sequence in the following two subsections.

2.2.1 Aerosol dynamics

To provide an initial focus for this overview of mesoscale aerosol dynamics, it is appropriate to show some typical examples of measured continental-tropospheric aerosol composition. Figure 2 provides some overview information in this regard, showing data from composite measurements of relative aerosol abundance for two different particle-size ranges.

The compositional attributes indicated here are particularly important in a mechanistic context, because particle-growth pathways differ appreciably among the various constituents. Particle development through deposition of sulfate ion via pathways such as reaction (2), for example, is normally presumed to be a one-way, terminal process. Deposition of ammonium and nitrate ions as well as semi-volatile organic compounds, on the other hand, is typically reversible in character. Particles associated with these species may either shrink or grow,

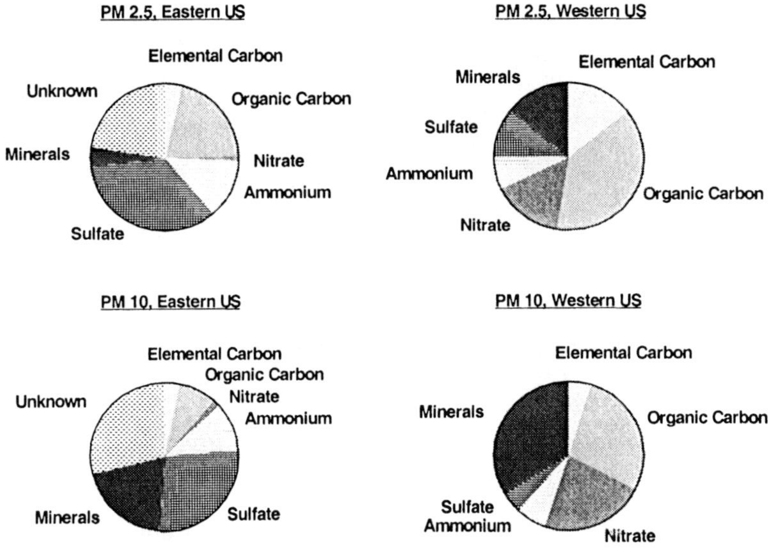

Figure 2: Measured relative abundance of aerosol constituents. (Adapted from EPA [9].)

depending on prevailing thermodynamic driving forces. Judging from the chemical differences shown in Figure 2, it is apparent that significant differences in mechanistic emphasis typically occur for different regions and particle-size ranges.

Within this context, it is important also to note the complexity associated with characterizing aerosol chemistry and its associated pathway processes, when compared to purely gas-phase phenomena. This can be appreciated by observing that purely gas-phase phenomena can be considered, locally, in terms of a discrete vector of chemical-species concentrations, e.g., $\vec{c} = \{c_{O_3}, c_{NO}, \ldots, c_i, \ldots\}$, whose interactive elements grow and diminish as dictated by stipulated transformation relationships. In contrast, however, one may reasonably expect that any collection of similar-sized particles sampled from a local air parcel will exhibit substantial (particle-to-particle) differences in composition. This feature combines with the size-distributed nature of aerosols to result in bimodal distributions for each chemical species, such as shown in Figure 3. This substantial increase in complexity typically necessitates the use of expedients such as parameter lumping, strong assumptions regarding chemical distribution, and the like, and results in considerable levels of predictive uncertainty.

As noted above, aerosol growth and decay occur through a variety of pathways. These can be categorized into the following general areas, pertaining to the generation (g) and loss (l) of a given chemical species (denoted here by i) onto particles in the size-range m, m+dm:

- $g1_i$: condensation of gaseous material to the surfaces of existing particles,
- $g2_i$: chemical production of i on or in the particles,
- $g3_i$: formation of i-containing particles in the range m, m+dm through coagulation of smaller i-containing particles,
- $g4_i$: formation of i-containing particles in the range m, m+dm through breakup of larger i-containing particles, and
- $g5_i$: homogeneous nucleation of i-containing particles from the gas phase,
- $l1_i$: evaporation of condensed material from existing particles,
- $l2_i$: chemical loss of i on or in the particles,

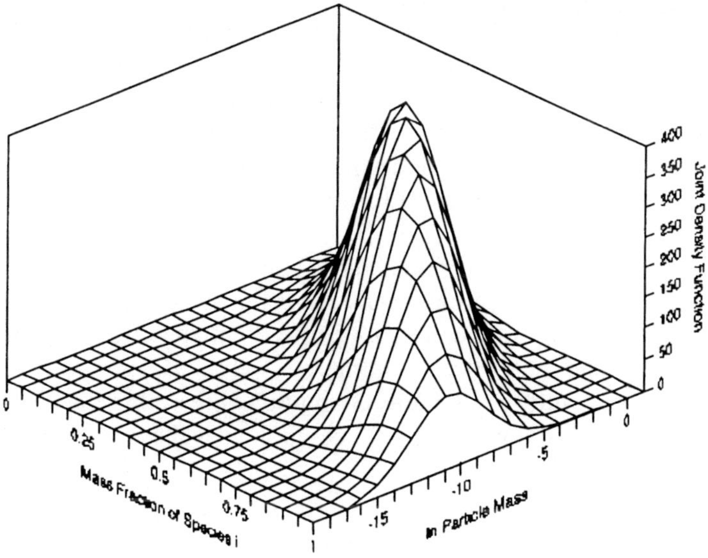

Figure 3: Hypothetical bimodal distribution for a selected chemical species in a size-distributed aerosol.

$l3_i$: loss of i-containing particles in the range m, m+dm through scavenging and/or coagulation to form larger particles, and

$l4_i$: loss of i-containing particles in the range m, m+dm through breakup into smaller particles.

Similar relationships can be set forth for all other chemical species as well as for aerosol total mass. Various mathematical expressions of these summations, corresponding to a variety of constraints and assumptions, are available in the extended literature (e.g., Hidy [10], Williams and Loyalka [11]). These are too voluminous to be described adequately in any brief discussion, and thus will not be elaborated in this overview.

It is worth noting, however, that under most conditions the quantification and mathematical processing of these terms poses a substantial challenge. Extensive effort has been focussed on processes g1 and l1 (cf. Friedlander [12], Williams and Loyalka [11]) particularly over the past several years for aerosols containing nitrate, ammonium, and sulfate ions, as well as a number of organic species (e.g., Seinfeld [13],

Bowman, et al. [14]). Owing to their complexity, processes g2 and l2 have received relatively little attention until recently. This situation has changed, however, with the growing recognition that heterogeneous processes can be significant determinants of atmospheric chemical behavior. Although the most dramatic example of this has been the contribution of ice and nitric acid trihydrate aerosols to ozone depletion in polar regions (WMO [15]), there is growing evidence that aerosol-based transformation phenomena are significant elsewhere in the stratosphere and troposphere as well (cf., Ravishankara [16], WMO [15], Jacob [17]). The importance of this pathway in specific cases of course depends on the chemical species of interest.

Coagulation pathways g3 and l3 can be expressed mathematically in terms of integral equations. For total aerosol mass these take the form

$$r_m \big|_{coagulation} = \frac{1}{2}\int_0^m K(\mu, m-\mu) n(\mu) n(m-\mu) d\mu - n(m)\int_0^\infty K(\mu, m) n(\mu) d\mu \quad (4)$$

where n is the density function for particle number (a function of particle mass), μ is a dummy integration variable, and K is a coagulation coefficient. Corresponding rate equations may be written for each chemical species i. Conventional numerical quadrature methods are unsatisfactory for solving such equations owing to mass-consistency problems and, combined with their general numerical complexity, this renders this aspect a particularly challenging feature of aerosol modeling. A number of specialized and rather elaborate techniques have been developed for this purpose (cf. Williams and Loyalka [11]).

Aerosol breakup processes g4 and l4 are usually presumed to be unimportant as an atmospheric mechanism. Homogeneous nucleation process g5 has been the subject of some rather elaborate and sometimes elegant theories, and can be an important source of ultrafine particles under conditions involving relatively sparse populations of pre-existing particulate material (cf. Friedlander [12]). Recent comparisons of measurements and theory of homogeneous sulfuric acid nucleation (Eisele and McMurry [18]), however, indicate a present inability to quantify such processes on a theoretical basis, at least in the real atmosphere. It seems apparent that much remains to be resolved before this issue can be treated quantitatively in any satisfactory manner.

2.2.2 Heterogeneous chemistry

As can be observed from the recent reviews by Ravishankara [16], and Jacob [17], our knowledge of the heterogeneous chemistry associated with aerosols is, with few exceptions, qualitative, at best. This unsatisfactory condition results from measurement and interpretive difficulties, rather from lack of alleged importance to atmospheric chemistry; indeed, there seems to be ample evidence that aerosol chemistry significantly affects the atmosphere's chemical state by operating in a number of areas, including:

- trapping and/or sequestration of radicals and other reactive species, thus removing them from the gas-phase reaction system;
- serving as host sites for catalysis of sorbed species, which may subsequently re-enter the gas phase;
- serving as reservoir sites for gaseous pollutant species, which condense on the particles and subsequently re-escape to the gas phase;
- providing reactive materials for chemical interaction with depositing gases, and possibly off-gassing reaction products as a result.

Some examples of pathways within these categories that have been observed, or at least speculated upon, include the following:

a. Trapping of peroxy and alkylperoxy radicals: This pathway depletes radicals from the homogeneous system, with obvious implications. Moreover these radicals may proceed to react in/on the condensed phase to produce further reactants and products.
b. Reduction of nitric acid to NO_x by reaction with carbonyls in sulfate aerosols: As can be noted from Figure 1, this process could recycle NO_x, which was formerly consumed in the photochemical sequence, to return for active participation in that sequence. Again, the implications are obvious.
c. Conversion of methanol to methyl nitrate by reaction with nitric acid in sulfate aerosols: As noted by Ravishankara [16], this process is a potential explanation for the higher-than-expected measurements of methyl nitrate in the atmosphere.

d. Catalytic reduction of hydrogen peroxide and ozone by dust particles:
e. Uptake of nitrogen oxides by sea-salt particles to liberate chlorine gas:
f. Nitric acid reduction on soot particles: Similar to b., above, this process has the potential to recycle NO_x to the homogeneous photochemical sequence.
g. Ozone reaction with moist sea-salt particles to produce chlorine gas: Recently Findlayson-Pitts and her co-workers (DeHaan et al. [3]) have detected parts-per-billion levels of molecular chlorine produced upon irradiation, at visible wavelengths, of moist sodium chloride aerosols in the presence of ozone. This is a potential explanation of observations of boundary-layer and free-tropospheric chlorine concentrations in this range by Spicer and his co-workers [8].
h. Nitrous acid sequestration: Stemming partly from observations of smog-chamber wall effects, the sorption and desorption of nitrous acid to and from aerosol particles has been speculated to be a potentially significant reservoir effect.
i. Nitric acid and ammonia exchange: aerosol "equilibrium": The reversible interactions between nitrate and ammonium ions and their gaseous counterparts, nitric acid and ammonia, have been demonstrated to be important determinants of aerosol composition, particularly in Western US environments where these species are typically abundant. Extensive research has been performed on this topic. In contrast to most of the pathways noted above, this phenomenon can be considered to have approached a quantitative level of understanding.

3 Chemical transport models

3.1 Homogeneous gas-phase processes

Dodge [19] provides a modern and quite thorough review of homogeneous-chemistry condensation schemes currently being used for chemical-transport model calculations. As noted in her paper, there have been a few attempts to develop comprehensive reaction descriptions corresponding to limited VOC subsets (e.g., Kerr and Calvert [20], Andersson-Sköld [21], Derwent, et al. [22], and Jenkin et al. [23]). Even though these "master" descriptions are themselves bounded and consolidated in various fashions, they nevertheless contain of the order of one thousand chemical species and several thousand chemical reactions. When one observes that one thousand species implies the need to solve roughly one thousand versions of eqn. (1) simultaneously, the computational need for mechanistic consolidation is readily apparent.

Dodge observes also that parameterization schemes used for this purpose can be classified, roughly into three general areas, as follows:

1. **Lumped structure parameterizations**. Within such schemes, molecules having similar types of carbon bonds are consolidated into single representative entities, which are assigned appropriately averaged rate parameters. Product yields are also represented as lumped entities.
2. **Lumped molecule parameterizations**. This parameterization category is similar to the one above, except that lumping is performed on a molecular-structure basis, rather than strictly with respect to carbon bonds.
3. **Morphecule parameterizations**. The term "morphecule" pertains to a fictitious ensemble of molecules, which are grouped on the basis of chemical similarity and assigned consolidated rate parameters based on the relative abundance of individual constituents. Under the morphecule system, versions of equation (1) corresponding to each morphecule class are solved for a limited time period. Subsequently the morphecules are decomposed, and their component species accounted algebraically on the basis of their individual rate parameters. Following this, new morphecules are composed, assigned revised rate parameters based on their updated makeup, and subjected to the next integration step. The morphecule concept is quite new, and has not seen extensive application to date.

It is noteworthy that all of the conventional parameterization schemes treat the inorganic chemistry represented in Figure 1 in a roughly identical fashion; thus the major differences between schemes lie mainly in the manner in which they represent the organic chemistry and those inorganic interactions not covered by the "core" system represented here. Usually these schemes are evolutionary, with updated versions appearing periodically as improvements are made. The paragraphs immediately following describe some of the more prominent parameterizations applied for urban and regional-continental modeling purposes during recent years.

Carbon bond mechanism (Gery [24]): As its name implies, this parameterization scheme applies a Type 1 approach. Single- and double-bonded straight- and branched-chain hydrocarbons are consolidated into respective single- and double-bond parameters, and singly and doubly substituted aromatic compounds are consolidated into respective single parameters as well. Extensive lumping is performed to consolidate the representation of $RO_x^{\bullet}$ reactions and products. The current version of this scheme contains 32 lumped species and 81 reactions.

Statewide Air Pollution Research Center (SAPRC) mechanism (Carter et al. [4]): The SAPRC scheme is a Type-2 parameterization which provides a somewhat more explicit characterization of organic reaction chemistry and product yields than the Carbon Bond Mechanism described above. This scheme also allows some flexibility in product-yield description, by incorporating product-yield parameters, which may be user-specified on the basis of the chemical environment. These parameters are established in the initialization process and remain fixed throughout the simulation. The current version, SAPRC-97, includes roughly 55 species and 160 reactions and may be downloaded from the SAPRC Web site: <http://helium.ucr.edu/~carter/saprc97.htm>.

Regional Atmospheric Chemistry Mechanism (RACM) (Stockwell et al. [25]): An updated version of the older (Type-2) RADM parameterization (Stockwell et al. [26]), this scheme applies substantially different lumping procedures for VOC reactants and provides a more explicit treatment of $RO^{\bullet}_x$ chemistry. Currently the RACM includes 69 species and 236 reactions.

Morphecule Mechanism: As noted above, this class of parameterization scheme is currently under development and, other than the previously cited reference by Dodge [19], little descriptive material is available in the published literature. We mention it here primarily because of the probability that it will emerge as a viable competitor with Type-1 and Type-2 schemes in the near future owing to its promised flexibility, adaptability, and computational efficiency. The reader is encouraged to consult Dodge's review for a more detailed description of this and the more standard approaches.

3.2 Aerosol dynamics and heterogeneous processes

As might be expected from the preceding text, the mathematical characterization of aerosol processes — from a "microphysicochemical" standpoint, at least — is impeded by the complexity and computational demands of the relevant rate equations as well as by uncertainties in many of the process parameters and mechanisms. The first of these impediments typically necessitates the stipulation of (often rather strong) simplifying assumptions, in addition to lumping, or at least discretization, of the distributed aerosol properties. An extreme example of this, applied in many of the earlier regional and airshed models, is to lump the total distribution of a given aerosol-bound chemical into a single variable, to make no distinction with regard to the overall chemical makeup of the aerosol particles, and to blanket the total mechanistic generation process by presuming that aerosol formation occurs immediately upon oxidation of the chemical's gaseous precursor.

Such practices, however, must be viewed largely as acts of convenience, rather than acts of science. In addition to being conceptually unsatisfying, this approach provides no information whatsoever with regard to aerosol size distribution. Furthermore, and while it may have some credence for low-volatility materials such as sulfuric acid, it becomes totally unsatisfactory for substances exerting higher vapor pressures such as, for example, nitric acid, ammonia and semivolatile organics.

There are several examples of urban and regional modeling efforts that apply less confining and more realistic assumptions to the modeling process. In their extensive review Seigneur et al. [27] argue that coagulation processes are relatively unimportant in typical urban and regional atmospheres, and thus can be neglected for most practical purposes. Moreover, nucleation processes produce only short-lived superfine particles, leading Seigneur to maintain that these can be "treated as a boundary condition at the lower end of the distribution" rather than modeled explicitly. This effectively leaves condensation/revolatilization as the sole operative microphysical process and, as Seigneur points out, several computer modules based on this assumption have been developed and operated to describe the processes in conjunction with comprehensive chemical-transport models. In general these modules share a number of similarities and differ mainly in the various assumptions regarding mass-transfer rates of the semivolatiles to and from the particle surfaces. The reader is referred to Seigneur's report for more specific information on these modules.

The importance of atmospheric water should be noted in this context as well. Saxena et al. [28] have noted that condensed organics can lead to significant changes in particle hygroscopicity, further leading to increases in particle size by water condensation. The implications of this to visibility impairment and cloud nucleation are obvious. Moreover, modifications to aerosol composition, mass and size-distribution by interaction with condensed water in cloud environments are significant determinants of aerosol features whenever an aerosol population has experienced a cloudy environment during its life history. This leads to profound complexity and a substantial challenge to the aerosol modeling community.

Finally, it should be noted that although the pragmatic suggestion of Seigneur et al. that processes such as coagulation and nucleation be (essentially) ignored for modeling purposes has a great deal of practical appeal, it is probably unwise to preclude these mechanisms from possible future consideration. The physiological mechanisms associated with fine-particle insult are presently far from clear, and it may turn out that — at least under some circumstances — the distributions and properties of superfine particles may have significant biological consequences. The emission of such particles by road vehicles, combined with their almost immediate subsequent exposure to human receptors, is an excellent example of a potentially important circumstance of this type (cf. Kittelson [29]). Modeling efforts where multi-mechanism phenomena are being simulated appear to be gaining momentum at this time (e.g., McGraw [30]).

4 Including chemistry in atmospheric models

As the chemistry models become more sophisticated and treat a plethora of chemical processes, it becomes imperative that the changes in the atmospheric dynamics and physics be fed to the chemistry model at a frequency that is appropriate to the scales of the atmospheric circulations as well as the chemical mechanisms modeled. An example would be the modeling of the effluents from a point pollutant source and their downwind transport. These pollutant plumes are usually emitted into the planetary boundary layer (PBL) from the industrial stacks and are transported and mixed by the dynamical and diffusive nature of the PBL as well as the flow in the free troposphere. These processes determine the concentrations of the pollutants in the plume, which in turn affect the chemical transformations that occur in the plume.

Traditionally, the chemical transport models and the atmospheric models were linked externally. The meteorological environments modeled by the atmospheric model were made available to the chemistry models via files written at appropriate intervals. In most cases this interval was about an hour or longer. To effectively model plumes that are affected by PBL processes, this data exchange interval has to be made much smaller (order of a few minutes). This increases the input/output (I/O) requirements of the simulation system, especially as the models are also run at an increased spatial and temporal resolution. For example, if the model resolution is increased by a factor of two in all three directions, the model time step has to be reduced by a factor of two to retain numerical stability. To account for higher frequency features that are thus resolved, the model output has to be processed at twice the frequency. Thus the I/O requirements have increased by a factor of 16. Significant advances have been achieved by some organizations which somewhat circumvent this problem. For example, EPA in its Models-3 program, uses sophisticated chemistry parameterizations to represent the chemical transformations that may be happening in the portions of the atmosphere that are not explicitly resolved (Byun and Ching [31]).

The ultimate in the interfacing of an atmospheric simulation system with a chemical model is to embed the chemical model into the atmospheric model and run them together in lock step. This overcomes the problem of heavy I/O requirements. To resolve and maintain concentration gradients in the pollutant fields, the atmospheric simulation has to be performed at the requisite high resolution in the region of the plume (or plumes). Traditionally this is achieved by a nesting approach as in PSU/NCAR MM5 (Grell et al. [32]) and the Regional Atmospheric Modeling System (RAMS) (Pielke et al. [33]). However, the nesting approach requires the a priori knowledge of the location of the plume so that the higher resolution nests can be strategically placed.

Scientists at the Science Applications International Corporation (SAIC, McLean, Virginia) have developed a novel atmospheric simulation system by combining the advances in the computational fluid dynamics arena with the advances in the atmospheric physics parameterizations. The resultant model, Operational Multiscale Environment model with Grid Adaptivity (OMEGA), is built upon an unstructured adaptive triangular grid in which the computational grid adapts to chosen features of the solution (Bacon et al. [34]). Unstructured adaptive grids have been used for more than a decade in the aerospace industry as well as in shock hydrodynamics modeling and this is a proven technique which yields high resolution results at great savings in computational cost (Baum and Löhner [35]). The adaptive grid technique achieves this by providing high-resolution only where the solution requires it and doing the computations at a lower acceptable resolution over the rest of the computational domain. The refinement and coarsening of the grid are triggered and controlled by a set of user-defined criteria known as "adaptivity criteria".

Figure 4 shows an example in which the OMEGA grid adapts to a plume represented by a collection of particles transported by the Lagrangian Particle Dispersion Model embedded in OMEGA. In this simulation the grid adaptivity criteria were set to particle locations; the grid refines to provide higher resolution at and around particle locations. The adaptation was performed once every hour of the simulation. The adaptation scheme provides a smooth transition from low to high resolution in the region where the plume is located. The adaptive grid allows for the better resolution of the pollutant concentrations as well as the ambient meteorology. As the adaptation refines the grid around the particle location, the grid is already refined in an area before the plume gets there. This allows for better resolution of terrain properties in advance of the arrival of the plume. The resolution of the ambient meteorology is significant, as the plume properties are very much

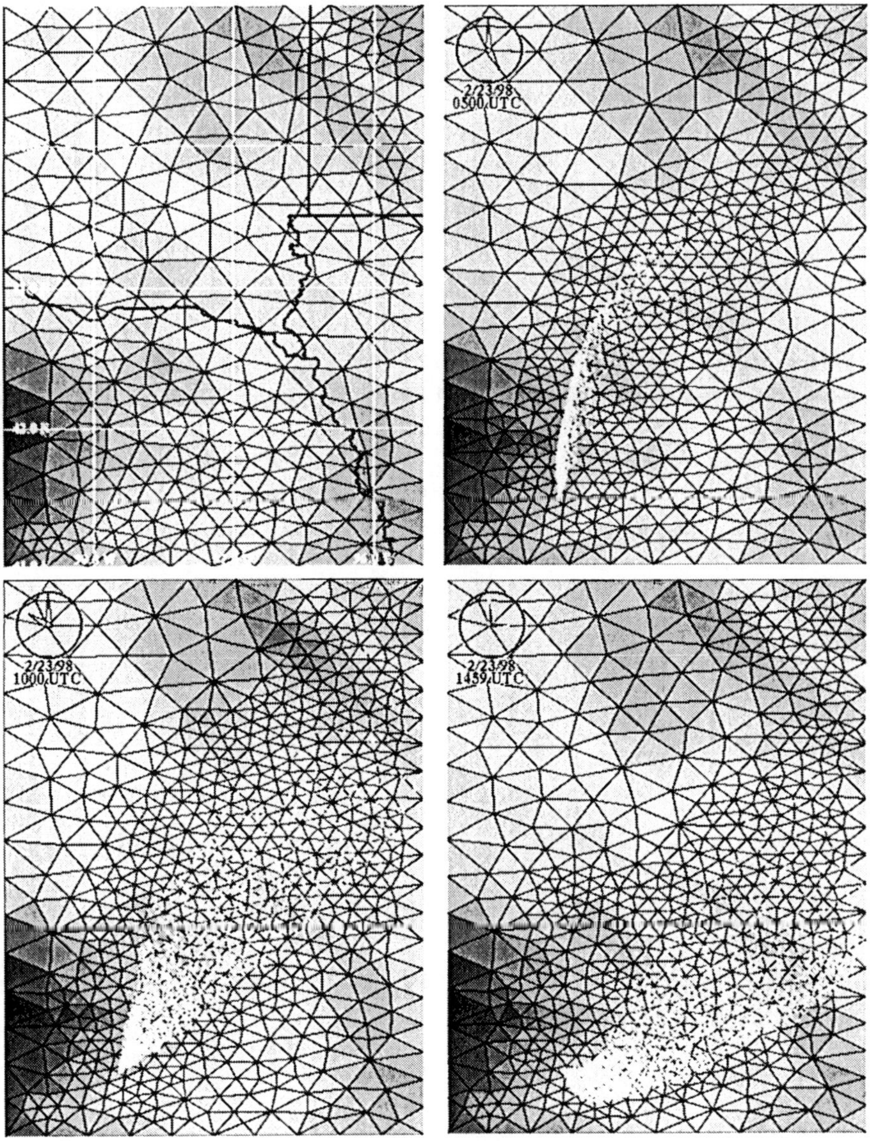

Figure 4: The unstructured grid in OMEGA adapts to an evolving plume. The adaptivity criteria used the particle locations to guide the grid refinement. The panels shown are five hours apart. The shading represents terrain elevation, with dark being the highest.

dependent on the ambient properties such as advective speed and turbulent mixing. The embedded nature of the particle model makes it possible for the plume to "feel" the turbulent fluctuations explicitly.

Figure 5 shows a comparison of the adaptive grid results with one in which the resolution was held high everywhere. The plumes from both these simulations are almost identical. However, the adaptive grid solution was achieved at less than half the computational cost of the high-resolution simulation.

Sarma et al. [36] took the concept of dynamic grid adaptation one step further to look at the potential benefit of using the adaptive grid methodology to atmospheric chemistry calculations. The plume dispersion simulation discussed in the previous paragraph was augmented with a very simple chemistry module. This module simulated a single chemical reaction which converted SO_2 to H_2SO_4 by the OH radical (Lurmann et al. [37]) and is given by the equation

$$OH + SO_2 \xrightarrow{M,O_2,H_2O} H_2SO_4 + HO_2 \tag{5}$$

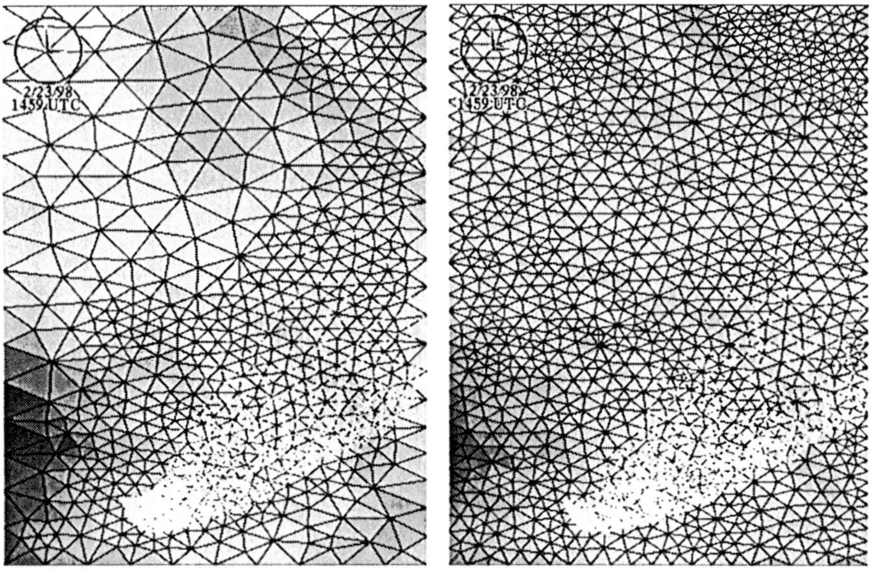

Figure 5: Comparison of the plume after fifteen hours of simulation in the adaptive OMEGA simulation (left) to that in a high-resolution static grid simulation (right).

An iterative scheme with OH and HO_2 radical concentration convergence as termination criteria was used to calculate pseudo-steady state concentrations of the gas phase species. In this experiment a continuous source of SO_2 was introduced in the domain with a source strength of 21 tons per day, which roughly corresponds to the emission from a 1000MW coal-fired power plant (Burns and McDonnell Engr. Co. [38]). Along with the gaseous source, a pseudo-particle source was also introduced at the same location, which injected massless particles at a fixed time interval. These particles were transported using the built-in Lagrangian particle dispersion model and served as the adaptivity criteria for dynamic adaptation.

To explore the impact of adaptive grid refinement on atmospheric chemistry calculations they performed three simulations. The first simulation used no adaptation; the grid was not changed during the simulation. The grid resolution varied from 3 km near the source to 30 km away from the source. The second simulation allowed the grid to adapt around particle locations. The grid resolution again varied from 3 to 30 km but the high resolution was determined by the location of the plume. The third simulation held the grid resolution uniformly high at about 3 km everywhere. Note that the resolution used in the third simulation was equivalent to the highest value set in the adaptation simulation. The grid was not allowed to change in this simulation. The results from the high-resolution simulation were assumed to provide the ground truth to compare against the other two simulations. Figure 6 shows the grids for these three simulations and the imposed wind field. Note that higher resolution was provided around the source for better source definition. The terrain was kept flat for these simulations.

Figure 7 shows the plume concentration contours after twelve hours of transport for the three simulations. A horizontal section as well as a vertical cross section is shown for each of the simulations. The adaptive grid simulation (Figure 7b) was able to capture most of the features of the plume as present in the high-resolution case (Figure 7c), while the non-adaptive solution diffused the plume too much (Figure 7a). This is further evidenced by the plume centerline concentrations shown in Figure 8. The changes in the concentration of SO_2 affect the production of sulfate. The total amount of sulfate produced within the computational domain as a function of time is shown for the three cases in Figure 9. The simulation with no adaptation produced only 60% of the total sulfate

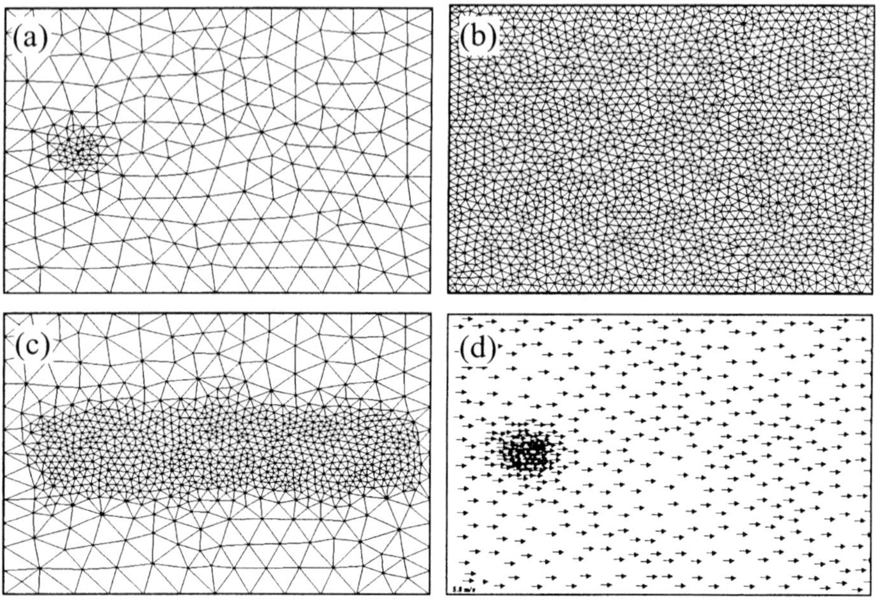

Figure 6: The grids used in a) the baseline, b) the high-resolution and c) the adaptive grid simulations. For the adaptive simulation, the grid shown is at the end of the simulation when the plume extends across the domain. The imposed wind field is shown in (d). The source was placed at the center of the high-resolution region in (a).

when compared with the high-resolution run while the adaptation was able to increase this to 85%.

The principal motivation behind using an adaptive grid is the efficient utilization of computer resources by providing high resolutions only where they are needed. The adaptivity criteria are very flexible and can be changed depending on the problem at hand. In the example discussed above, the savings in computer time was significant. The high-resolution case took twelve times as much CPU time as the baseline case. The adaptive grid case was able to achieve a significantly improved solution with only a factor of 2.5 increase in the CPU time when compared to the baseline case (a factor of 5.3 lower than the high-resolution case).

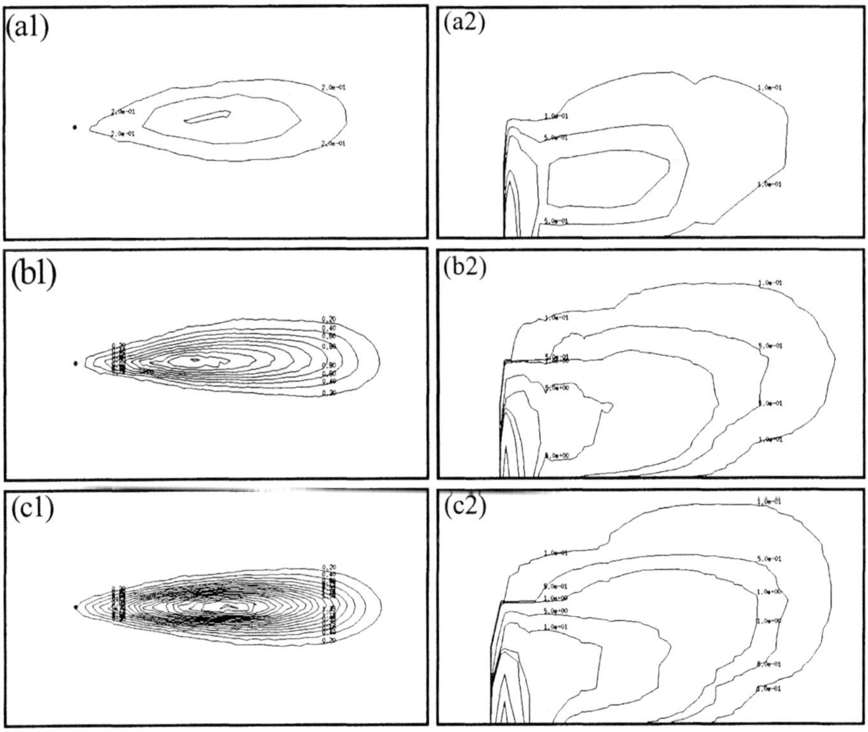

Figure 7: Plume concentration distributions for (a) the no-adaptation, (b) the adaptation, and (c) the high-resolution cases. The figures on the left show horizontal slices at an altitude of about 100m (AGL). The contours are plotted at intervals of 0.2 starting with 0.2 as the outermost contour. The figures on the right show cross sections along the centerline of the plume. The vertical dimension has been exaggerated and represents a total height of 250m. The contours represent values of 0.1 (outermost contour), 0.5, 1, 5, 10, 50, 100, 500, and 1000 (arbitrary units).

The work by Sarma et al. [36] paves the way for the next generation of atmospheric chemistry models, which will work synergistically with atmospheric multi-scale models. Thus the advanced atmospheric chemistry modules can benefit from the high-resolution (temporal and spatial) weather information making possible the simulation of the effects

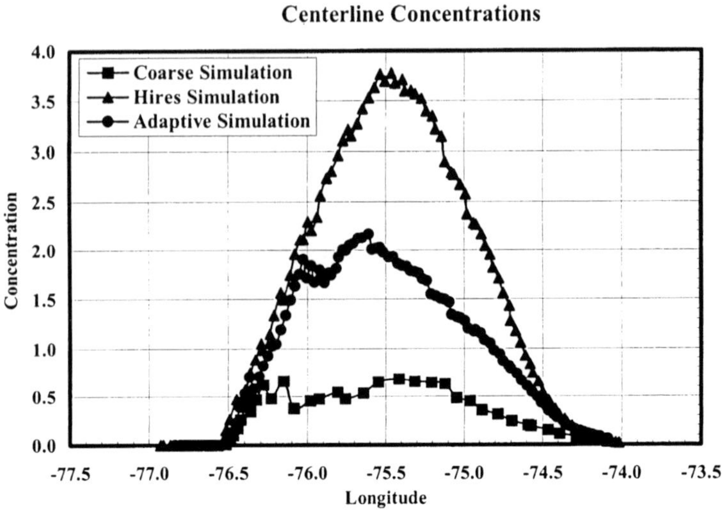

Figure 8: Plume concentration along the plume centerline for the baseline, high-resolution and adaptive grid cases.

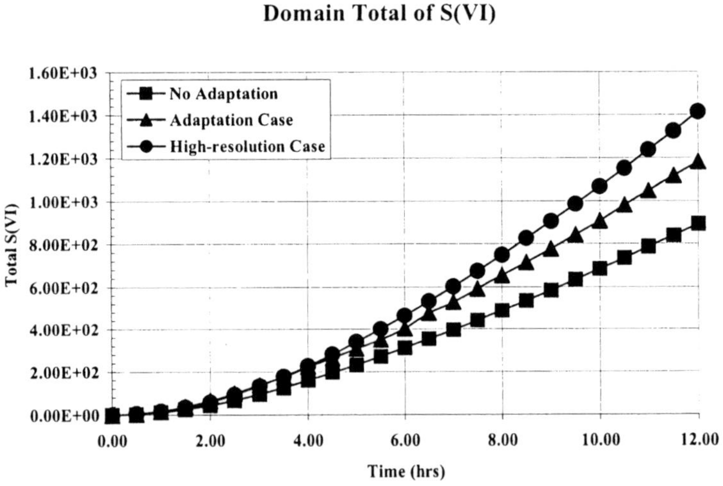

Figure 9: Total sulfate formed within the domain for the baseline, high-resolution and grid adaptive cases.

of small scale environmental features on pollutant chemistry. The atmospheric physics routines such as the cloud microphysics and radiation can be enhanced with the atmospheric chemistry information such as aerosol composition and concentration. Such synergy is also crucial in dealing with pollutant removal processes such as precipitation scavenging and deposition processes. However, the high-resolution simulations will require initial and boundary condition data of commensurate resolution describing the various chemical species modeled. This will continue to pose a challenge to the researchers working in the atmospheric measurement and remote sensing fields.

5 Summary

In this chapter we have presented some of the issues that relate to atmospheric chemistry from a mesoscale perspective. Current generation numerical weather prediction models use a horizontal grid resolution down to a kilometer. At these scales these models will be able to resolve explicitly some of the reactions that were previously deemed too fast and were represented by some equilibrium state. Even though our state of knowledge on some of these reactions has improved, an issue that poses a significant challenge is the collection of data describing the spatial variations of the key species at a resolution that will be appropriate to the high-resolution meteorological models. Emission inventories do not exist yet at a resolution of O(1km), except for some specialized data sets over some regions. It would also be important to improve the parameterizations of some of the removal processes such as dry and wet depositions, especially when high-resolution surface property data sets (land use, vegetation, soil moisture, albedo, etc.) are used. Finally, advances in the field of numerical weather prediction will allow us to explore in great detail the chemical and physical behavior of urban and industrial plumes.

References

[1] Atkinson, R., Kinetics and mechanisms of the gas-phase reactions of the NO3 radical with organic compounds. *J. Phys. Chem. Ref. Data*, **20**, pp. 459-507, 1999.

[2] Dransfield, T. J., Perkins, K. K., Donahue, N. M., Anderson, J. G., Sprengnether, M. M. and Demerjian, L., Temperature and Pressure Dependent Kinetics of the Gas-Phase Reaction of the Hydroxyl Radical with Nitrogen Dioxide. *Geophys. Res. Ltrs.*, **26**, pp. 687-690, 1999.

[3] DeHaan, D. O., Brauers, T., Oum, K., Stutz, J., Nordmeyer, T. and Finlayson-Pitts, B. J., Heterogeneous Chemistry in the Troposhere: Experimental Approaches and Applications to the Chemistry of Sea-Salt Particles. *Ont. Rev. Phys. Chem*. (in press), 1999.

[4] Carter, W. P. L., Luo, D. and Malkin, I. L., *Environmental Chamber Studies for Development of an updated Photochemical Mechanism for VOC Reactivity Assessment,* Final Report to California Air Resources Board, Contract 92-3455, 1997.

[5] Atkinson, R., Atmospheric Chemistry of VOCs and NO_x. *Atmos. Environ.*, (in press), 1999.

[6] Finlayson-Pitts, B. J. and Pitts, J. N., *Atmospheric Chemistry*, Wiley Interscience, New York, 1986.

[7] Finlayson-Pitts, B. J. and Pitts, J. N., Chemistry of the Upper and Lower Atmosphere: Theory, Experiments and Application, Academic Press, New York, 2000.

[8] Spicer, C. W., Chapman, E. G., Finlayson-Pitts, B. J., Pastridge, R. A., Hubbe, J. M., Fast, J. D. and Berkowitz, C. M., First Observations of Cl_2 and Br_2 in the Marine Troposphere. *Nature*, **394**, pp. 353-356, 1998.

[9] EPA, *Air Quality Criteria for Particulate Matter.* EPA/600/P-95/001af. Environmental Protection Agency Office of Research and Development, Washington, D. C., 1996.

[10] Hidy, G., Aerosols: An Industrial and Environmental Science. Academic Press, New York, 1984.

[11] Williams, M. and Loyalka, S., *Aerosol Science: Theory and Practice.* Pergamon Press, Oxford, U. K., 1991.

[12] Friedlander, S. K., *Smoke, Dust and Haze: Fundamentals of Aerosol Behavior,* Wiley Interscience, New York, 1977.

[13] Seinfeld, J. H., Atmospheric Chemistry and Physics of Air Pollution. Wiley, New York, 1986.

[14] Bowman, F., Odum, A., Seinfeld, J. and Pandis, J., Mathematical Model for Gas-Particle Partitioning of Secondary Organic Particles. *Atmos. Environ.,* **23**, pp. 3921-3931, 1997.

[15] WMO, *Scientific Assessment of Ozone Depletion: 1994,* World Meteorological Organization, 1995.

[16] Ravishankara, A. R., Heterogeneous and Multiphase Chemistry in the Troposphere. *Science,* **276**, pp. 1058-1065, 1997.

[17] Jacob, D. J., Heterogeneous Chemistry and Tropospheric Ozone, *Atmos. Env.,* (in press), 1999.

[18] Eisele, F. L. and McMurry, P. H., Recent Progress in Understanding Particle Nucleation and Growth. *Phil. Trans. Roy. Soc. London,* **B 352**, pp. 191-201, 1997.

[19] Dodge, M. C., Chemical Oxidant Mechanisms for Air-Quality Modeling. *Atmos. Env.,* (in press), 1999.

[20] Kerr, J. A. and Calvert, J. G., *Chemical Transformation Modules for Eulerian Acid Deposition Models,* Volume I: The Gas Phase Chemistry., Report submitted to the US Environmental Protection Agency, EPA/600/3-85-015, 1985.

[21] Andersson-Sköld, Y., *Updating the Chemical Scheme for the IVL Photochemistry Trajectory Model,* IVL Report B 1151, Swedish Environmental Research Institute, Götegorg, Sweden, 1995.

[22] Derwent, R. G., Jenkin, M. E. and Saunders, S. M., Photochemical Ozone Creation Potentials for a Large Number of Reactive Hydrocarbons under European Conditions. *Atmos. Environ.,* **30**, pp. 189-199, 1996.

[23] Jenkin, M. E., Saunders, S. M. and Pilling, M. J., *Constructing a Detailed Chemical Mechanism for Use in Tropospheric Chemistry Models.* Report AEA/RAMP/18360008/003, AEA Technology, Culham Laboratory, Oxfordshire, U.K., 1996.

[24] Gery, M. W., Whitten, G. Z., Killus, J. P. and Dodge, M. C., A

Photochemical Kinetics Mechanism for Urban and Regional Scale Computer Modeling. *J. Geophys. Res.* **94**, pp. 12925-12956, 1989.

[25] Stockwell, W. R., Kirchner, F., Kuhn, M. and Seefeld, S., A New Mechanism for Regional Atmospheric Chemistry Modeling. *J. Geophys. Res.* **102**, pp. 25847-25880, 1997.

[26] Stockwell, W. R., Middleton, P., Chang, J. S. and Tang, X., The Second Generation Regional Acid Deposition Model Chemical Mechanism for Regional Air Quality Modeling. *J. Geophys. Res.* **95**, pp. 16343-16367, 1990.

[27] Seigneur, C, Hopke, P. and Grosjean, D., *Review of Air Quality Models for Particulate Matter.* Document CP015-97-1a, American Petroleum Institute, Washington, D.C., 1998.

[28] Saxena, P., Hildemann, L., McMurry, P. and Seinfeld, J., Organics Alter Hygroscopic Behavior of Atmospheric Particles. *J. Geophys. Res.*, **100**, pp. 18685-18700, 1995.

[29] Kittelson, D., *Characterization of Diesel Particles in the Atmosphere*. Report to the Coordinating Research Council, Atlanta, G.A., 1988.

[30] McGraw, R., Description of Aerosol Dynamics by the Quadrature Method of Moments. *Aerosol Sci. and Technol.*, **27**, pp. 255-265, 1997.

[31] Byun, D. W. and Ching, J. K. S., *Science Algorithms of the EPA Models-3 Community Multiscale Air Quality (CMAQ) Modeling System*, EPA Report EPA/600/R-99/030, U.S. Environmental Protection Agency, Washington, D.C., 1999.

[32] Grell, G. A., Dudhia, J. and Stauffer, D. R., *A description of the Fifth-Generation Penn State/NCAR Mesoscale Model (MM5).* NCAR Technical Note, NCAT/TN-398+IA, 1993.

[33] Pielke, R. A., Cotton, W. R., Walko, R. L., Tremback, C. J., Nicholls, M. E., Moran, M. D., Wesley, D. A., Lee, T. J. and Copeland, J. H., A comprehensive meteorological modeling system – RAMS. *Meteor. Atmos. Phys.*, **49**, pp. 69-91, 1992.

[34] Bacon, D. P., Ahmad, N. N, Boybeyi, Z, Dunn, T. J., Hall, M. S.,

Lee, P. C. S., Sarma, R. A. and Turner, M. D., A Dynamically Adapting Weather and Dispersion Model: The Operational Multi-scale Environment model with Grid Adaptivity (OMEGA)., *Mon. Wea. Rev.*, (in press), 2000.

[35] Baum, J. D. and Löhner, R., Numerical simulation of shock-elevated box interaction using an adaptive finite element shock capturing scheme. *Proc. of the 27th Aerospace Science Meeting, AAIA-89-0653*, 1989.

[36] Sarma, A., Ahmad, N., Bacon, D. P., Boybeyi, Z., Dunn, T., Hall, M. and Lee, P., Application of Adaptive Grid Refinement to Plume Modeling, *Air Pollution VII*, WIT Press, Southampton, U.K., pp. 59-68, 1999.

[37] Lurmann, F. W., Lloyd, A. C. and Atkinson, R., A chemical mechanism for use in long range transport / acid deposition computer modeling, *Journal of Geophysical Research*, **91**, pp. 10905–10936.

[38] Burns and McDonnell Engr. Co., Report on the environmental analysis for the Laramie River Station Steam Generating facility of the Missouri Basin power project. Report No. 74-012-4-001, 1975.

Chapter 9

Numerical modeling of gas deposition and bi-directional surface–atmosphere exchanges in mesoscale air pollution systems

Devdutta S. Niyogi & Sethu Raman
North Carolina State University
Raleigh, NC 27695–7236, USA

Abstract

Accurate representation of surface processes such as vegetation has a significant role in air pollution models. In a variety of situations, the surface acts as a sink for the pollutants. Using pristine relations developed on fluid mechanical concepts, different formulations are discussed in this chapter to develop deposition flux estimates in air pollution models. An interesting scenario also develops when the soil and vegetation, in particular, acts as a source, in addition to being a sink for the gaseous material. Hence as a generalized framework in air pollution systems, the ability of the surface to generate bi-directional fluxes needs to be represented. Accordingly, different modeling techniques are presented, ranging from regression equations to modifications in the resistance pathways, and a detailed eco-physiological leaf scaling approach. Finally, of particular relevance to mesoscale applications is the area averaging and regional mapping of the bi-directional fluxes. Accordingly different methods are discussed based on a combination of surface measurements, remote sensing and model parameterizations.

1 Introduction

One of the important considerations in designing an air pollution modeling system is the accurate and efficient representation of the surface processes. Mathematically, surface processes form a boundary condition in the atmospheric analysis thus becoming a pivotal component of the modeling system for both surface energy balance as well as mass transfer. As discussed in Niyogi and Raman [1], surface processes manifest changes through soil vegetation atmosphere transfer (SVAT) processes in the boundary layer. At the local scale, surface features modulate humidity, temperature, and surface energy balance. At a mesoscale, heterogeneity in surface fluxes leads to mesoscale circulation, which affects transport and diffusion characteristics of the atmosphere. The presence of vegetation thus leads to humidity exchange potential [2]. In addition to its impact on the local as well as the regional scale thermodynamic structure, vegetation has a dominant role in air pollution models principally as a sink or a depositing surface [3], [4]. Until the mid 1970s most deposition research assumed gases deposit to bare ground with only a marginal impact due to vegetation [5], [6]. Subsequent studies, however, have shown that vegetative processes are a dominant pathway for surface–atmosphere exchanges [4], [7], [8], [9], [10].

In this chapter, we will discuss the issues pertaining to mathematically representing surface effects with particular emphasis on pollutant deposition. In the classical sense, such exchange processes have been evaluated principally considering homogeneity. It is being increasingly recognized that one of the largest sources of uncertainty lie in representing surface couplings within the air pollution simulation framework. Although in general gaseous pollutants are deposited to the surface, in some instances the vegetation interface can form a source for the biogenic emissions. Hence the biospheric atmospheric interactions have a dominant role in the atmospheric environmental mass balance, for a variety of applications ranging from regulatory purposes to ecosystem synthesis (see Mooney [11]).

Accordingly, this chapter is structured in the following manner. In the following section, the role of multimedia couplings is discussed both at the micro and the regional scale. Different gas deposition parameterizations are discussed, with particular emphasis on mesoscale models. In Section 3, the land–atmosphere coupling is discussed with particular emphasis on the bi-directional (source as well as sink) exchanges. Here both the modifications in the traditional resistance approach as well as more mechanistic plant physiological relations are discussed. One of the pertinent issues for mesoscale models is scaling the parameterizations and the model results to a regional scale. Section 4, therefore, outlines the procedures and approaches for regional analysis. Finally, Section 5 presents the conclusions and a discussion for future research in this area.

1.1 Surface processes in environmental analysis

Recent years have witnessed phenomenal growth in the computational resources. Numerical models are being run with higher and higher grid resolution. With increasing horizontal grid resolution, the impact of surface processes becomes increasingly dominant in numerical simulations (see for example, Pielke [12]). Hence, there is growing concern regarding a realistic representation of the continuum between the surface atmosphere exchanges [13], [14], [15]. However, most models that consider the coupling between land, air and water treat the continuum more for energy balance than for mass balance. Hence parameterizations need to be developed for air pollution models and environmental assessment systems in which mass exchange or more correctly gas exchange is explicitly accounted for to provide source/sink representation. We will discuss formulations and development of such parameterizations in this chapter.

In creating a continuum, it is important to develop the coupling by linking the heterogeneous media dynamically. As a general approach, it will be beneficial in developing such couplings so as to integrate different parameterizations and then link them via different conservation equations. Such a system can thus accommodate more than one process and media. This not only allows representation of processes other than linear relations but also provides opportunity for checks and validations of the exchanges. It can also replicate the reality as much as possible in terms of what dominates the pathway and the exchange. Ideally, as discussed earlier, these conservation equations would involve explicit energy as well as mass exchanges. It is important that the system has to be designed with interactive couplings. Also, different processes involved in the coupling should be sensitive to direct as well as second–order or indirect changes and thus show variations for reaction as well as the stimuli. Consequently, the system has to be generalized so as to show valid spatial and temporal variations. Hence the equations that form coupling interfaces in the system need to have prognostic variables and have some form of landscape or regional factors embedded in their specification. Finally the coupled system has to be tested with observations made under different conditions for verification.

Consider the example of a coastal watershed. These are important regions of socio economic activities for which the environmental issues (air and water quality) have significant implications. The multi media (land, air, water) exchanges form critical pathways for pollutant as well as nutrient transfer. The pollutants released on the land would be either detained on the soil surface, or deposited over the vegetation. This material can subsequently transport through runoff or percolate into the soil and roots and affect the ground water. A portion of the pollutant, which is not detained on the landmasses, is passed over the water body. The terrestrial loading can thus induce changes in the water nutrient status and affect flora and fauna. An additional feature of such a watershed is the mesoscale variability in the surface and subsurface, and the meteorological features which can affect the hydro meteorological exchanges. These features need to be resolved through a mesoscale model with the multimedia systems coupled to it.

1.2 Developing the mathematical framework

In developing the simulation framework, the first principle approach can be adopted. Thus, pristine laws relating to the laminar and turbulent fluid flows are modified to represent the boundary conditions. Typical examples of the extension of such principles are the wind profile 'laws', development of the mixing length (see Holt and Raman [16]), turbulent kinetic energy closure schemes, and the atmospheric boundary layer similarity theories (see Stull [17]). Such fundamental fluid assumptions have been extended in almost every numerical as well as analytical model developed for environmental simulations. The equations and models for simulating the atmospheric or oceanic phenomena have some fundamental assumptions at their core. One such assumption is surface homogeneity. However, in recent years, there have been significant improvements in the modeling strategies. More and more realistic features are being incorporated into the model equations, with emphasis on linking causal behavior within heterogeneous systems.

Interfacing heterogeneous media in its simplest form involves the development of equations, which pass on the base state at System 1 boundary condition as the initial state for the System 2 boundary. However, this transition is not easy as the two systems cannot be treated in a homogeneous state. There has to be a buffer or interface layer that can act as a moderator for the two boundary conditions to be in congruence with the prevailing environments (without shocking or destabilizing the local equilibrium). This intrinsic concept will be discussed for applications involving water vapor and other gas exchanges and deposition over vegetation (canopy, leaf, stomata and intercellular conduits) and atmospheric boundary.

In the following section, we review some of the underlying principles and assumptions of gas exchange for differential media, including the approximations made in experimental field studies for toxic deposition and exchange. Then, we discuss the techniques and approaches for vapor exchange at vegetation–atmosphere interface. Finally, we present a discussion summary and conclusions regarding some of the limitations and developments for next generation models.

2 Developing gas deposition relations

Understanding the transport and fate of gaseous pollutants is important for diverse applications. For example, it is now widely recognized that plants emit volatile organic compounds (VOCs) such as monoterpene and isoprene (cf. Arey et al. [18]). These biogenic emissions can undergo photolysis and other chemical transformation and generate pollutants such as ozone in the lower troposphere [19]. Results from both special observational campaigns (e.g., Monson and Fall [20]) as well as numerical modeling studies (e.g., Langford and Fehsenfeld [21]; Guenther et al. [22]) confirm this feature at diverse scales. For instance, Fitzgerald [23] estimated that of the 6000 to 7500 tons of mercury emitted into the atmosphere, 25–50 % is released from natural surfaces.

In addition to being a source of biogenic emission and other primary as well as secondary pollutants, the terrestrial biosphere is known to be a significant sink (see Wesley [24]). The gaseous pollutants can either deposit to the vegetated surface or be absorbed into the leaf cells. This exchange between the terrestrial biosphere and the atmosphere thus results in a net reduction of the available pollutant in a mass balance analysis. Hence, the roles of the vegetative surfaces both as a source as well as a sink are not isolated from each other and are in fact interactive (see Fowler and Unsworth [25], Fuentes and Gillispie [26], and Gao et al. [27]). Such an issue has to be addressed as a continuum problem. The release is therefore from what can be referred to as the VOC rich surface. In the atmosphere, it undergoes transformation as well transport and diffusion in the air medium. Indeed the gases (and the transformed aerosols) subsequently deposit back to the surface. Atmospheric deposition thus plays a pivotal role in determining both the air as well as the water and the airshed/watershed quality at a regional scale [28]. The deposited toxin, over land or water, through the continuum impacts regional hydrology, and the air quality both at diurnal as well as climatic scales, depending on its residence time in the ecosystem. Such a scenario, for instance, has been considered to be a major contributor of pollutant recycling in various North American lakes and watersheds [29]. Hence it is critical to introduce the bi-directional role of vegetation and natural surfaces in an environmental analysis.

In order to develop such couplings regarding the source/sink characteristics of natural surfaces, several different approaches have been addressed in the literature. We will review some of the underlying concepts and formulations here.

2.1 Modeling surface deposition flux

For different pollutants including VOCs, the concentration flux can be represented as:

$$F_i = - (D_i + K_i).\ dC/dz, \qquad (1)$$

where D_i is the molecular diffusivity, K_i is the respective eddy diffusivity, and dC/dz is the vertical gradient of the concentration. Based on results from several empirical formulations such as Businger [30], and Droppo [31], K_i is taken to be equal to the eddy diffusivity of heat, K_h. Integrating eqn. (1), assuming a constant flux layer near the surface, yields:

$$F_i = - (R_a + R_b)^{-1} (C_r - C_s), \tag{2}$$

where, R_a and R_b are the turbulent aerodynamic and molecular bulk boundary layer resistance for gas exchange, and C_r, C_s are gas concentrations at the reference level (model lowest level) and the surface. The turbulent resistance is the total resistance across the turbulent layer and can be estimated by integrating the changes in the vertical eddy diffusivity for heat between a thickness comprising δ and z_r, that is the layer above the molecular exchange and the model reference level. Using similarity equations,

$$R_a = Pr .(u_*.k)^{-1}. [\ln (z_r - d)/ (\delta - d) - \Psi_h]. \tag{3a}$$

In the above, in addition to the terms described before, *Pr* is the Prandtl number (0.923 in neutral conditions), u_* is friction velocity, and Ψ_h is the non-dimensional stability function for heat [32], [33]. Under neutral conditions [34], [35] eqn. (3a) becomes,

$$R_a = Pr. u / u_*^2. \tag{3b}$$

The molecular bulk resistance, on the other hand, is obtained by integrating the total molecular as well as eddy diffusivity between the receptor surface, z_s, and the molecular layer, δ. Using similarity relations the boundary layer resistance can be represented as

$$R_b = (u_*.k)^{-1}. [\ln (z_o/ \delta)]. \tag{4a}$$

Since it is difficult to prescribe or determine the molecular layer thickness, empirical approaches are adopted. A simplification can be adopted for vegetation canopies, yielding [36] $R_b=A.(l/ u)^{0.5}$, where l is the characteristic length scale of the leaf canopy (order of 0.05 m), and A is a constant (~90 $s^{1/2}$ m^{-1}). Similarly, following Owen and Thompson [37], eqn. (4a) can be represented as

$$R_b = (u_*.B)^{-1}, \tag{4b}$$

where *B* is the interfacial sublayer Stanton number. However, estimations involving *B* are still largely uncertain (see a discussion by Kramm et al. [38]). With increasing turbulence, values of *B* range from 1 to 1.25. Over vegetation canopies, under turbulent conditions, Parlange et al. [39] obtain *B*=2.5, which have been applied in studies such as Leuning et al. [40]. Following theoretical considerations of Owen and Thompson [37],

Schlichting [41] and Brutsaert [42] for example, the Stanton number can be estimated as

$$B^{-1} = a.\ Sc^{b}.\eta^{c}\ . \tag{4c}$$

In the above equation, following Kramm et al. [38], a=0.52, b=0.8, and c=0.45, and η is the roughness Reynolds number. For an aerodynamically rough surface, $\eta = u_*.(z_o+d)/\nu$, while for an aerodynamically smooth surface, eddy diffusivity can be solved for the Reynolds number:

$$K_m / \nu = \kappa\ (\eta - 11 \tanh(\eta/11))\ . \tag{4d}$$

Another approach for calculating R_b [43], [44],

$$R_b = Pr\ (2 / k.u_*).(Sc^{\ 0.67})\ , \text{ or} \tag{5a}$$

$$R_b = 1.47\ Sc^{0.67}\ (l/uv)^{0.5}\ B^{-1}\ , \tag{5b}$$

where Sc is the Schmidt number, which is the ratio of the kinematic viscosity and the diffusivity of the gas of interest. Thus, $Sc = \nu/D$, and D is calculated through Graham's law as,

$$D = \nu\,(Wa / Wp)^{0.5}\ . \tag{5c}$$

At the core of these surface exchange parameterizations is an exchange velocity often referred to as the 'deposition velocity' (V_d). In the bi-directional atmosphere–vegetation exchange, a positive deposition velocity indicates deposition from the atmosphere to the surface while a negative V_d can be interpreted as an emission from the surface to the atmosphere. Thus deposition velocity is a convenient way to parameterize surface–atmosphere exchange for use in analytical as well as numerical models at a regional scale. Deposition flux (F_d, which is the amount of deposited material per unit area for unit time) is defined as the product of V_d with concentration, C, of the depositing material, yielding

$$F_d = V_d\ .\ C\ . \tag{6}$$

Equation (6) is useful, as deposition velocity estimates are available in literature and are often known to vary over a small range under similar conditions. Knowing ambient pollutant concentrations through measurements (see for example, Padro [45]) or by using air quality models (e.g., Chang et al. [46]), and by prescribing deposition velocity from look-up tables (e.g., Sehmel [47], Voldner et al. [48]) or using dynamic formulations described later, deposition flux can be estimated.

2.2 Obtaining deposition velocity

One way of developing a look-up table for V_d values is by using observations based on micrometeorological considerations. Studies such as Duyzer and Fowler [49], Erisman et al. [50], and Erisman and Baldocchi [51] outline different measurement techniques adopted for measuring gaseous deposition. The different micrometeorological techniques are categorized as eddy correlation, gradient, variance, conditional sampling, and eddy accumulation methods. Additionally, surface analysis methods are adopted particularly for complex terrain, in which snow analysis, throughfall, and leaf wash techniques are included. Finally, for specie specific studies, chamber techniques are often used. Each of these techniques, however, has significant limitations. Erisman et al. [50] discusses these limitations and uncertainties in detail. However, for a homogeneous surface, both F_d and C can be measured with a fair degree of certainty using different approaches (cf. Arya [52], [53], Pleim et al. [54]). If the measurements are made over an extended period (a few weeks) and during different seasons as well as under different environmental conditions (such as wetness, radiation and temperature) a sufficiently realistic range of V_d values can be obtained. These values can be prescribed for similar landscapes in air pollution models for developing deposition flux estimates. Typically, the concentration estimates are made using a simple gaussian plume model, and the deposition flux is obtained from multiplying concentrations with V_d value [52], [53]. However, measurements have significant uncertainties and are limited in the range of data and variability in the atmospheric conditions (see Brook et al. [55], Niyogi et al. [56]). Hence there is a continued need for parameterizations that can accurately estimate deposition velocities both for observational studies and for numerical/diagnostic models [57].

In parameterizing deposition velocity, often the resistance pathway described earlier is further extended (e.g., Garland [58]). As can be inferred, the higher the resistance for the material to deposit, the lower will be the deposition flux. Also, following eqn. (6), the deposition flux is directly proportional to deposition velocity. Combining these two features, one obtains deposition velocity as the inverse of total resistance (similar to the approach obtained in eqns. (1) through (6)). For a natural vegetative landscape, the resistance can be offered through different pathways. First, the material has to be dispersed to the receptor surface, then it has to penetrate the quasi laminar surface boundary layer, and then it has to be captured by the vegetative surface. At the vegetated surface the material encounters resistance from leaf cuticles, stomata, mesophyll, and the soil surface [1]. These resistance pathways can be represented by a total resistance (R_T), comprising turbulent aerodynamic resistance [*Ra*], surface quasi laminar boundary [*Rb*], and surface resistance [*Rc*], and can be linked to deposition velocity as,

$$V_d = (R_T)^{-1} = (Ra + Rb + Rc)^{-1} \tag{7}$$

A schematic representation of the three resistances involved in surface deposition is shown in Fig. 1. Of the three resistances, *Ra,* calculations are independent of the type of gas (see Wesley [24]). *Rb* calculations, however, need to consider molecular diffusivity of the depositing gas, and *Rc* estimates show dynamic variations for biophysiochemical responses. We will, therefore, discuss parameterizations for *Rc* in more detail. Also, of the three resistances, the *Rc* term is generally dominant, typically by an order of magnitude in various cases (see Baldochi et al. [34], Lynn and Carlson [59]). Following Niyogi and Raman [1], and Niyogi et al. [60], [61], [62] the parameterization approaches can be conveniently classified as 'environmental' and 'physiological'.

In the 'environmental' approach, the resistance pathways are modeled as a function of environmental variables such as humidity, surface radiation and temperature, treating the biospheric components in a parametric form (either through constants or functional entities). In the 'physiological' approach, on the other hand, both the environmental and biospheric components are treated as interactive variables. Mathematically, in the environmental approach, the vegetation is treated as a parameter and its role in the mass/energy balance is considered only implicitly, while in the physiological representation, vegetative processes are treated more causally and as a component of the conservation and mass/energy balance equations explicitly [63]. Figure 2, based on Niyogi et al. [60] shows the schematic representation of the two approaches. A similar control between external (environmental) and internal (physiological) parameters on the canopy resistance was developed by Lynn and Carlson [59] using water stress as the driver.

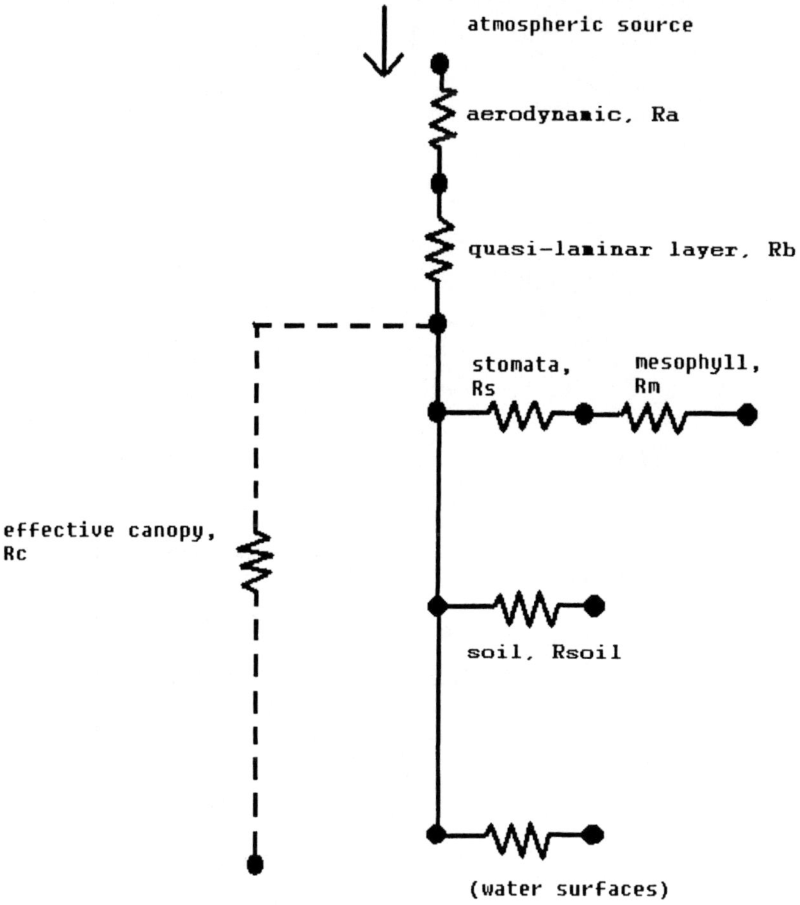

Figure 1: Resistance pathways for depositing material. *Ra* is the aerodynamic resistance, *Rb* is the resistance offered by the quasi laminar layer close to the depositing surface. *Rc* is the effective canopy or the surface resistance. *Rc* is generally dominant and can comprise various different resistances before the material deposits. *Rc* dynamically responds to gas characteristics, environmental and meteorological feedback as well as changes in the vegetation canopy.

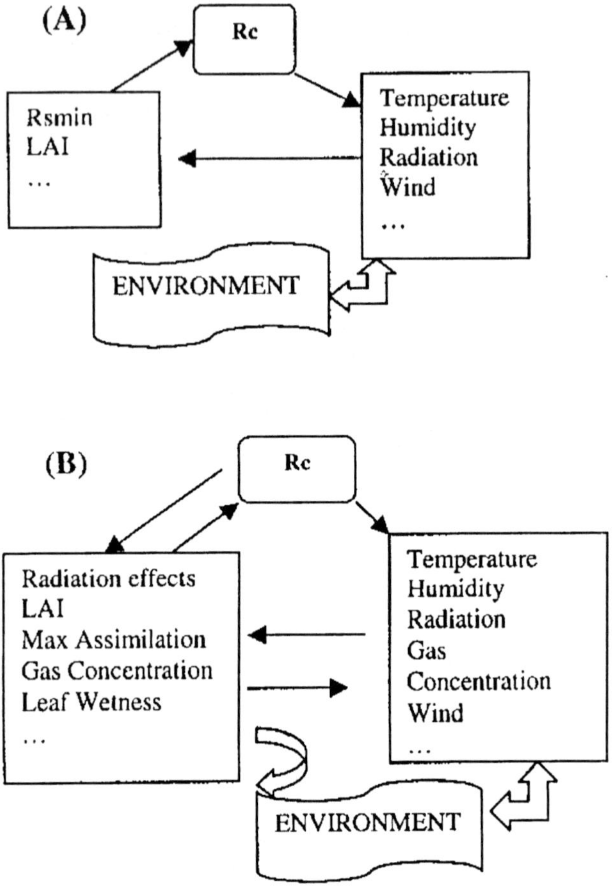

Figure 2: Schematic representation of the (A) meteorological, and the (B) physiological pathways for parameterizing the canopy/surface resistance (*Rc*) term for estimating depositing velocity (*Vd*). In the physiological parameterizations there is explicit feedback interactions between the environment and the depositing surface (foliage).

2.3 Environmental approach for parameterizing V_d

Following eqn. (6), to parameterize V_d, one needs to develop sub models for the *Rc* term. In estimating *Rc*, all the additional surface resistances (as in eqn. 8) need to be integrated. A central approach of the environmental method for estimating *Rc* was proposed in a seminal work by Jarvis [64]. The surface stomatal resistance was parameterized as a function of the so called 'minimum stomatal resistance' (R_{smin}), modulated through atmospheric variables such as temperature, radiation, water stress and humidity. R_{smin} is the minimum resistance the vegetation offers for water vapor exchange. Typically, afternoon values of stomatal resistance in the absence of moisture stress can be considered to be 'minimum' (see Niyogi et al. [65], Avissar et al. [66]). R_{smin} varies according to the vegetation and the season and a typical value of R_{smin} for grassy type vegetation is around 50 s m^{-1}. A detailed review regarding the R_{smin} values for various plant species can be found in Schulze et al. [67] as well as in Kelliher et al. [68]. The Jarvis–type approach has thus become a primary evapotranspiration stomatal resistance parameterization at all scales. [See for example, Alapaty et al. [69], Wetzel and Chang [70] and Dickinson et al. [71], for modeling studies from micro to global scales.]

The principal equation for the Jarvis type scheme can be stated as follows (see Noilhan and Planton [72]):

$$R_s = \frac{R_{s\min}}{LAI.F_1.F_2.F_3.F_4}, \tag{8}$$

where R_{smin} and leaf area index (LAI) are prescribed variables, while

$$F_1 = \frac{1+f}{f + R_{s\,min}/R_{s\,max}}, where, f = .55\frac{G}{G_l}\frac{2}{LAI}, \tag{9a}$$

$$F_2 = \begin{cases} 1, if\,(w_2 > 0.75 w_{sat}) \\ \dfrac{w_2 - w_{wilt}}{0.75 w_{sat} - w_{wilt}}, if\,(w_{wilt} \le w_2 \le 0.75 w_{sat}) \\ 0, if\,(w_2 < w_{wilt}) \end{cases}, \tag{9b}$$

$$F_3 = 1 - 0.025.D\,, \tag{9c}$$

$$F_4=1-1.6X10^{-3}(298.0-T_a)^2\,. \tag{9d}$$

In the above, G is the net radiation reaching the foliage, G_l is the radiation limit at which photosynthesis is assumed to start (about 100 W m^{-2} for crop like vegetation). Similar to R_{smin}, it is necessary to prescribe the maximum stomatal resistance (R_{smax}) the foliage can practically offer, and it is generally set at a constant value of 5000 s m^{-1} (see Niyogi and Raman [1], Schulze et al. [67], Kelliher et al. [68]). In the F_2 term (eqn. 9b), w_2 is the deep soil moisture (at 1m below the surface) and w_{wilt} and w_{sat} are the wilting and saturated soil moisture values for the soil (see Clapp and Hornberger [73], Cosby et al. [74], and Noilhan and Planton [72]). Of the remaining terms, D is the vapor pressure deficit given by $[e_{sat}(T_s) - e_a]$; where $e_{sat}(T_s)$ is the saturated vapor pressure at the surface temperature Ts, and e_a is the vapor pressure at the ambient temperature T_a.

Several other forms have been proposed. Wesley [24] adopts one such. In this formulation, it is assumed that stomatal resistance is principally controlled by solar radiation, giving

$$R_s = R_{smin}\,.\,D_{hx}\,[1 + \{200/\,(G+0.1)\}^2\,]\,[400\,/\,T_a\,(40 - T_a))] \tag{10}$$

where D_{hx} is the ratio of the diffusivity of water vapor to that of the gaseous pollutant in air. Typical values of D_{hx} for some pollutants are 1.9 for sulfur dioxide and vaporous nitric acid, 1.6 for ozone, carbon dioxide, nitrogen dioxide and nitrous acid, 1.4 for hydrogen peroxide, and 1 for ammonia.

In addition to the R_s calculations, other resistance pathways (such as cuticular, mesophyll) still need to be estimated. A popular choice for developing the different resistances for vegetated surfaces is given by Wesley [24] and Walmsley and Wesley [75]. Rc is considered to be a combination of stomatal (s), mesophyll (m), upper canopy leaf cuticle (lu), gas phase transfer through convection (dc), lower canopy (cl), canopy height and density (ac), and ground surface (gs) resistances. Thus,

$$Rc = [(R_s + R_m)^{-1} + R_{lu}^{-1} + (R_{dc} + R_{cl})^{-1} + (R_{ac} + R_{gs})^{-1}]^{-1} \tag{11}$$

In the above, Rs can be calculated following eqns. (8, 9 a-d, and 10). The mesophyll acts as an interface for the gases to react with the foliage humidity and the resistance of this exchange is dependent on the gas kinetics alone. Hence a gas reactivity based resistance pathway is adopted, as the gases pass through mesophyll:

$$R_m = (H / 3000 + 100 . f)^{-1} . \tag{12}$$

In eqn. (12), H is the Henry's law constant and f is the reactivity factor for the gas. For relatively inert gases such as sulfur dioxide, ammonia and nitric oxide $f = 0$, for moderately reactive gases such as nitrogen dioxide and nitrous oxide, $f = 0.1$, and for reactive gases such as ozone and hydrogen peroxide, Wesley [24] suggests a value of $f = 1$. The cuticular resistance term (R_{lu}) is calculated in a manner similar to the mesophyll resistance. For a dry surface,

$$R_{lu} = R'_{lu} . (10^{-5} . H + f) . \tag{13a}$$

In this calculation, R'_{lu} is a default resistance factor assigned as a function of landscape (land use) and the season (see Wesley [24]). Typical values range from 2000 s m^{-1} in midsummer for green vegetation, to about 9000 s m^{-1} during winter. The cuticles due to their functional role in plant transpiration are highly responsive to moisture availability. Hence the cuticle resistance also changes significantly due to variations in surface wetness (see Fuentes and Gillespie [26]). Under humid conditions, the cuticle resistance is largely controlled by the hygroscopic properties of the exchanging gas. Wesley [24] suggests an adjustment as $R_{lu}=[1/W+1/(3R_{lu})]^{-1}$. Considering ozone for instance, the resistance value (referred to as R_{luo}) under moderately wet conditions, such as dew formation, can be calculated with $W = 3000$, and with $W = 1000$ for rain wetted conditions. Wesley [24] provides another limit specifically for sulfur dioxide transfer. For this, he suggests R_{lu} as 100 s m^{-1} for dew wetted conditions and 50 s m^{-1} for urban regions. Additionally, W is assigned a value of 5000 for the rain event. For any gas other than sulfur dioxide and ozone, in wet conditions, use of the following equation is suggested:

$$R_{lu,\,wet} = [1 / (3\, R_{lu,dry}) + 10^{-7}\, H + f / R_{luo}]^{-1} . \tag{13b}$$

Sutton et al. [76] propose an alternative general formula, based on different laboratory and field measurements. They consider relative humidity (rh) control on the resistance, which is more convenient to model rather than dew and rain. They propose:

$$R_{lu} = 2 \exp([100 - rh]/12) . \qquad (13c)$$

The depositing matter has to encounter two additional resistances, namely the lower canopy (R_{cl}) and the ground surface (R_{gs}). Typically, the resistance values for both ozone and sulfur dioxide are of the order of 200 s m^{-1} for the ground surface and of the order of 1000 s m^{-1} for the lower canopy resistance. Using these resistance values for sulfur dioxide and ozone, more generalized formulae for the resistances are derived, yielding:

$$R_{clx} = [\, H/(10^5 . R_{cls}) + f / R_{clo}]^{-1} , \qquad (14a)$$

$$R_{gsx} = [H / 10^5 R_{gss}) + f / R_{gso}]^{-1} . \qquad (14b)$$

Based on several observations, it is recognized that surface temperature has a significant control over the cuticular, lower canopy, and ground surface response to depositing matter. For lower temperatures, Walmsley and Wesley [75] suggest a correction term $R_{low}=10^3.exp(-Ta-4)$, to be added to the resistance such as R_{lu}, R_{cl}, and R_{gs}.

Finally from eqn. (10), the buoyant–convection resistance is a function of radiation and the slope of the terrain (ϑ, in radians) and is estimated as $R_{dc}=100[1+1000(G+10)^{-1}].(1+1000\vartheta)^{-1}$. The slope can be conveniently assumed to be zero in a regional scale analysis [43], [44]. Knowing the different resistances, surface–atmosphere exchange fluxes can be obtained (cf. Kramm et al. [38]), by explicitly representing the vegetation–atmosphere exchange (F_f) and the ground surface and the atmosphere flux (F_g):

$$F_f = - [(c_f - H.c_{int}). (R_s + H.R_{int})^{-1}] - [(c_f - c_{cu}).(R_{cu})^{-1}]$$
$$- [(H.R_{wf})^{-1} . (c_f - H. C_{wf})] , \qquad (15a)$$

$$F_g = - [(c_g - H.c_{wsl}). (H.R_{wsl})^{-1}] - [(c_g - c_{sl}).(R_{sl})^{-1}] . \qquad (15b)$$

In the above equations, *c* and *C* refer to surface and aqueous state concentrations, and subscripts *f*, *int*, *cu*, and *g* correspond to the foliage, intercellular cells, cuticle and ground surface levels respectively. *H* is the Henry's law constant, *R* is the resistance offered to the exchange, while the additional subscripts, namely *wf* and *wsl* refer to the wetness at foliage and the soil surface. An alternative deposition flux parameterization involving K-theory along with a vegetation canopy–atmosphere interface can be found in Underwood [77]. Note that since the resistance is a function of the surface it represents (vegetation or ground), it has to be normalized by the surface area. For a grid, this can be normalized by vegetal cover fraction (*veg*). Thus for the foliage the effective resistance will be obtained by normalizing through *veg*, and the soil resistances by (1-*veg*).

An important aspect for operational models is that it is only the *net effect* (combined resistances of all the vegetative and surface parameters) that is required (as R_c) in estimating V_d. Hence, while designing the gas exchange system, one can either approach it in the detailed analysis as eqn. (11), or represent the R_c term by scaling the R_s over the entire canopy through LAI. Thus most models adopt

$$R_c = \Sigma\, R_s \, . \, LAI. \tag{16a}$$

Further, to account for the differences in the surface characteristics due to vegetal cover, the effective R_c can be represented as

$$1 / R_{c,eff} = veg / R_s + (1\text{-}veg) / R_{soil} \, . \tag{16b}$$

Knowing R_a, R_b and R_c, gas deposition fluxes can be obtained.

One of the effects of the material deposition is that it reduces the material to be dispersed further downwind. This can be adjusted through a net reduction in the mass transfer over the grid boundaries or as an effective reduction in the emission source strength for subsequent time steps (see Horst [78], Arya [53]).

3 Terrestrial biosphere as a source and sink

One of the limitations of the approaches and formulations presented up to now is that they do not explicitly account for the changes in the source/sink relationship discussed earlier. As observed with the biogenic VOCs, for example, the resistance pathways need to be designed for bi–directional exchanges so as to be generalized. To address this issue, recent efforts have been directed towards the development of modified flux emission–deposition approaches that would account for bi-directional exchanges. We will discuss this issue, using two recently published studies as examples: Xu et al. [29] and Sutton et al. [79]. The Sutton et al. [79] study focuses on the traditional bi-directional issue, namely that of the 'ammonia compensation point' (see Farquhar et al. [80], [81]). Ammonia can be efficiently deposited to moist surface and get converted into ammonium. However, the leaves themselves generate ammonium radicals in plant tissues. This ammonium can be subsequently dissociated resulting in ammonia release in the substomatal cavity. Such a compensating scenario can lead to an equilibrium in the direction of the mass flow (emission or deposition), depending on the ambient pollutant concentration. We will review this issue in more detail because of its importance for the development of bi-directional exchange models. The Xu et al. [29] work, on the other hand, is of interest because it deals with a regional scale application of mercury emission and deposition and is based on regression equations-based flux assumptions. Thus it differs from the more mechanistic 'compensation point'-based studies.

Hanson et al. [82] provide observational evidence of the bi-directional flow of the mercury mass exchange in the terrestrial ecosystem. The starting assumption of the regression-based models, such as that of Xu et al. [29], is that there is a non-causal relation between evapotranspiration and surface emission. Following this assumption, a biogenic/surface pollutant (mercury) emission rate (Fc) can be assumed, as a product of surface evapotranspiration (Ec), and pollutant concentration (Cs) at the surface soil solution, $Fc = Ec.Cs$. Interestingly, a similar approach was recently successfully adopted by Pleim et al. [54] to develop deposition estimates from field observations. Under wet and unstressed conditions, the evapotranspiration is assumed to be small and the vegetation is conducive for deposition.

Ec can be estimated from routine climatic information (temperature, humidity and rainfall, see Mahfouf [83]) or from detailed micrometeorological observations (see for example, Pleim et al. [54]). Additionally, Ec can be estimated using detailed soil-vegetation-atmosphere transfer (SVAT) models, such as those of Deardorff [84], Noilhan and Planton [72], Alapaty et al. [69], Wetzel and Chang [70, and Xue et al. [85], that are embedded in different mesoscale and regional scale models. Alternatively, empirical relations such as the Penman–Monteith equation (see Acs [86] and Monteith and Unsworth [87]) can be used to develop Ec estimates. Xu et al. [29], for instance, used the Penman–Monteith relation for estimating Ec along with a soil water deficit factor of Raupach [88]. They estimate Rc as (R_{smin} / f_m), where f_m

is a function of soil water deficit (M). For values of M between 0 and 5 cm of water, f_m is 1; for M greater than 20 cm of water, f_m is 0. For intermediate ranges of M, f_m is linearly scaled as $(20-M)/15$. In addition to this transpirative control of pollutant emission, the regression based models also relate soil surface emissions using surface variables such as soil temperature, radiation and soil moisture status. In their study, Xu et al. [29] adopt a regression model based on Carpi and Lindberg's [89] observations. They develop an equation of the form, $log(F_{soil})=a.(T_{soil})+b$. The coefficients a and b are hypothesized to be a function of surface characteristics such as soil wetness. In the case of mercury, for instance, Xu et al. [29] use values of a = 0.057 and 0.064, and b = –1.7 and –2.03. Concurrently, the deposition flux is obtained by estimating V_d as discussed earlier. Their results show some typical features that are of relevance to mesoscale models. First, there is a strong diurnal variability in both the emission as well as the deposition rates. Second, the transfer velocities for emission as well as deposition are coupled with land use pattern. This difference is principally due to the variations in the Rc term as a function of the vegetation/surface changes. The deposition rate was about 10% of the emission for agriculture landscape. This increased to about 15 to 20% for urban areas and about 20% for forests. As discussed earlier, the availability of surface moisture has a profound impact on deposition estimates (see Harley et al. [90]). For mercury, Xu et al. [29] estimate a 11% change in the area averaged dry deposition for a 1% wet surface.

Although in this study, the model processes are not causal in their formulation, they serve as an example of how data from field experiments can be adopted in developing regional scale scenario evaluation for air–surface exchanges (see Lemon and van Houtte [91]). Indeed it is possible to develop statistical-dynamical couplings for other terrestrial emissions such as isoprene, monoterpenes and different nitrogenous compounds. These models are also computationally efficient due to their analytical form and simplicity. An additional feature is that such models allow an easier interpretation of the cause-effect relationship often sought for policy development [65].

However, there are some limitations when dealing with such statistical-dynamical relations. One, typically, is that they cannot help develop a general scenario, as the range over which they can be applied could be fairly limited. The range over which observations are available for developing such regression relations generally sets these limits. Two, these relations are developed from observations which can have a significant error bar (sometimes of the same order of magnitude as the mean observed value itself see for example, Fowler et al. [92]). This introduces significant uncertainty in the results. Dynamical processes can be fairly compensatory over a wide range, while in the statistical relations the uncertainty persists (cf. Niyogi et al. [56]). Third, the functional form used in developing the relations can make the system biased towards emission or deposition, and thus create a difference in the direction of the mass flow. As an example, the Xu et al. [29] study adopts the use of logarithm in the emission term, making it biased for positive values, namely emissions as against deposition. Fourth, unless explicitly specified, the statistical models often cannot account for effects of co varying or second order interactions in the environmental system [93]. For instance changes in the basal characteristics of the biogen such as the emission rate (e.g., Monson et al. [94]), or seemingly contrasting changes in the physical characteristics (such as increased radiation as well as humidity, increased air temperature and lowered foliage temperature) can lead to very different results in the modeled outcome and reality (see Nikolov et al. [95]).

For such reasons, the 'compensation point' approach [80], [81], [82] appears more encouraging for developing bi-directional models. A study by Sutton et al. [79] which develops and applies the bi-directional exchange for ammonia is discussed here. As discussed earlier, due to the physiological emission of the ammonium radicals and their subsequent dissociation from ammonia, the concentration flux can vary its direction and magnitude significantly [96]. Indeed a similar case can be made for mercury and several other VOCs.

Following eqns. (1) and (7), gradients in the surface concentrations can be used for estimating surface fluxes. As a first approximation, Sutton et al. [79] assume that the surface concentration is zero and that the gas concentration at reference level can yield a gradient for calculating the fluxes. These fluxes can be linked with total resistance (R_T) following eqn. (7). Then *Ra*, and *Rb* can be calculated following eqns. (3), (4) and (5). Knowing R_T, *Ra* and *Rb*, *Rc* is obtained as a residual resistance term. This *Rc* is then used with other meteorological and chemistry datasets (observed or modeled) to generate corresponding flux values. Meyers and Paw [97], as well as Meyers and Baldocchi [98], used a similar approach both with observations as well as with vegetation models. However, in their study for nitric acid deposition, they assume a zero resistance from the canopy for exchanges between the surface and the overlying atmosphere.

Indeed such approaches do not provide a mechanistic feedback between physiological changes and the environment. However, they can provide a better analysis mode for bi-directional fluxes for specialized fields experiments in which detailed observations are available against which robust regression models can be generated. To introduce the physiological feedback, the resistance terms can be linked to yield a compensation concentration point (C_{cp}),

$$C_{cp} = Cr + F_i \,.\, (Ra + Rb)\,. \tag{17}$$

Using the compensation point as a central parameter rather than *Rc*, a flux-resistance approach can be developed.

Total flux has to be conserved between cuticle (F_{lu}) and the stomata (F_s). This yields $F_w = C_{cp}/R_{lu}$ and $F_s = [C_{cps} - C_{cp}]/R_s$, where C_{cps} is the compensation point for the stomata (rather than the canopy), and is calculated similar to eqn. (17) as $[Cr + F_i.(Ra + Rb + Rs)]$. The cuticle resistance can be estimated from eqn. (13 a, b, c), and a generalized compensation point can be obtained as

$$C_{cp} = \frac{\left[C_r/(R_a + R_b) + C_{cps}/R_s \right]}{\left[(R_a + R_b)^{-1} + R_s^{-1} + R_w^{-1} \right]}. \tag{18}$$

In eqn. (18), the effect of soil emission can be introduced by simply adding one more soil flux term (F_{soil}) in the numerator (see Sutton et al. [99]). It is important to note that the soil flux term added thus will not be a constant background type additive correction to the bi-directional flux model (see Carpi and Lindberg [89]). Since the flux is introduced in a manner similar to altering the resistance, there will be feedback in the leaf surface/stomata and the soil exchanges. Indeed such an additive flux emission is a simplification and more explicit and detailed representation may be necessary for regions in which soil surface emissions are dominant [89]. One example of such a case could be an animal waste site in agricultural utility regions. Since by definition, at compensation point there will not be any mass exchange, eqn. (18) can be simplified to yield:

$$C_{cp} = \frac{C_{cps}/R_s}{R_s^{-1} - R_w^{-1}}. \qquad (19)$$

Using eqn. (18), Sutton et al. [79] obtain results that agree well with observations for bi–directional flux transport over a wheat canopy. Though the above example is not unique, and other researchers have proposed similar exchange models even within a gaussian plume framework (see for example, Asman [100]), two features deserve additional mention. First, the example demonstrates a method in which the existing deposition velocity–resistance relations (eqn. 6) can be conveniently modified to configure the bi-directional exchange within the modeling system. This can be achieved at a landscape as well as at a regional scale. Second, leaf and canopy scale processes have profound impact on the landscape flux characteristics. Related to the leaf scale control issue, several studies such as Rondon and Granat [101], Monson et al. [94], and Schjoerring et al. [93] for instance, provide observations on the physiological control of plant-atmosphere exchange. They find convincing evidence that stomatal conductance (inverse of resistance) controls the exchanges; and physiological variables such as foliage temperature, for instance, couples tightly with the exchange fluxes. The models discussed so far have an advantage in that they represent the modeling process within the traditional meteorological/environmental approach itself (see Fig. 2). For bi–directional exchanges, due to their very nature, models are needed explicitly to account for the physiological responses. Efforts are under way to couple biological processes with meteorological models, thus giving an opportunity to develop realistic surface–atmosphere exchange modules. In the following section, we will discuss these model formulations in more detail.

3.1 Biosphere atmosphere exchanges by physiological approach

As described in Niyogi and Raman [1] and Niyogi et al. [60], for the environmental Jarvis type schemes, it is the ambient environment or the meteorology that is a dominant factor in determining the stomatal resistance. Lack of physiological feedback may be valid for various routine field situations. However, there is increasing evidence that the physiological feedbacks in the form of biochemical stimuli or plant photosynthesis are requisite causal indicators of the stomatal activity terrestrial exchange. Here we discuss the physiological gas exchange and surface evapotranspiration model (GEM) developed by Niyogi et al. [28], [61], [62], [103] and Niyogi and Raman [102].

GEM considers *Rc* as a combination of various resistances at the surface (vegetated or bare soil), to yield an areal average for each grid point in the numerical model,

$$\text{Rc} = (1\text{-veg})\ \text{Rsoil} + (\text{veg})\ \text{Rs}\ . \tag{20}$$

Unlike the meteorological/environmental formulations, *Rs* is taken as a continuum between soil thermal and hydraulic feedback in terms of the surface energy and soil moisture balance, and the canopy processes. The model considers mechanistic leaf scale gas assimilation/photosynthesis, evapotranspiration and energy balance, along with a dynamic link with the atmosphere through mass and energy transfer as well as gas exchange (cf. Cowan [63], Elssworth [104], Niyogi and Raman [1]). The logical flow in GEM is as follows: stomatal resistance (r_s) is inverse of stomatal conductance (g_s); conductance is known to correlate with carbon assimilation or photosynthesis (A_n) (cf. Wong et al. [105]); and the photosynthesis is dependent on the biophysiochemical state of the foliage and its surrounding environment such as carbon dioxide at surface (C_s), within the leaf cells (C_i), and humidity at leaf surface (rh_s) (cf. Farquhar et al. [81], [106], Farquhar and Sharkey [107]).

In GEM, stomatal conductance (g_s) is estimated following the Ball Woodrow Berry model (see Ball [108], Ball et al. [109], Niyogi and Raman [1] as well as a critique by Aphalo and Jarvis [110], Mott and Parkhurst [111], and Monteith [112], [113]). Accordingly,

$$g_s = [m\ .\ A_n.\ rh_s\ /\ C_s] + b\ . \tag{21}$$

Photosynthesis or carbon assimilation [11], [114], [115], is taken as the residue of gross carbon assimilation (A_g) and loss due to respiration (R_d). A_g is taken as the weighted minimum of three limiting functions, namely, assimilating efficiency of photosynthetic enzymes (Rubisco limited; w_c), radiation (w_e), and carbon dioxide (w_s) (see Collatz et al. [116], [117]). Giving

$$w_c = V_m\ , \tag{22a}$$

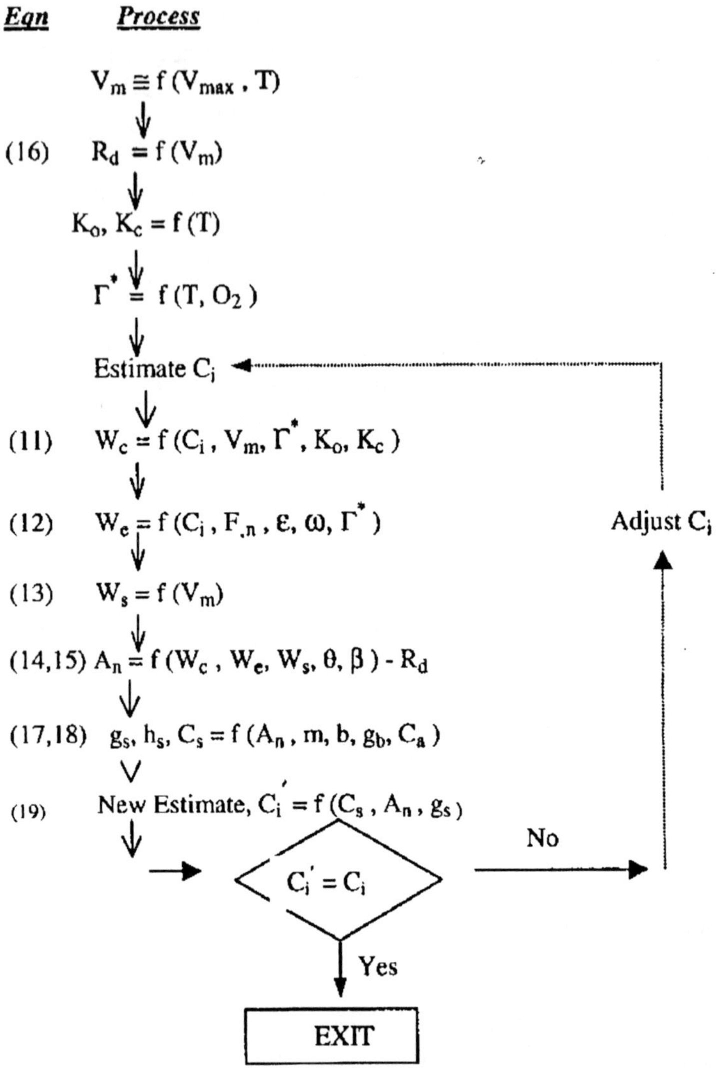

Figure 3. Flowchart showing the decision process in solving the iterative $g_s - A_n$ equations in GEM. (Based on Sellers et al. [125], Su et al. [127], Niyogi et al. [103])

$$w_e = PAR.\varepsilon.(1 - \omega_\pi) \ , \tag{22b}$$

$$w_s = \frac{20000.V_m.C_i}{P} \tag{22c}$$

where ε is the intrinsic quantum efficiency for carbon dioxide uptake, and ω_π is the leaf scattering coefficient for photosynthetically active radiation (PAR, see Sellers et al. [118], Niyogi et al. [61], [62]). V_m is estimated as a function of soil moisture and ambient temperature [117], [119]. P is the atmospheric pressure, which is prescribed or obtained as a tendency term in the meteorological model, and PAR is calculated following Noilhan and Planton [72] as a function of net radiation. Knowing w_c, w_e and w_s, corresponding smoothed minima (cf. Collatz et al. [116]) gives the gross primary productivity (A_g). Knowing A_g, photosynthesis (A_n) can be calculated by taking off the loss in resources due to respirative processes (R_d). Following Collatz et al. [116], R_d can be taken as a certain percentage (1.5 to 2.5%) of V_m. Another approach developed and presented in Goudriaan et al. [120], Kim and Verma [121], van Heemst [122], Jacobs et al. [123] and Calvet et al. [119] is also available. Accordingly, R_d is parameterized as 0.11 A_m, where, A_m (or $A_{m,max}$) is the maximum assimilation rate [67] limited through carbon dioxide deficit (see Jacobs et al. [123], Jacobs [124]), and mesophyllic conductance (g_m) as,

$$A_m = A_{m,max}\,[1 - exp(-g_m(C_i - \Gamma) / A_{m,max})] \ , \tag{23}$$

g_m is parameterized following Calvet et al. [119], and Niyogi et al. [61], [62],

$$g_m = g_{m,max}(2^{Q_t})\left[\frac{1 + exp(0.3(T_c - S_2))}{1 + exp(0.3(S_1 - T_c))}\right] \cdot \left[\frac{(w_2 - w_{wilt})}{(w_{sat} - w_{wilt})}\right] \tag{24}$$

In the above, $g_{m,max}$ is typically of the order of 17.5 x 10^{-3} m s^{-1}, S_1 and S_2 are land use based temperature coefficients. Sellers et al. [125], [126], provide look-up tables for these coefficients as a function of vegetation type. Tc is the surface temperature, and w_2, w_{wilt} and w_{sat} are the deep (root level) soil moisture, and the wilting and saturation capacity of the soil. All the variables are either specified as a function of the land use or determined prognostically in the SVAT module in GEM. Obtaining g_b, the carbon dioxide concentration at the leaf surface (C_s) can be estimated following Su et al. [127] as

$$C_s = C_a - \frac{A_n}{g_b} . \tag{25}$$

From eqn. (25) a first estimate of g_s is made. Closure is obtained by estimating C_i as

$$C_i = C_s - \frac{D_{hx} \cdot A_n P}{g_s} . \tag{26}$$

D_{hx} is the diffusivity ratio to accommodate for different gases, as discussed earlier [24]. The set of equations is solved in an iterative mode until a convergence is achieved. Finally in GEM, a lag is introduced for the changes between (from or to) the physiological and the ambient environment (see Jones [128] for observations and Su et al. [127] for large eddy simulation results). Accordingly the temporal stomatal response $g_s(t)$ for the steady state stomatal conductance, g_s, is introduced based on the stomatal conductance value for the prior time step $g_s(t\text{-}1)$ giving,

$$g_s(t) = g_s(t-1) + [g_s - g_s(t-1)] \times [1 - exp(-k(\Delta t/\tau))] . \tag{27}$$

GEM and similar modules (SiB2 [125], CANVEG [129], [130], IBIS [119]) can be efficiently coupled to any PBL or mesoscale model through surface energy balance and SVAT [1], [69], [72], to achieve complete integration. Figure 3 shows the process flow chart for the physiological terrestrial atmosphere exchange. In GEM, the Ball–Berry model for the g_s–A_n relation is used. Niyogi and Raman [1] review additional physiological approaches, one of which is vapor pressure deficit (D)-stomatal conductance (g_s) and – carbon assimilation (A_n) analytically relation of Jacobs et al. [123]. Calvet et al. [119] describe a recent application of such a model, and a summary is presented here.

The gas (water vapor) exchange parameterization is efficiently linked with a stomatal conductance model through gas concentrations and a canopy scaled conductance. They use the following equation:

$$G_s = g_s^* + E\, M_a / M_v \,.\, (C_s + C_i) / [2\,(C_s - C_i)]\,, \tag{28a}$$

$$g_s^* = 1.6\, g_{sc}^* + g_c\,, \tag{28b}$$

$$g_{sc}^* = \frac{A_n - \left(\frac{D_s}{D_{max}} \times \frac{A_n + R_d}{A_m + R_d}\right) + R_d\left(1 - \frac{A_n + R_d}{A_m + R_d}\right)}{C_s - C_i}\,, \tag{28c}$$

$$C_i / C_s = f + [(1 - f)\ (\Gamma / C_s)]\,, \tag{28d}$$

$$f = f_o\,(1 - D/D_{max}) + f_{min}\,(D/D_{max})\,, \tag{28e}$$

where g_c is the cuticular conductance, M_a, and M_v are molecular masses of air and water vapor, Γ is the carbon dioxide compensation point. This is the CO_2 leaf intercellular concentration, below which a leaf is unable to carry out photosynthesis, due to photorespiration. The compensation point is about 45 ppm for woody trees, and 3 ppm for grassy landscape. Leaf transpiration E is estimated as $\rho_a g_s D_s$. f_o is value of f for D_s=0, and is of the order of 0.85 for woody trees, and 0.5 for grass. Following Choudhury and Monteith [131], D_{max} is often taken as 45 g kg^{-1}. Using this formulation Calvet et al. [119] find good agreement between observations and predictions for different cases. Indeed some studies [110], [111], [112], [113] question the causality of relative humidity based g_s–A_n relations such as the Ball–Berry model, and prefer the vapor pressure deficit approach (see also Dougherty et al. [132]). In reviewing different g_s, formulations Niyogi and Raman [1] and Niyogi et al. [60] conclude that at short time scales (a few hours), the two approaches indeed provide different values. Qualitatively both the humidity representations were similar in their functional form and responses, but different in terms of their numerical outcome. These results suggest that it is important to choose the canopy resistance scheme appropriately depending on the trace species and their humidity dependence. Further, it may often be necessary to independently evaluate the resistance pathways for the trace gas and pollutant, and their successful validation for surface energy balance and meteorological models may not be sufficient criteria for choosing one formulation over another.

4 Regional mapping

Up to now, we have discussed different modeling approaches in the development of deposition and bi-directional fluxes. One of the major issues in including such an analysis for a mesoscale model, is the scaling of the deposition estimates to a regional scale. In this section, we will discuss the application of the various formulations and the techniques used for such scaling.

Studies involving regional mapping of deposition and regional mass balance recognize the limitations of point observations [133]. Most observations are limited in time and space and cannot be efficiently used for generating spatial distributions or maps. This issue is discussed following a recent deposition mapping exercise adopted for the United Kingdom using observational data alone [49].

A literature review is generally necessary for various deposition velocity estimates as a function of measurement techniques, landscape characteristics as well as the pollutant. The reported V_d values for different gases include: nitric oxide ranging from 10^{-4} to 2 mm s^{-1} over soil and about 1 mm s^{-1} over vegetation [134], [135]; nitrogen dioxide from 2 mm s^{-1} over grass [92], 3 to 7 mm s^{-1} for forest surface, as well as sand and clay soils [136], [137], to 13 mm s^{-1} for alfa-alfa grass [4]; peroxy acetyl nitrate (PAN) from 2 mm s^{-1} over grass [9] to 8 mm s^{-1} over alfa-alfa [4], 0.9 mm s^{-1} for acidic moorland to 2 mm s^{-1} over calcareous soil/patchy grass [138]. Erisman and Baldocchi [51] and Erisman et al. [139] summarized recent sulfur dioxide deposition estimates from various studies. They report values ranging from 1 mm s^{-1} to 20 mm s^{-1}, for bare soil to coniferous forests, with the higher values for forest landscape (see also Rondon et al. [140]). Various such measurements are available from field programs though mostly in the U.S. and in Europe. Using these values of V_d, with representative concentration values, deposition fluxes are calculated. These fluxes are then mapped using geographical information system (GIS) based approaches.

Models are still widely used due to lack of observations in many cases. For example, Duyzer and Fowler [49] report absence of PAN observations over forest regions, and nitrous acid and resort to using Wesley's [24] model (which yielded a value of 5 mm s^{-1} for PAN and 10 mm s^{-1} for nitrous acid) for the forest land use. Similarly there is a need to generate diverse climatic scenarios for different landscapes since moisture has a major effect on the deposition potential. A typical approach involves the following steps. First, high resolution surface features, and an emission inventory of the pollutant to be analyzed is generated. The emissions are fed into an environmental/air pollution–deposition model. The model also requires climatological or prognosticated humidity, precipitation, temperature, winds and radiation fields [141], [142]. A high resolution surface data set comprising vegetation type, soil type, leaf area index, roughness length, and vegetation characteristics is required. The accuracy of the surface data will determine, to a large extent, the appropriateness of analyzed and predicted deposition fields. The resulting distributions are generated hourly or integrated over month, season, and year. The generated estimates can be used for mapping the zonal susceptibility (see, for example, Singles et al. [143]).

In addition to surface observations, use of satellite data will be beneficial for providing input values for generating mesoscale model based deposition output. While developing the bi-directional models, we discussed the relations between evapotranspiration and the gas exchange fluxes. Taconet et al. [144] present a canopy resistance model based on thermal infrared remote sensed data:

$$R_S = R_{s\,\min}\left[\frac{R_{n\,\max}}{c_1\,R_{n\,\max}+Q}+\left(\frac{1.2w_{wilt}}{w_s}\right)^2\right].\frac{1+0.5LAI}{LAI}\,. \tag{29}$$

Here, c_l is of the order of 0.03, and other variables are similar as described in eqns. (8) and (9). Similarly, Nemani and Running [145] successfully tested the potential of using spectral index *NDVI* (normalized differential vegetation index) for evapotranspiration and hence *Rc* estimates. Studies such as Asrar et al. [146], for instance, have established the correlation between leaf area coverage and the infrared/red band values [147], [148]. Green leaves, due to the presence of chlorophyll, absorb radiation in red wavelengths. Hence red reflectance (*RED*, 0.55–0.68μ) is negatively correlated with chlorophyll and green vegetation, while the near infrared (*NIR*, 0.73–1.1μ) is scattered by the physiological characteristics of the leaves. *NDVI* is calculated as:

$$\mathrm{NDVI} = [(\mathrm{NIR} - \mathrm{RED}) / (\mathrm{NIR} + \mathrm{RED})]. \tag{30}$$

Nemani and Running [149] obtain a relation between *LAI* and *NDVI* as:

$$\mathrm{NDVI} = [\ln (\mathrm{LAI} / 0.65)] \text{ X } 34 . \tag{31}$$

Using this as a starting point, Nemani and Running [145], represent a relation between surface temperature (*Ts*) and NDVI of the form $T_s{=}a{+}b$ *NDVI*. For their data they obtained a ranging from 28 to 60 and b from 24 to 45. They refer to the slope of the T_s *NDVI* best fit as σ, and deduce a relation between σ and R_c as: $R_c{=}{-}ln(\sigma/48)X(-10)$.

Additional details regarding such an approach in mesoscale–regional landscape perspective can be found in studies such as Cihlar et al. [150], Seevers and Ottmann [151], and Szilagyi et al. [152]. Indeed, development of R_c estimates allows regional deposition assessment for models. The above approach is attractive as it has direct relevance for biospheric models that seek information regarding surface characteristics from satellite data [126]. In the final section, therefore, we will describe such a methodology, for R_c estimation with both the Jarvis type and the physiological g_s–A_n models in perspective [118], [126], [153], [154]. For instance, Sellers et al. [118] proposed

$$g_s = \left[\frac{b_1 + F.n}{a_1 + b_1 c_1 + c_i F.n}\right]\left[f(T).f(w_2).f(e)\right], \tag{32}$$

where a_1, b_1, c_1 are specie dependent constants. Using different parameterizations (e.g., Charles–Edwards and Ludwig [155], Jarvis [64], Farquhar et al. [106], and Collatz et al. [116], [117]), Sellers et al. [118] estimate the constants to be 13966, 0.1, and 28 respectively. $F.n$ is the normal component of the vector flux for *PAR*, and $f(T)$, $f(w2)$, $f(e)$ are the temperature, soil moisture, and vapor pressure deficit, as explained in eqn. (9). The canopy conductance (inverse of resistance), which is used for estimating surface deposition, can be obtained by integrating the conductance over canopy *LAI*. In developing radiometric data based biophysical parameters, in addition to *NDVI* (eqn. 29), Sellers [153] uses the simple ratio (*SR*) (defined as ratio of *NIR* to *RED*). Under conditions such as uniform canopy, dark soil and light to moderate water stress, the following relation can be established:

$$\frac{\partial g_c}{\partial(LAI)} \alpha \frac{\partial(SR)}{\partial(LAI)}. \tag{33}$$

However for physiologically intensive photosynthesis–based schemes, Sellers et al. [118] extend eqn. (32) to yield

$$\langle g_c \rangle = \frac{1}{S} \int_0^s g_c ds \, \alpha < SVI >, \tag{34}$$

where < > refers to area–integral and spatial averages (*S* is the area of the region), and *SVI* is spectral vegetation index such as *SR* and *NDVI*. Thus the adaptation of the spectral indices, as obtained through satellite imagery for instance, into mesoscale models will be governed by the type of resistance formulation adopted. In summary, spatial averages of the spectral indices can yield spatial averages of canopy conductance (inverse of resistance). A functional form can be deduced based on leaf biophysical properties (such as radiation limit and V_{max}, see Goudriaan [156]), photosynthetically active radiation (*PAR*) usage over the canopy, and environmental feedback through forcings such as temperatures, humidity, and moisture availability:

$$g_c \equiv [V_{max}, F_o].[\Pi].[F_i]$$

$$g_c \equiv [Leaf\ Properties].[PARUsage].[EnvironmentalFeedback] \tag{35}$$

Additional description of the use of spectral indices for biophysical processes can be found in Asrar [157]. Significant uncertainty has to be assigned due to limitations in spatial resolution, and the heterogeneity of bio physiochemical as well as radiative properties within the model landscape. However, the radiance based approach for a mesoscale air pollution modeling appears to be quite promising for mesoscale deposition analysis.

5 Summary and concluding discussion

We have discussed the role of vegetation and terrestrial biosphere in the mesoscale air pollution systems. Generally the surface acts as a sink and aids gaseous deposition. Accordingly, we discussed the development of different parameterization schemes using a resistance pathway between the aerodynamic, surface boundary layer, and the depositing surface. Such schemes can be categorized under Jarvis–type environmental or meteorological exchange schemes in which the biospheric characteristics are not dominantly interactive. In addition to being a sink, the terrestrial biosphere can also be a significant source for what is referred to as biogenic emissions. Examples include gases such as isoprene, monoterpene (which can be photochemically transformed into ozone), nitrogen compounds and mercury. Such scenarios provide an impetus for implementation of generalized bi-directional surface–atmosphere exchange processes in air pollution models. The bi-directional exchange is parameterized principally through development and modification of surface resistance schemes.

Two developments were discussed in which the Jarvis type environmental approach could be utilized. The first involves adopting a statistical regression equation based estimation of the exchange flux. The second method is based on a 'compensation point' approach. In this, the differences in gas concentration within the canopy (surface) and the air around it determine the direction of flux transport. In addition to the modifications to the Jarvis type schemes, a physiologically intensive surface–atmosphere exchange was also discussed. In this the exchange is not only controlled by the environmental forcing, but also by the within canopy bio-physiochemical reactions. A methodology for coupling such exchanges in mesoscale air pollution systems was discussed. Finally, a critical component of adopting such surface–atmosphere deposition or exchange modules in the mesoscale perspective is the regional scaling. Here we presented use of the parameterizations discussed earlier using both observations as well as model simulated exchanges at a regional scale. In addition to the surface observations, application of remote sensed radiometric data and the spectral indices was also linked (see Gao et al. [27]).

One of the critical issues for future research into mesoscale deposition estimates is the propagation of uncertainty in the analysis. Despite efforts such as the National Acid Deposition Program (NADP), there is still a paucity of experimental data to validate the regional scale deposition velocity estimates under diverse atmospheric conditions. Wesley and Lesht [158] estimate an uncertainty in estimates of deposition velocity of about ±30% even with surface homogeneity. In addition to the effect of inhomogeneity, a feature almost always encountered in reality, mesoscale deposition estimates will have an uncertainty due to the interactive role of the reactive nature of depositing material. Further, most deposition and mass balance studies have been performed over continental United States and Europe, and the tropics still remain largely unexplored. The vegetative feedback in the tropics is dominated by soil moisture availability [159] and hence the variations in the deposition velocity over an extended period would be of particular interest. Overall, surface wetness and temperature significantly control the bi-directional fluxes in the tropics as well as in other regions of the world (see Sharkey and Singsaas [160]).

Hence, mesoscale models that do not have explicit parameterizations for surface moisture and surface temperature will have an inherently high uncertainty in estimating V_d values (see Niyogi et al. [28]). Further, the resistance parameterizations are a function of land use type, which lead to a simplification in the mass balance procedures at a regional scale. The parameterizations are developed and often validated using observations at micrometeorological scale (few minutes, and few meters), while the estimates are applied for meso to regional scale (Walcek et al. [161]). Wesley [24] suggests that V_d estimates can be considered most appropriate for long term averages and time scales ranging over several weeks and large regions. Thus, the effects of diverse surface characteristics need to be averaged to 'median' values.

Further, the role of varying vegetative clusters, differences in vegetation characteristics (senescence or greening), and the chemical activity of the soil (alkalinity or acidic, see Finlayson Pitts and Pitts [162]) are still largely unknown. Duyzer and Fowler [49] outline three additional deficiencies in the regional deposition studies. Firstly, since deposition flux is a product of ground level concentration and deposition velocity, depending on the model vertical resolution, the average boundary layer concentration has to be converted to ground level concentration. This can be difficult for chemically active species and for topographically and physiologically variable domains [18], [163]. Secondly, the edges of major discontinuities such as forests and water bodies act as significant sinks [140], [164]. Special routines may be thus necessary for such complex terrains with regression based interpolations for vegetative and topographical features. Thirdly, for mesoscale eco-physiological studies in which issues such as nutrient loading to watershed or acidic deposition to a sensitive biota are under consideration, subgrid scale processes become critical. Duyzer and Fowler [49] suggest statistical corrections for dynamical models for critical loading as a subgrid scale parameterization. In conjunction with such uncertainties, Lovblad and Erisman [165] and Erisman and Baldocchi [51] have developed interesting summary tables of the uncertainty in deposition estimates for Europe. At a regional scale, they consider deposition to be modulated by different variables such as: emission strength, wind speed, roughness length, surface wetness, orography, co deposition, canopy resistance, dry deposition, wet deposition, cloud and fog.

Of these factors, for the entirety of Europe, the highest uncertainty is estimated for dry deposition, in addition to factors such as surface/canopy resistance calculations, surface wetness and the role of fog and clouds that influence the deposition process. It is pertinent to recognize that most experiments undertaken for validation of the V_d modules are under fair weather conditions which may not yield a range of values valid for diverse situations often possible within model grid cells. Additionally, the role of mechanistic models, such as the eco-physiological models described previously, at a regional scale need to be further evaluated. Also the role of water films on vegetation [18], co deposition and canopy leaching is still unresolved [51], [79]. Indeed there are large gaps in our understanding of the processes related to deposition and bi-directional fluxes, as well as translation of these effects to a regional grid structure. However, the issue is of pivotal importance for regional climate change, and socio economic policies. Thus efforts in mesoscale air pollution studies should focus on developing modules for point to area (for converting microscale observations to mesoscale grids), as well as area to point (for satellite datasets to link with tower measurements) conversions. Use of bi-directional eco physiological models coupled with dynamical models appears to be the most promising approach in developing universal terrestrial biosphere exchange scaling relations.

Acknowledgements

This work was funded in part by the National Oceanic and Atmospheric Administration (NOAA) and the State Climate Office of North Carolina. The authors would like to acknowledge fruitful discussions with Dr. Ellen Cooter, NOAA, during the course of this study. The State Climate Office of North Carolina is a public service center for climate–environment interactions at the North Carolina State University.

References

[1] Niyogi D. S. and Raman S., Comparison of four different stomatal resistance schemes using FIFE data. *J. Appl. Meteor.*, **36**, pp. 903–917, 1997.

[2] Grantz, D. and Meinzer, F., Stomatal responses to humidity in a sugarcane field: Simultaneous porometric and micrometeorological measurements. *Plant, Cell and Environ.*, **13**, pp. 27-37, 1990.

[3] Monteith, J., Gas exchange in plant communities. In *Environmental Control of Plant Growth*, ed. L.T. Evans, Academic Press, New York, 1963.

[4] Hill, A., Vegetation: a sink for atmospheric pollutants. *J. Air Poll. Control Assoc.*, pp. 341–346, 1971.

[5] van der Hoven, I., Deposition of particles and gases. In *Meteorology and Atomic Energy*, ed. D. Slade, U. S. Atmospheric Energy Commission, Office of information Services, Oak Ridge TN, USA, pp. 202–208, 1968.

[6] Turner, N., Rich, S. and Waggoner, P., Removal of ozone by soil. *J. Environ. Quality,* **2**, pp. 259–264, 1973.

[7] Turner, N., Waggoner, P. and Rich, S., Removal of ozone by soil and vegetation. *Nature,* **250**, pp. 486–489, 1974.

[8] Bennett, J. and Hill, A., Absorption of gaseous air pollutant by a

standardized plant canopy. *J. Air Pollut. Control Assoc.,* **23**, pp. 89–98, 1973.

[9] Garland, J. and Penkett, S., Absorption of peroxy acetyl nitrate and ozone by natural surfaces. *Atmos. Environ.,* **10**, pp. 1127–1131, 1976.

[10] Wesley, M. and Hicks, B., Some factors that affect the deposition rates of sulfur dioxide and similar gases on vegetation. *J. Air Pollut. Control. Assoc.,* **27**, pp. 1110–1116, 1977.

[11] Mooney, H., The carbon balance of plants. *Ann. Rev. Ecol. Syst.,* **3**, pp. 315–346, 1972.

[12] Pielke, R., Climate prediction as an initial value problem. *Bull. Amer. Meteor. Soc.,* **79**, pp. 2743–2746, 1998.

[13] Randall, D., Dazlich, D., Zhang, C., Denning, A., Sellers, P., Tucker, C., Bounoua, L., Berry, J., Collatz, G., Field, C., Los, S., Justice, C. and Fung, I., A revised land surface parameterization (SiB2) for GCMs. Part III: The greening of the Colorado State University General Circulation Model. *J. Clim.*, **9**, pp. 738–763, 1996.

[14] Foley, J., Prentice, C., Ramankutty, N., Levis, S., Pollard, D., Stich, S. and Haxeltine, A., An integrated biosphere model of land surface processes, terrestrial carbon balance, and vegetation dynamics. *Glob. Biogeochem. Cyc.*, **10**, pp.603–628, 1996.

[15] Cox, P., Betts, R., Bunton, C., Essery, R., Rowntree, P. and Smith, J., The impact of new land surface physics on the GCM simulation on climate and climate sensitivity. *Clim. Dynamics*, **15**, pp. 183–203, 1999.

[16] Holt, T. and Raman S., A review and comparative evaluation of multi-level boundary layer parameterizations for first order and turbulent kinetic energy closure schemes. *Rev. Geophys.*, **26**, pp. 761-780, 1988.

[17] Stull, R., *An introduction to boundary layer meteorology*, Kluwer Academic Publishers, Dordrecht, pp. 666, 1988.

[18] Arey, J., Winer A., Atkinson, R., Aschman, S., Long, W., Marrison, C. and Olszyk, D., Terpenes emitted from agricultural species found in California's central valley. *J. Geophys. Res.,* **96**, pp. 9329–9336, 1991.

[19] Chameides, W., Lindsay, R., Richardson, J. and Kiang, C., The role of biogenic hydrocarbons in urban photochemical smog: Atlanta as a case study. *Science,* **241**, pp. 1473–1475, 1990.

[20] Monson, R. and Fall, R., Isoprene emissions from Aspen leaves. The influence of environment and relationship to photosynthesis and photorespiration. *Plant Physiol.,* **90**, pp. 267–274, 1989.

[21] Langford, A. and Fehsenfeld, F., Natural vegetation as a source or sink for atmospheric ammonia: a case study. *Science*, **255**, pp. 581–583, 1992.

[22] Guenther, A., Zimmerman, P. and Harley, P., Isoprene and Monoterpene emission rate variability: Model evaluations and sensitivity analyses. *J. Geophys. Res.,* **98**, pp. 12609–12617, 1993.

[23] Fitzgerald, W., Is mercury increasing in the atmosphere? The need for an atmospheric mercury network. *Water, Air, and Soil Pollution,* **80**, pp. 245–254, 1995.

[24] Wesely, M., Parameterization of surface resistance to gaseous dry deposition in regional scale numerical models. *Atmos. Environ.,* **23**, pp. 1293–1304, 1989.

[25] Fowler, D. and Unsworth, M., Turbulent transfer of sulfur dioxide to a wheat canopy. *Q. J. Roy. Meteor. Soc.,* **105**, pp. 767–783, 1979.

[26] Fuentes, J. and Gillespie, T., A gas exchange system to study the effects of leaf surface wetness on deposition of ozone. *Atmos. Environ.,* **26**, pp. 1165–1173, 1991.

[27] Gao., W., Wesley, M., Cook, D., and Hart, R., Air surface exchange of H_20, CO_2, and O_3 at a tall grass prairie in relation to remotely sensed vegetation indices. *J. Geophys. Res.,* **97**, pp. 18663–18671, 1992.

[28] Niyogi, D., Raman, S. and Alapaty, K., A dry deposition model that includes vegetation atmosphere interactions through photosynthesis /gas assimilation. *Atmos. Environ.,* in review, 1999.

[29] Xu, X., Yang, X., Miller D., Helble, J. and Carley, R., Formulation of bi-directional atmosphere surface exchanges of elemental mercury. *Atmos. Environ.,* **33**, pp. 4345–4355, 1999.

[30] Businger, J., Evaluation of the accuracy with which dry deposition can be measured with current micrometeorological techniques. *J. Appl. Meteor.,* **25**, pp. 1100–1124, 1986.

[31] Droppo, J., Concurrent measurements of ozone dry deposition using eddy correlation and flux profile methods. *J. Geophys. Res.,* **90**, pp. 2111–2118, 1985.

[32] Beljaars, A. and Holstlag, A., Flux parameterization over land surfaces for atmospheric models. *J. Appl. Meteor.,* **30**, pp. 327–341, 1991.

[33] Monin, A. S. and A.M. Yaglom, *Statistical Fluid Mechanics. Vol. I,* MIT Press, Cambridge, pp. 468-504, 1971.

[34] Baldocchi, D., Hicks, B. and Camara, P., A canopy stomatal resistance model for gaseous deposition to vegetated surfaces. *Atmos. Environ.,* **21**, pp. 91–101, 1987.

[35] Baldocchi, D. and Meyers, T., On using eco-physiological, micrometeorological, and biogeochemical theory to evaluate carbon dioxide, water vapor, and trace gas fluxes over vegetation: Synthesis and application. *Agric. Forest Meteorol.,* **90**, pp. 1–25, 1995.

[36] Kustas, W., Estimated values of evapotranpiration with one and to layer model of heat transfer over partial canopy cover. *J. Appl. Meteor.,* **29**, pp. 704–715, 1990.

[37] Owen, P. and Thompson, W., Heat transfer across rough surfaces. *J. Fluid Mech.,* **15**, pp. 321–334, 1963.

[38] Kramm, G., Beier, N., Foken, T., Muller, H., Schroder, P. and Seiler, W., A SVAT for NO, NO2, and O3 - Model description and test results. *Meteor. Atmos. Physics,* **61**, pp. 89–106, 1996.

[39] Parlange, Y., Waggoner, P. and Heichel, G., Boundary layer resistance and temperature distribution on still and flapping leaves. I. Theory and Laboratory experiments. *Plant Physiol.,* **48**, pp. 437–442, 1971.

[40] Leuning, R., Neumann, H. and Thurtell, G., Ozone uptake by corn: a general approach. *Agric. Meteorol.,* **20**, pp. 115–136, 1979.

[41] Schlichting, H., *Boundary Layer Theory,* translated by J. Kestin,

McGraw-Hill, New York, pp. 647, 1960.

[42] Brutsaert, W., *Evaporation into the Atmosphere*, D. Reidel Publishing Company, Dordrecht, 1982.

[43] Draxler, R. and Hess, G., *Description of the HySplit-4 modeling system*, NOAA Tech. Mem. ERL/ ARL –224, 1997. [Available from http://www.arl.noaa.gov/READY]

[44] Draxler, R. and Hess, G., An Overview of the Hysplit_4 Modeling System for Trajectories, Dispersion, and Deposition. *Aust. Met. Mag.,* **47**, pp. 295-308, 1998.

[45] Padro, J., Observed characteristics of the dry deposition velocity of O_3 and SO_2 above a wet deciduous forest. *Sci. Tot. Environ.,* **146/147**, pp. 395–400, 1994.

[46] Chang, J., Brost, R., Isaksen, I., Madronich, S., Middleton, P., Stockwell, W. and Walcek, W., A three dimensional Eulerian acid deposition model: physical concepts and formulation. *J. Geophys. Res.,* **92**, pp. 14681–14700, 1987.

[47] Sehmel, G., Particle and gas dry deposition: A review. *Atmos. Environ.,* **14**, pp. 983–1012, 1980.

[48] Voldner, E., Barrie, L. and Sirois, A., A literature review of dry deposition of oxides of sulphur and nitrogen with emphasis on long range transport modeling in North America. *Atmos. Environ.,* **20**, pp. 2101–2123, 1986.

[49] Duyzer, J. and Fowler, D., Modeling land atmosphere exchange of gaseous oxides of nitrogen in Europe. *Tellus,* **46B**, pp. 353–372, 1994.

[50] Erisman, J., Beier, C., Draaijers, G. and Lindberg, S., Review of deposition monitoring methods. *Tellus,* **46**, pp. 79–93, 1994.

[51] Erisman, J. and Baldocchi, D., Modeling dry deposition of SO_2. *Tellus*, **46B**, pp. 159–171, 1994.

[52] Arya, S., *Introduction to Micrometeorology,* Academic Press, San Diego, pp. 307, 1988.

[53] Arya, S., *Air pollution meteorology and dispersion*, Oxford University Press, New York, pp. 310, 1999.

[54] Pleim, J. E., Finkelstein, P. and Ellestad, T. G., A technique for estimating dry deposition velocities based on similarity with latent heat flux. *Atmos. Environ.*, **33**, pp. 2257–2265, 1999.

[55] Brook, J., Di-Giovanni, F. and Meyers, T., Estimation of dry deposition velocity using inferential models and site specific meteorology- Uncertainty due to siting of meteorological towers. *Atmos. Environ.,* **31**, pp. 3911–3920, 1997.

[56] Niyogi, D., Raman, S. and Alapaty K., Uncertainty in specification of surface characteristics, Part 2: Hierarchy of interaction explicit statistical analysis. *Boundary-Layer Meteorol.*, **91**, pp. 341–366, 1999.

[57] Travis, C. and Hattemer--Frey, H., Uptake of organics by aerial plant parts: A call for research. *Chemosphere*, **17**, pp. 277–283, 1988.

[58] Garland, J., The dry deposition of sulfur dioxide to land and water surfaces. *Proc. R. Soc. Lond,*. **12**, pp. 245-268, 1977.

[59] Lynn, B. and Carlson, T., A stomatal resistance model illustrating plant vs. external control of transpiration. *Agric. For. Meteorol.*, **52**, pp.5-43, 1990.

[60] Niyogi, D., Raman, S. and Alapaty, K., Comparison of four different stomatal resistance schemes using FIFE observations, Part 2: Analysis of terrestrial biospheric-atmospheric interactions. *J. Appl. Meteorol.,* **37**, pp. 1301–1320, 1998.

[61] Niyogi D., Alapaty K., Raman S. (1999a) An advanced carbon assimilation B surface evapotranspiration scheme for mesoscale models, MM5 Workshop on Land Surface Modeling and its Application to Mesoscale Models, 24-25 June 1999, Boulder CO.

[62] Niyogi, D., Alapaty, K. and Raman, S., A gas exchange / assimilation surface evapotranspiration model (GEM) for mesoscale applications. *J. Appl. Meteorol.,* in review, 1999.

[63] Cowan, I., Regulation of water use in relation to carbon gain in higher plants', in *Encylopedia of Plant Physiology, New Series, Vol. 12B, Physiological Plant Ecology II,* eds. Lange O., Nobel P., Osmond C. and Zeigler H., Springer Verlag, Berlin, pp.589-615, 1982.

[64] Jarvis, P., The interpretation of leaf water potential and stomatal

conductance found in canopies in field. *Phil. Trans. R. Soc. Lond. B.*, **273**, pp. 593–610, 1976.

[65] Niyogi, D., Raman, S., Alapaty, K. and Han J., A Dynamic Statistical Experiment for Atmospheric Interactions. *Environ. Model. Assess. (A)*, **2,** pp. 307–322, 1997.

[66] Avissar, R., Avissar, P., Mahrer, Y. and Bravdo, B., A model to simulate response of plant stomata to environmental conditions'. *Agric. For. Meteorol.*, **34**, pp. 21-29, 1985.

[67] Schulze, E., Kelliher, F., Korner, C., Lloyd, J. and Leuning R., Relationships among maximum stomatal conductance, ecosystem surface conductance, carbon assimilation rate, and plant nitrogen nutrition: A global ecology scaling exercise'. *Ann. Rev. Ecol. Syst.*, **25**, pp. 629-660, 1994.

[68] Kelliher, F., Leuning, R., Raupach, M. and Schulze, E., Maximum conductances for evaporation from global vegetation types. *Agric. For. Meteor.,* **73**, pp. 1–16, 1995.

[69] Alapaty, K., Pleim, J., Raman, S., Niyogi, D. S. and Byun, D., Simulation of Atmospheric Boundary Layer Processes using Local - and Nonlocal-Closure Schemes. *J. Appl. Meteorol.*, **36**, pp. 214–233, 1997.

[70] Wetzel P. and Chang J., Evapotranspiration from non-uniform surface: A first approach for short-term numerical weather prediction. *Mon. Wea. Rev.,* **116**, pp. 600-621, 1988.

[71] Dickinson, R., Henderson-Sellers, A., Kennedy, P. and Wilson M., *Biosphere-atmosphere transfer scheme (BATS) for the NCAR community climate model,* NCAR / TN - 275 + STR, 1986. [Available from the National Center for Atmospheric Research, Boulder, CO.]

[72] Noilhan, J. and Planton, S., A simple parameterization of land surface processes for meteorological models'. *Mon. Wea. Rev.*, **117**, pp. 536-549, 1989.

[73] Clapp, R. and Hornberger, G., Empirical equations for some soil hydraulic properties. *Wat. Resources Res.,* **14**, pp. 601–604, 1978.

[74] Cosby, B., Hornberger, G., Clapp, R. and Ginn, T., A statistical exploration of the relationships of soil moisture characteristics to the physical properties of soil. *Wat. Resources Res.,* **20**, 682 – 690, 1984.

[75] Walmsley, P. and Wesely, M., Modification of coded parametrizations of surface resistances to gaseous dry deposition. *Atmos. Environ.*, **30**, pp. 1181–1196, 1996.

[76] Sutton, M., Pitcairn, C. and Fowler, D., The exchange of ammonia between the atmosphere and the plant communities. *Advances Ecol. Res.*, **24**, pp. 301–393, 1993.

[77] Underwood, B., Dry deposition to a uniform canopy: evaluation of a first order closure mathematical model. *Atmos. Environ.,* **21**, pp. 1573–1585, 1987.

[78] Horst, T., A surface depletion model for deposition from a Gaussian plume. *Atmos. Environ.,* **11**, pp. 41–46, 1977.

[79] Sutton, M., Burkhardt, J., Guerin, D., Nemitz, E. and Fowler, D., Development of resistance models to describe measurements of bi-directional ammonia surface - atmosphere exchange. *Atmos. Environ.,* **32**, pp. 473–480, 1998.

[80] Farquhar, G., Firth, P., Wetselaar, R. and Weir, B., On the gaseous exchange of ammonia between leaves and the environment: determination of the ammonia compensation point. *Plant Physiol.,* **66**, pp. 710–714, 1980.

[81] Farquhar, G., Wetselaar, R. and Weir, B., Gaseous nitrogen losses from plants, in *Gaseous Loss of Nitrogen from Plant – Soil Systems*, ed. J. R. Freeny, and J. R. Simpson, Nijhoff/Dr. W. Junk Publishers, The Hague, pp. 159–180, 1983.

[82] Hanson, P., Lindberg, S., Tabberer, T., Owens, J. and Kim, K., Foliar exchange of mercury vapor: evidence for a compensation point. *Water, Air, Pollution,* **80**, pp. 373–382, 1995.

[83] Mahfouf, J., Analysis of soil moisture from near-surface parameters. A feasibility study. *J. Appl. Meteor.*, **30**, pp. 461–467, 1991.

[84] Deardorff, J., Efficient prediction of ground surface temperature and moisture with inclusion of a layer of vegetation. *J. Geophy. Res.*,

20, pp. 1889-1903, 1978.

[85] Xue Y., Sellers P., Kinter J. and Shukla J., A simplified biosphere model for climate studies'. *J. Clim.*, **4**, pp. 345-364, 1991.

[86] Acs, F., A Coupled Soil-Vegetation Scheme: Description, Parameters, Validation and Sensitivity Studies. *J. Appl. Meteor.*, **33**, pp. 268-284, 1994.

[87] Monteith, J. and Unsworth, M., *Principles of Environmental Physics,* Edward Arnold, London, pp. 291, 1990.

[88] Raupach, M., Vegetation-atmosphere interaction in homogeneous and heterogeneous terrain: some implications of mixed-layer dynamics'. *Vegetation*, **91**, pp. 105-120, 1991.

[89] Carpi, A. and Lindberg, S., Application of a Teflon dynamic flux chamber for quantifying soil mercury flux: tests and results over background soil. *Atmos. Environ.,* **32**, pp. 873–882, 1998.

[90] Harley, P., Guenther., A. and Zimmerman, P., Environmental controls over isoprene emission in deciduous oak canopies. *Tree Physiol.,* **17**, pp. 705–714, 1997.

[91] Lemon, E. and van Houtte, R., Ammonia exchange at the land surface. *Agron. J.*, **72**, pp. 876–883, 1980.

[92] Fowler, D., Flechard, C., Sutton, M. and Storeton-West, R., Long term measurements of the land - atmosphere exchange of ammonia over moorland. *Atmos. Eviron.*, **32**, pp. 453–459, 1998.

[93] Schjoerring, J., Husted, S. and Mattsson, M., Physiological parameters controlling plant atmosphere ammonia exchange. *Atmos. Environ.,* **32**, pp. 491–498, 1998.

[94] Monson, R., Lerdau, M., Sharkey, T., Schimel, D. and Fall, R., Biological aspects of constructing VOC emission inventories. *Atmos. Environ.,* **29**, pp. 2989–3002, 1995.

[95] Nikolov, N., Massman, W., Schoettle, A., Coupling biochemical and biophysical processes at the leaf level: an equilibrium photosynthesis model for leaves of C3 plants'. *Ecol. Modeling*, **80**, pp. 205-235, 1995.

[96] Sutton, M., Schjoerring, J. and Wyers, P., Plant atmosphere exchange of ammonia, *Phil. Trans. Roy. Soc. London,* **351**, pp. 261–278, 1995.

[97] Meyers, T. and Paw U., K., Modeling the plant canopy micrometeorology with higher-order closure principles. *Agri. For. Meteor.*, **41**, pp. 143-163, 1987.

[98] Meyers, T. and Baldocchi, D., A comparison of models for deriving dry deposition fluxes of O_3 and SO_2 to forest canopy. *Tellus,* **40B**, pp. 270–284, 1988.

[99] Sutton, M., Asman, W. and Schjorring, J., Dry deposition of reduced nitrogen. *Tellus,* **46B**, pp. 255–273, 1994.

[100] Asman, W., Factors influencing local dry deposition of gases with special reference to ammonia. *Atmos. Environ.,* **32**, pp. 41–421, 1998.

[101] Rondon, A. and Granat, L., Studies on the dry deposition of NO_2 to coniferous species at low NO_2 concentrations. *Tellus*, **46B**, pp. 339–352, 1994.

[102] Niyogi, D. and Raman, S., *Dynamical Coupling of Heterogeneous Fluid Media for an Environmental Interface, Proc. 1st Intrnl. Conf. Fluid Mech. Fluid Power*, Dec. 15 - 17, Delhi, IIT Delhi, New Delhi, pp. 107-110, 1998.

[103] Niyogi D., Alapaty K. and Raman S., Developing multi-media couplings to link ambient meteorological information with depositing surface environment. *Workshop on Atmospheric Nitrogen Compounds: Emissions Transport, Transformation, Deposition, and Assessment*, June 7 B 9 1999, Chapel Hill, NC, 1999.

[104] Elssworth, D.,Water relations and gas exchange of *Acer saccharum* seedlings in contrasting natural light and water regime. *Tree Physiol.*, **10**, pp. 1-20, 1992.

[105] Wong, S., Cowan, I. and Farquhar G., Stomatal conductance correlates with photosynthetic capacity, *Nature*, **282**, pp. 424–426, 1979.

[106] Farquhar, G., von Caemmerer, S. and Berry, J., A biochemical model of photosynthetic CO_2 assimilation in leaves of C3 species'. *Planta*, **149**, pp. 78-90, 1980.

[107] Farquhar, G. and Sharkey, T., 'Stomatal Conductance and Photosynthesis'. *Ann. Rev. Plant Physiol.*, **33**, pp. 317-345, 1982.

[108] Ball, J., Calculations related to gas-exchange. In *Stomatal Function,* ed. E. Zeiger, G.D. Farquhar and I. R. Cowan, Stanford University Press, Stanford, pp. 445–476, 1987.

[109] Ball J., Woodrow I. and Berry J., A model predicting stomatal conductance and its contribution to the control of photosynthesis under different environmental conditions, In *Progress in Photosynthetic Research*, ed. J. Biggins Vol IV, Martinus Nijhoff Publishers, Dordrecht, 1987.

[110] Aphalo, P. and Jarvis, P., Do stomata respond to relative humidity?, *Plant, Cell and Environ.*, **14**, pp. 127-132, 1991.

[111] Mott, K. and Parkhurst, D., Stomatal response to humidity in air and elox. *Plant, Cell, Environ.,* **14**, pp. 509–515, 1991.

[112] Monteith, J., A reinterpretation of stomatal responses to humidity'. *Plant, Cell Environ.,* **18**, pp. 357–364, 1995.

[113] Monteith, J., Accommodation between transpiring vegetation and the convective boundary layer'. *J. Hydrol.*, **166**, pp. 251-254, 1995.

[114] Mooney, H., Vitousek, P. and Matson, P., Exchange of materials between terrestrial ecosystems and the atmosphere. *Science,* **238**, pp. 926–932, 1987.

[115] Hosker, R. and Lindberg, S., Review: Atmospheric deposition and plant assimilation of gases and particles. *Atmos. Environ.,* **16**, pp. 889–910, 1982.

[116] Collatz, G., Ball, J., Grivet, C. and Berry, J., Physiological and environmental regulation of stomatal conductance, photosynthesis and transpiration: a model that includes a laminar boundary layer'. *Agric. For. Meteorol.*, **54**, pp. 107-136, 1991.

[117] Collatz, G., Ribas-Carbo, M. and Berry, J., Coupled photosynthesis-stomatal conductance model for leaves of C4 plants. *Aust. J. Plant Physiol.*, **19**, pp. 519-538, 1992.

[118] Sellers, P., Berry, J., Collatz, G., Field, C. and Hall, F., Canopy

reflectance, photosynthesis and transpiration, III. A reanalysis using enzyme – kinetics – electron transport models of leaf physiology. *Remote Sens. Environ.*, **42**, pp. 187–216, 1992.

[119] Calvet, J., Noilhan, J., Roujean, J., Bessemoulin, P., Cabelguenne, M., Olioso, A. and Wigneron, J., An interactive vegetation SVAT model tested against data from six contrasting sites. *Agric. For. Meteor.*, **92**, pp. 73–95, 1998.

[120] Goudriaan, J., van Laar, H., van Keulen, H. and Louwerse, W., Photosynthesis, CO_2, and plant production, *Wheat growing and modeling, NATO ASI Series, Series A, Vol 86,* ed. W. Day, and R. Atkin, Plenum Press, New York, pp. 107–122, 1985.

[121] Kim J. and Verma S., Modeling canopy photosynthesis: scaling up from a leaf to canopy in a temperate grassland ecosystem. *Agri. For. Meteorol.*, **57**, pp. 187 – 208, 1991.

[122] van Heemst, Potential crop production. In *Modeling of agricultural production: weather, soil, and crops*, ed. H. van Keulen and J. Wolf, Simulation Monograph, Pudoc, Wageningen, 1986.

[123] Jacobs, C., van den Hurk, B. and de Bruin, H., Stomatal behavior and photosynthetic rate of unstressed grapevines in semi-arid conditions. *Agric. For. Meteor.*, **80**, pp. 111–134, 1996.

[124] Jacobs, C., Direct impact of atmospheric CO_2 enrichment on regional transpiration'. Ph.D. thesis, Wageningen Agricultural University, The Netherlands, pp. 1-179, 1994. [Available from the Department of Meteorology, Wageningen Agricultural University, Duivendaal 2, 6701 AP Wageningen, The Netherlands.]

[125] Sellers, P., Randall, D., Collatz, J., Berry, J., Field, C., Dazlich, D., Zhang, C., Collelo, G. and Bounous A., A revised land surface parameterization (SiB2) for atmospheric GCMs: Model formulation. *J. Clim.,* **9**, pp. 676-705, 1996.

[126] Sellers P., Los, S., Compton, T., Justice, C., Dazlich, D., Collatz J., Randall, D., A revised land surface parameterization (SiB2) for atmospheric GCMs. Part II: The generation of global fields of terrestrial biophysical parameters from satellite data. *J. Clim.,* **9**, pp. 706-736, 1996.

[127] Su, H. B., Paw U, K. T. and Shaw, R., Development of a coupled leaf and canopy model for the simulation of plant-atmosphere interaction. *J. Appl. Meteor.*, **35**, pp. 734-748, 1996.

[128] Jones, H., *Plants and Microclimate, A quantitative approach to environmental plant physiology,* Cambridge University Press, Cambridge U.K. and New York USA, pp. 428, 1992.

[129] Baldocchi, D., A Lagrangian Random-Walk model for simulating water vapor, CO_2, and sensible heat flux densities and scalar profiles over and within a soybean canopy'. *Bound Layer Meteorol.*, **61**, pp. 113-144, 1992.

[130] Baldocchi, D., A comparative study of mass and energy exchange over a closed C3 (Wheat) and an open C4 (Corn) canopy: The partitioning of available energy into latent and sensible heat exchange. *Agric. For. Meteorol.,* **67**, pp. 191-220, 1994.

[131] Choudhury, B. and Monteith, J., Implications of stomatal response to saturation deficit for the heat balance of vegetation. *Agric. For. Meteorol.*, **36**, pp. 215-225, 1986.

[132] Dougherty, R., Bradford, J., Coyne, P. and Sims, P., Applying an empirical model of stomatal conductance to three C-4 grasses. *Agric. For. Meteorol.*, **67**, pp. 269-290, 1994.

[133] Sandnes H. and Styve H., *Calculated budgets for airborne acidifying component in Europe, 1985, 1987, 1988, 1989, 1990, and 1991.* EMEP Report 1/92. Norwegian Meteor. Inst. Oslo, Norway. (ISSN 0332-9879), 1992.

[134] van Aalst, R., Dry deposition of NO_x. In *Air Pollution by Nitrogen Oxides,* ed. T. Schneider and L. Grant, Elsevier, Amsterdam, pp. 263–270, 1982.

[135] Hanson, P. and Lindberg, S., Dry deposition of reactive nitrogen compounds: A review of leaf, canopy and non-foliar measurements. *Atmos. Environ.,* **25A**, pp. 1615–1634, 1991.

[136] Hanson, P., Rott, K., Taylor, G., Gunderson, C., Lindberg, S. and Ross–Todd, B., NO_2 deposition to elements representative of a forest landscape. *Atmos. Environ.,* **23**, pp. 1783–1794, 1989.

[137] Judeikis, H. and Wren, A., Laboratory measurements of NO and NO2 deposition onto soil and cement surfaces. *Atmos. Environ.,* **12**, pp. 2315–2319, 1978.

[138] Niyogi, D., Raman, S., Prabhu, A., Kumar, A. and Joshi, S., Direct estimation of stomatal resistance for meteorological applications. *Geophys. Res. Lett.,* **24**, pp. 1771–1774, 1997.

[139] Erisman, J., Bleeker, A. and Jaarsveld, H., Atmospheric deposition of ammonia to semi-natural vegetation in the Netherlands - Methods for mapping and evaluation. *Atmos. Environ.*, **32**, pp. 481–489, 1998.

[140] Rondon, A., Johansson, C. and Granat, L., Dry deposition of nitrogen dioxide and ozone to coniferous forests. *J. Geophys. Res.,* **98D**, pp. 5159–5172, 1993.

[141] Cohen Y. and Ryan P., Mass transfer across wind-sheared interfaces. *Int. Comm. Heat Mass Transfer*, **12**, pp. 139–148, 1985.

[142] Pasquill, F. and Smith, F., Atmospheric Diffusion; the dispersion of windborne material from industrial and other sources, Ellis Horwood, Chichester, UK, pp. 297, 1983.

[143] Singles, R., Sutton, M. and Weston, K., A multi-layer model to describe the atmospheric transport and deposition of ammonia in Great Britain. *Atmos. Environ.,* **32**, pp. 393–399, 1998.

[144] Taconet, O., Bernard, R. and Vidal–Madjar, D., Evapotranspiration over an agriculture region using a surface flux / temperature model based on NOAA – AVHRR data. *J. Clim. Appl. Meteor.,* **25**, pp. 284–307, 1986.

[145] Nemani, R. and Running, S., Testing a theoretical climate – soil-leaf area hydrological equilibrium of forests using satellite data and ecosystem simulation. *Agric. For. Meteor.,* **44**, pp. 245–260, 1989.

[146] Asrar, G., Fuchs, M., Kanemasu, E. and Hatfield, J., Estimating absorbed photosynthetic radiation and leaf area index from spectral reflectance in wheat. *Agron. J.,* **76**, pp. 300–306, 1984.

[147] Tarpley, J., Schneider, S. and Money, R., Global vegetation indices from the NOAA-7 meteorological satellite. *J. Clim. Appl. Meteor.,* **23**, pp. 491–494, 1984.

[148] Tucker, C., Vanpraet, C., Sharman, M. and van Itersum, G., Satellite remote sensing of total herbaceous biomass production in the Senegalese Sahel: 1980–1984. *Remote Sens. Environ.,* **17**, pp. 233–249, 1985.

[149] Nemani, R. and Running, S., Estimation of regional surface resistance to evapotranspiration from NDVI and Thermal –IR AVHRR data. *J. Appl. Meteor.,* **28**, pp. 276–284, 1989.

[150] Cihlar, J., St. Laurent, L. and Dyer, J., Relation between the normalized difference vegetation index and ecological variables. *Remote Sens. Environ.,* **8**, pp. 279–298, 1991.

[151] Seevers, P. and Ottmann, R., Evapotranspiration estimation using a normalized difference vegetation index transformation of satellite data. *Hydrol. Sciences,* **39**, pp. 333–345, 1994.

[152] Szilagyi, J., Rundquist, D., Gosselin, D. and Parlange, M., NDVI relationship to monthly evaporation. *Geophys. Res. Let.,* **25**, pp. 1753–1756, 1998.

[153] Sellers, P., Canopy reflectance, photosynthesis and transpiration. *Int. J. Remote Sens.,* **6**, pp. 1335–1372, 1985.

[154] Sellers, P., Canopy reflectance, photosynthesis and transpiration Part II: The role of biophysics in the linearity of their interdependence. *Remote Sens. Environ.,* **21**, pp. 143–183, 1987.

[155] Charles–Edwards, D. and Ludwig, L., A model for leaf photosynthesis by C3 plant species. *Ann. Bot.,* **38**, pp. 921–934, 1974.

[156] Goudriaan, J., The bare bones of leaf-angle distribution in radiation models for canopy photosynthesis and energy exchange'. *Agric. For. Meteorol.*, **43**, pp. 155-169, 1988.

[157] Asrar, G., *Theory and Applications of Optical Remote Sensing,* Wiley Interscience, New York, pp. 734, 1990.

[158] Wesley, M. and Lesht, B., *Comparison of the RADM dry deposition module with site – specific routines for inferring dry deposition,* EPA/600 / S4-88/027, U.S. Environmental Protection Agency (available as PB88-238191/AS from NTIS, Springfield VA), 1988.

[159] Raman, S., Niyogi, D., Prabhu, A., Ameenullah S., Nagaraj, S., Jayyanna, K. and Udai Kumar, VEBEX: A Vegetation and Energy Balance Experiment for the Tropics., *Proc. Ind. Acad. Sci. (Earth and Planetary Sciences)*, **107**, pp. 97–105, 1997.

[160] Sharkey, T. and Singsaas, E., Why plants emit isoprene?. *Nature,* **374**, pp. 769, 1995.

[161] Walcek, C., Brost, R., Chang, J. and Wesley, M., SO_2, sulfate, and HNO_3 deposition velocities computed using regional landuse and meteorological data. *Atmos. Environ.,* **21**, pp. 1759–1763, 1986.

[162] Finlayson–Pitts, B. and Pitts, J., *Atmospheric Chemistry: Fundamentals and Experimental Techniques,* John Wiley and Sons, New York, pp. 1098, 1986.

[163] Finnigan, J. and Raupach, M., Transfer processes in plant canopies in relation to stomatal characteristics, In *Stomatal Function*, ed. E. Zeiger, G. Farquhar, I. Cowan,. Stanford University Press, Stanford, CA, pp. 385-429, 1987.

[164] Baldocchi, D. and Rao, K.S., Intra-field variability of scalar flux densities across a transition between a desert and an irrigated potato field'. *Bound.Layer Meteorol.*, **76**, pp.109-121, 1995.

[165] Lovblad, G. and Erisman, J., Deposition of nitrogen in Europe, *Critical loads for Nitrogen,* ed. P. Grennfelt, and E. Thornelof, Lokeberg, 6–10 April 1992, Sweden, Report No. Nord 1992:41, Nordic Council of Ministers, Copenhagen, Denmark, pp. 239–286, 1992.

WITPRESS

Air Pollution Emissions Inventory

Editors: ***H. POWER****, Wessex Institute of Technology, UK and* ***J.M. BALDASANO****, Universitat Politécnica de Catalunya, Spain*

An indispensable tool for all researchers in the field, this book not only provides an idea of pollution sources but also gives essential information for future pollution modelling studies. It contains a selection of invited review articles covering various topics on emissions inventory and modelling and places particular emphasis on emission data in urban settings.

The following chapters from distinguished contributors are featured: Guidelines and Formulation of an Upgrade Source Emission Model for Atmospheric Pollutants; Spatial and Temporal Disaggregation of Emission Inventories; Modelling the Emissions of Road Vehicles at Macroscale and Microscale; The Construction and Use of an Emissions Inventory for Local Air Quality Management; Validation and Verification of Emission Inventory Data; On the Forecasting of SO_2 Time Series.

Series: Advances in Air Pollution, Vol 3

ISBN: 1-85312-517-2 1998

256pp

£73.00/US$117.00

Air Pollution VIII

Editors: ***C.A. BREBBIA****, Wessex Institute of Technology, UK,* ***J.W.S. LONGHURST****, University of the West of England, UK and* ***H. POWER****, Wessex Institute of Technology, UK*

This book contains state-of-the-art contributions from scientists working on various aspects of air pollution. Over 70 papers are included and these were first presented at the eighth in a highly successful series of international conferences focusing on this problem.

The topics covered are as follows: Air Pollution Modelling; Air Quality Management; Urban Air Pollution; Urban and Suburban Transport Emissions; Monitoring and Laboratory Studies; Global Studies; Comparison of Modelling with Experiments; Indoor Pollution; Pollution Engineering; Fluid Mechanics for Environmental Problems; Chemistry of Air Pollution; Aerosols and Particles; and Health Problems.

Series: Advances in Air Pollution

ISBN: 1-85312-822-8 2000

808pp

WIT*Press*

Ashurst Lodge, Ashurst, Southampton, SO40 7AA, UK.

Tel: 44 (0)23 8029 3223

Fax: 44 (0)23 8029 2853

E-Mail: witpress@witpress